Mysteries of Modern Physics: Time

Sean Carroll, Ph.D.

PUBLISHED BY:

THE GREAT COURSES
Corporate Headquarters
4840 Westfields Boulevard, Suite 500
Chantilly, Virginia 20151-2299
Phone: 1-800-832-2412
Fax: 703-378-3819
www.thegreatcourses.com

Copyright © The Teaching Company, 2012

Printed in the United States of America

This book is in copyright. All rights reserved.

Without limiting the rights under copyright reserved above,
no part of this publication may be reproduced, stored in
or introduced into a retrieval system, or transmitted,
in any form, or by any means
(electronic, mechanical, photocopying, recording, or otherwise),
without the prior written permission of
The Teaching Company.

Sean Carroll, Ph.D.
Senior Research Associate in Physics
California Institute of Technology

Professor Sean Carroll is a Senior Research Associate in Physics at the California Institute of Technology. He did his undergraduate work at Villanova University and received his Ph.D. in Astrophysics from Harvard in 1993. His research involves theoretical physics and astrophysics, with a focus on issues in cosmology, field theory, and gravitation.

Prior to arriving at Caltech, Professor Carroll taught and did research at the Massachusetts Institute of Technology; the Kavli Institute for Theoretical Physics at the University of California, Santa Barbara; and the University of Chicago. His major contributions have included models of interactions among dark matter, dark energy, and ordinary matter; alternative theories of gravity; and violations of fundamental symmetries. His current research involves the foundations of quantum mechanics, the physics of inflationary cosmology, and the origin of time asymmetry.

While at MIT, Professor Carroll won the Graduate Student Council Teaching Award for his course on general relativity, the lecture notes of which were expanded into the textbook *Spacetime and Geometry: An Introduction to General Relativity*, published in 2003. In 2006, he received the College of Liberal Arts and Sciences Alumni Medallion from Villanova University, and in 2010, he was elected a Fellow of the American Physical Society.

Professor Carroll is the author of *From Eternity to Here: The Quest for the Ultimate Theory of Time*, a popular book on cosmology and time. His next book is *The Particle at the End of the Universe*, about the Higgs boson and the Large Hadron Collider. He is active in education and outreach, having taught more than 200 scientific seminars and colloquia and given more than 50 educational and popular talks. Professor Carroll has written for *Scientific*

American, *New Scientist*, *The Wall Street Journal*, and *Discover* magazine. His blog, *Cosmic Variance*, is hosted by *Discover*. He has been featured on such television shows as *The Colbert Report*, PBS's *NOVA*, and *Through the Wormhole with Morgan Freeman* and has acted as an informal science consultant for such movies as *Thor* and *TRON: Legacy*.

The first of Professor Carroll's Great Courses was *Dark Matter, Dark Energy: The Dark Side of the Universe*. ■

Table of Contents

INTRODUCTION

Professor Biography i
Course Scope 1

LECTURE GUIDES

LECTURE 1
Why Time Is a Mystery 3

LECTURE 2
What Is Time? 23

LECTURE 3
Keeping Time 42

LECTURE 4
Time's Arrow 63

LECTURE 5
The Second Law of Thermodynamics 84

LECTURE 6
Reversibility and the Laws of Physics 105

LECTURE 7
Time Reversal in Particle Physics 126

LECTURE 8
Time in Quantum Mechanics 146

LECTURE 9
Entropy and Counting 165

Table of Contents

LECTURE 10
Playing with Entropy 185

LECTURE 11
The Past Hypothesis 206

LECTURE 12
Memory, Causality, and Action 225

LECTURE 13
Boltzmann Brains 243

LECTURE 14
Complexity and Life 263

LECTURE 15
The Perception of Time 282

LECTURE 16
Memory and Consciousness 302

LECTURE 17
Time and Relativity 321

LECTURE 18
Curved Spacetime and Black Holes 341

LECTURE 19
Time Travel 360

LECTURE 20
Black Hole Entropy 380

LECTURE 21
Evolution of the Universe 399

Table of Contents

LECTURE 22
The Big Bang 421

LECTURE 23
The Multiverse 442

LECTURE 24
Approaches to the Arrow of Time 462

SUPPLEMENTAL MATERIAL

Bibliography 482

Mysteries of Modern Physics: Time

Scope:

This course addresses one of the most profound questions of modern physics: Why does time work the way it does? Time is as mysterious as it is familiar, and over the course of these lectures, we will explore where those mysteries come from and how they are addressed by physics, philosophy, biology, neuroscience, and cosmology.

We will start by exploring how time works at a basic level: what time is and how we measure it using clocks and calendars. But we will quickly come up against a central mystery: Why does time have a direction? The difference between past and future will be a primary concern throughout the course.

We will see that the fundamental laws, ever since Isaac Newton, have a profound feature: They do not distinguish between past and future. They are reversible; if we make a movie of the motion of a planet around the Sun or the back-and-forth rocking of a pendulum, we can play it backward and it seems perfectly sensible. But for systems with many moving pieces, there is a pronounced directionality to time. Many familiar processes are irreversible: scent dispersing into a room, cream mixing into coffee, the act of scrambling an egg. In the real world, these happen in one direction in time, never backward. That difference is the arrow of time.

Explaining why time has an arrow is a primary concern of modern physics. We will see that it does not arise from quantum mechanics or particle physics. Rather, it is due to the increase of entropy—a way of measuring how messy or disorderly a system is—as time passes. The increase of entropy is responsible for many deeply ingrained features of time, such as our ability to remember the past or make decisions that affect the future.

The question then becomes: Why does entropy increase? The increase of entropy toward the future is known as the second law of thermodynamics and was explained in modern terms by Ludwig Boltzmann in the 19th century. Boltzmann's insight is that entropy increases because there are more ways for a system to have high entropy than low entropy; thus, high entropy is a natural condition.

This raises a new question: Why was entropy lower in the past? That turns out to be a much harder problem, one that traces back to the very beginning of time. The low entropy of the past is ultimately due to the fact that our universe had low entropy 13.7 billion years ago, at the time of the Big Bang.

Cosmology would like to explain why the Big Bang had low entropy, but our best current models aren't up to the task. It's possible that the ultimate explanation might lie beyond our observable cosmos, in a larger multiverse. Even without knowing what that explanation will be, we can marvel at the deep connections between time in our everyday lives and the larger universe in which we live. ■

Why Time Is a Mystery
Lecture 1

Time is something we're all familiar with, but what's less clear is what science has to say about time. You probably know that the idea of time is important to physics; it's also important to biology, medicine, neuroscience, psychology, and the human sciences, such as history, politics, and economics. Time is absolutely central to every endeavor of humankind, and physics is a kind of bedrock for time. If we come to understand how time works from the point of view of physics, we can better understand how it works elsewhere in science and in our everyday lives.

Penetrating the Mysteries of Time

- Although we think about and use it every day, time has a reputation for being mysterious. As Saint Augustine wrote, "What is time? If no one asks me, I know. But if I wish to explain it to someone who asks, I know not."

- The difference between our everyday dealings with time and the attempt to understand time on a deeper level is similar to the difference between operating a microwave oven and understanding how one works. As with the microwave oven, even though time may appear to be mysterious and incomprehensible, it can be understood.

- Our goals in this course will be to understand time as physicists and then to connect that physical understanding to other ways that time manifests itself in science and in our everyday lives. Along the way, we'll discover that time has a manifestation both in our everyday lives and in a more cosmic perception and that there is a direct connection between how time works in the two realms.
    - The reason that time works the way it does in our everyday lives is ultimately because of how the universe works.

- Once we penetrate the mysteries of time, we'll find that the secret to those mysteries isn't in our everyday experience but in the creation of the universe itself.

Telling Time
- We have time-measuring devices—clocks—in almost everything around us—our phones, our computers, and so on. At a basic level, a clock is something that repeats itself in a predictable way. Both elements of that definition are important: repeatability and predictability.

- The classic idea of a clock is the Sun moving across the sky, rising in the east in the morning and setting in the west in the evening. That's the simplest possible timekeeper we have. Other clocks are based on that same basic principle: doing the same thing over and over again in a repeatable, predictable way.

- As we'll learn in this course, it's not easy to make a clock, and even though clocks are all around us, they're not very accurate. There is a great deal of challenge in building a clock that's as accurate as the modern world demands.

Time within Ourselves
- A clock measures the passage of time, but we also feel time within ourselves, at least partly because clocks—mechanisms that repeat themselves in predictable ways—exist within us, such as our breathing and the beating of our hearts.

- There are also things about us that don't repeat themselves but that progress as time passes: We age; we think; we make choices; we plan for the future; we remember the past. These different aspects of time are crucial to what it means to live our lives, to be human beings.

- Perhaps the most important aspect of the passage of time for our lives as human beings is the accumulation of experiences.

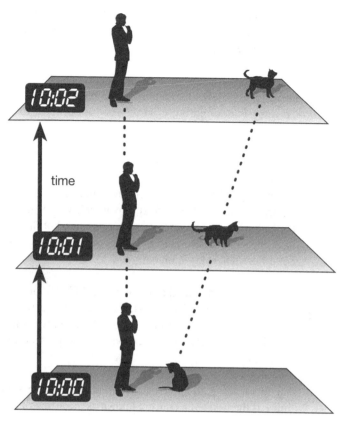

Both individual frames and the whole series of frames define what we think of as the universe; it's a four-dimensional thing—both space and time.

- It's not just that we experience something today, and then tomorrow, and then the next day, but we experience something today and we remember it tomorrow. As time passes, we have more life experience to draw on. There's a gradual buildup of our life spans through time.

- The past that we have experienced is important to who we are today. Even in the process of watching these 24 lectures, you'll learn things, you'll change, and you'll come out of this course a different person than you were when you came in. That's a central aspect of how time works.

Moving through Time

- When we look closely, the idea that time passes or that we move through time is not entirely clear. What does it mean to say that time passes?

- The notion of time passing has two different aspects: continuity and progress.
    - The passage of time isn't a story of complete rearrangement of reality; it's more like an uncut movie reel. We see moments in the history of the universe one after another, but the moments have the same basic "stuff." Things evolve, but everything is not different at every moment.

    - Time isn't only a label that we apply to different moments, but it's an ordered list of moments, moving in a direction. Just as a movie isn't a set of frames scattered randomly on the floor, likewise, the progress of time from past, to present, to future is crucial.

    - The fact that we all agree on the directions of the past and the future is the basis of a concept known as the arrow of time.

- How fast are we moving through time? What else could the speed of time be other than something like 1 second per second?
    - When we move through space and we have a velocity, what we're talking about is how much space we travel through per unit of time, such as how many miles per hour.

    - But if we're talking about the speed of time, that kind of measurement doesn't quite make sense. If there were something called the speed of time, it could only be 1 minute per minute, 1 year per year, and so on. Time moves at the rate time moves, or time doesn't move at all; it simply is.

- Time happens, whether we like it or not. Space is something that we can choose to move through, but time is inevitable. There is no velocity at which we move through time; we just experience time moment after moment.

The Direction of Time
- The fundamental distinction between time and space is the division of time into past, present, and future. The progression of time from the past, to the present, to the future only happens in one way, and as it happens, things occur always in the same sense.
  - Memory, for example, is something we have about the past; we don't have a memory of the future. When we age, we always are born young, we grow older, and then we die; that's how time works.

  - This is the arrow of time; the arrow points from the past, into the present, and toward the future. Time is absolutely, fundamentally directed in that way.

- Einstein taught us that time is related to space, but space doesn't have any fundamental directionality; it has a sort of nonfundamental directionality.
  - If you drop something in a room, it will always go down as opposed to up, but the difference between up and down in a room doesn't reflect any deep feature of reality. If you were in Australia instead of the United States, the downward direction would be something different; the "arrow of space" depends on where you are.

  - We don't even really think about an arrow of space because it's obviously a feature of our environment. There's an arrow of space because we live in the vicinity of an influential object called Earth.

  - The arrow of time, in contrast, seems to be a deep feature of reality; it seems to be the same everywhere.

- The ultimate explanation for the arrow of time is similar to the ultimate explanation for the arrow of space.
  - Just as the arrow of space is explained by the fact that we live in the vicinity of an influential object—the Earth—the arrow of time is explained by the fact that we live in the vicinity of an influential event—the Big Bang, the beginning of the universe.
  - At a deep level of reality, there is no arrow of time; there's no difference between the past and the future as far as the ultimate laws of physics are concerned.
  - The arrow of time has something to do with the macroscopic behavior of objects in the world, but it doesn't appear at the level of two particles bumping into each other and scattering off; that's something that treats the past and the future absolutely identically.
  - The arrow of time arises only when there are many, many particles—when we have a person or a cloud of gas in the solar system forming a new planet.

Entropy
- Entropy is a way of talking about the disorderliness of "stuff" in the universe. An egg is very orderly, but if we break the egg, it becomes disorderly; if we scramble the egg, it becomes even more disorderly. A scientist would say that the entropy of the egg is increasing.

- Left to their own devices, objects in the universe experience an increase of entropy; they become more disorderly as time passes. That, in fact, is the important distinction between the past and the future.

- There are many ways in which the arrow of time manifests itself, but there's one underlying feature that explains all those different manifestations: In a system left to itself, entropy increases. This is so important that it's literally a law of nature: the second law of thermodynamics.

- In the 19th century, scientists made great strides toward understanding how entropy works, why it tends to go up toward the future, and how it underlies the arrow of time in all of its manifestations. But that understanding left us with one unanswered question: Why was entropy lower in the past?
    - This question brings us all the way back to the Big Bang. Our universe started 13.7 billion years ago in a condition of very low entropy and very high organization. That's what got time started in the way we experience it in our everyday lives.
    - Ever since the Big Bang, we've been living out the process by which the universe increases in entropy. That's the influential event in the aftermath of which we live.
- In this course, we will come to understand what entropy is and why it tends to increase. Once we understand how entropy is responsible for the arrow of time and all of its forms, we will go back and explain why it's there; that's a matter of understanding relativity, spacetime, cosmology, the Big Bang, and how all these ideas fit together.

Suggested Reading

Carroll, *From Eternity to Here*, chapter 1.

Falk, *In Search of Time*.

Frank, *About Time*.

Greene, *Fabric of the Cosmos*, part II.

Questions to Consider

1. How would you define "time"? How do you use it in your everyday life? Do you think of yourself as moving through time or time passing through you?

2. In what ways is the past different from the future? Do those differences seem fundamentally important? How do they relate to each other?

Why Time Is a Mystery
Lecture 1—Transcript

Welcome to the mysteries of time. I should say right from the start that time is an unusual subject for one of The Great Courses. It's not like dark matter or medieval history where we have to be introduced to the subject matter; we all know about time. There was a study by the *Oxford English Dictionary* that said "time" is the most used noun in the English language. We're surrounded by time all the time. We have birthdays; we know how old we are; we know the typical lifespan of a human being. We have schedules; we have calendars; we have clocks all around us. It's amazing how many clocks there are in the everyday life of a typical person in this day in age. You have a clock in your phone; you have a clock on your microwave oven; you have a clock in the computer you use to type on. We literally cannot escape the phenomenon of time.

So why are we here? It's because when you think a little bit more carefully about time, mysteries do appear. Our goal in these lectures is to understand what science has to tell us about time. It's not quite as simple as building a clock or reading a clock. When we thing about time we realize that there is a lot going on. It's actually a very subtle subject, and science is going to have a lot to say. Our primary goal will be to understand the physics of time. In some sense, time belongs to physics. I'm sure you've heard of this guy named Albert Einstein. He was named the Person of the Century by *Time* magazine, and not just because the magazine is called *Time*. One of Einstein's major contributions was to understand that time and space are part of one thing, one four-dimensional universe that we call "spacetime." We're going to be talking about how spacetime works and how it relates to time itself. You may have heard of the book by Stephen Hawking called *A Brief History of Time*. Hawking's book was a phenomenal bestseller. It took by surprise both the publishing community and the science community. Part of that is Hawking's personal story which is very compelling. Another part is that the idea of time and the ideas next to it like spacetime and evolution and entropy are fascinating to all of us. They belong in some sense to physics. If you took a Physics 101 class, you must have talked about motion, speed, and acceleration. These are all different concepts that relate to how things evolve through time, how objects move, how they change their positions and their identities as time passes.

But of course, time also matters for other subjects. It matters for biology, very obviously; it matters for evolution—natural selection is a story about how species in our biosphere here on Earth have changed with time. The very word "evolution" brings in time to the equation. We also have psychology and neuroscience; we have the way that we think about time. What does it mean to remember something in the past, to predict something in the future, or to perceive time passing? Why is it that we're always late? We'll talk about these subjects and relate them back to the physics of time.

Time is part of language. Not only is the word "time" used a lot, it's essentially impossible to construct a sentence in the English language without somehow referring to the concept of time. We have verbs that can be past-, present-, or future-oriented. For that matter, time is part of almost every subject that you could imagine thinking about: history, culture. History itself is a story of what happens through time. Culture is a story about what happens at any one moment in time, but cultures evolve. Politics and economics talk about structures within human societies that change with time. And in some cases, especially in economics, how they change is absolutely central. Even the arts, music, and literature evolve with time in certain definite ways. When you read a novel, knowing when it was written and what the context is can be important; novels do not simply appear out of nowhere. Your personal experience: You age through time yourself. We all do. We feel time passing ourselves.

Physics is the bedrock. If you want to know what time is, we'll talk about all these different aspects of time, but they're all built on the physical understanding of how time works. Our strategy will be to understand the physics and then to apply it to other ways in which time appears in our everyday lives.

In particular, we're going to care about the mysteries of time. Time has a funny status, despite the fact that we use it every day, that it is literally around us in all these clocks, and in the back of our minds when we're thinking about what we do as time passes, time nevertheless has a reputation for being mysterious. This was put very elegantly by Saint Augustine, a thinker in north Africa back in the year 400. Now, Augustine was a theologian primarily—he worried about original sin and the city of God and his own personal sins—but he also had an interest in cosmology and metaphysics. He wrote several famous passages about the nature of time. In one of them he asks, "What is time? If no one asks me, I know. If I wish to explain it to one who asks, I know not."

He's basically summing up the dilemma here. We use time very easily. If someone says to us, "It's 5 o'clock. Please be somewhere at 6 o'clock," we are not seized by existential panic—"What does that mean, '6 o'clock'?" We all know what it means and how to use it, but if we think about it deeply, if someone says, "What is the definition of this concept, time, that you keep talking about?" it's actually very hard to define. I'm going to take the philosophy that the reputation that time has as mysterious is actually somewhat overblown. We will talk about the mysteries, but we'll say that many of these mysteries can be solved. I want you to first appreciate the superficial level of time, then the deeper level of which there are mysteries, but then an even deeper level than that, at which we can say, "You know what? We can understand what these mysteries mean and how to actually resolve them." At the end we'll be left with a few remaining mysteries lurking around. These are big research projects for scientists right now and into the future.

Think about how we use time all the time. If I say, "This is a 30-minute lecture," or if I say, "The TV show starts at 7 pm," you know what that means. Operationally, we know what time means because we know how to react. If you say something that contains an interval of time, I know how to respond to it in a sensible way. What we're going to be doing is to be looking deeper at it. We're going to be saying, why is that the case? Why is it the case that the future is a certain way and the past is a certain way? When we talk about an interval of time, what does that mean? Does everyone agree on the same intervals of time? Does everywhere in the universe agree? Could it have been different?

The central theme of this course is that time is more subtle and mysterious than you might think. But we can understand it; it is ultimately comprehensible.

And most interestingly, the process of understanding it connects the very mundane and everyday—the way we experience time in our day-to-day lives—to some of the most profound questions in modern physics. We will literally travel from the kitchen to the multiverse. It turns out that if you want to understand why you can mix cream into coffee, but you can't unmix it—in other words, why certain process move forward in time but never backward in time—you're ultimately led to ask questions about what happened at the Big Bang. How time works in our everyday lives is understandable to us, but understanding it at a deeper level means we need to understand the whole universe and maybe even beyond that to the multiverse.

On the everyday level, we measure time using clocks. As we said, clocks are everywhere these days—clocks, calendars, schedules. So what is a clock? We know how to read it; we know how to use it. If we were told to build one, what would we have to do? The simple answer is that a clock is something that does the same thing over and over again. A clock is something that repeats itself in a predictable way compared to other clocks in the universe. We're lucky enough to live in a world where there are many clocks. The simplest ones are the motions of the Sun and the stars in the sky. Those were the earliest ways that we kept time; in ancient times, you would use what you saw in the sky as a way of telling what time it was. It turns out that as a technology question, if you want to go beyond the stars and the Sun, if you want to build a clock, it's actually difficult to achieve. It was quite a technological challenge to build clocks that could keep accurate time. There are many, many things in the universe that do more or less the same thing over and over again, but it's making them absolutely predictable that is difficult. A pendulum rocking back and forth does basically the same thing over and over again. That's why a pendulum finds a role in a grandfather clock, for example. But put that pendulum on a boat where the deck is moving beneath you and suddenly it does not keep time very well anymore.

So one of the interesting aspects of time is simply how to measure it. Measuring time is a much more subtle operation than measuring space. There are rulers and there are clocks, but rulers are easy to make and clocks are difficult to make. On the other hand, there is more to time than just measuring it. In some sense, what the clock is doing is fairly simple. It is just moving forward at a very predictable pace; we know what the clock is going to do. Contrast that with our experience of time. Part of the way that we as human beings experience time is clock-like; we have processes in our own bodies that do the same thing over and over again in a more or less predictable way: We have the beat of our heart; we have our breath going back and forth.

But there are other aspects of time that are not repetitive: the aspects of change and evolution. In our personal lives that means aspects of aging. We are different now than we were when we were younger, and we are different now than we will be when we are older. It is not a repetitive cyclic kind of thing; it's motion in a certain direction. The way that we think in our brains is one of directedness: We process information; we make choices; we refer back to the past and we predict the future. The way that we actually think about time is not simply repetitive; every day is not a copy of the previous days. The important feature here is that as time passes, we accumulate. We change because have more time towards the past and less time towards our personal futures. We have more memories; we have more experiences. That's the personal experience of time.

As we go through this course, you will experience things. You will see and hear different lectures and you will know things at the end (hopefully) that you didn't know at the beginning. For a personal example, as I am speaking right now giving these lectures, I'm also engaged in writing a book about the Large Hadron Collider and the search for the Higgs boson. The Large Hadron Collider is a giant particle accelerator outside Geneva. One of the things it's trying to do is to find this hypothetical particle called the Higgs boson. The Higgs will help us understand why other particles like quarks and electrons have the properties they do, such as their mass. The problem is—or the fact is, whether it's a problem or not—as I am speaking, we have not yet discovered the Higgs boson. The LHC, the Large Hadron Collider is hard at work on it. Right now, the Higgs boson is a prediction. It is something we are predicting will happen in the future that this particle will appear in our experiment. But I don't know for sure. I have no memories of the Higgs boson. It hasn't existed in the past. At some future date, perhaps by the time you are listening to this lecture, we will know about the Higgs boson.

(And now, we do know about the Higgs boson. Greetings from the future, or at least the future of when I was originally recording the lectures, although it's still the past of when you were watching them. I'm just popping in very briefly to point out that a month after those original lectures were recorded, I was in Geneva to witness the announcement at CERN, at the great particle physics laboratory, that the Higgs boson had finally been discovered. So time has moved forward; we've learned a little bit more about the fundamental nature of reality.)

The point is that the questions we can ask and the attitudes we take toward the world depend on what time it is and how much knowledge and experience we have accumulated in that amount of time. As interesting as it is to think about clocks and repetition, the truly important and fascinating aspect of time is that things change and that they don't change back.

We have a feeling because of this change that time passes. We use verbs to express the workings of time (we were here, we are here, we will be over there). So what does that mean? Why are there verbs? What does it mean to have a past, a present, and a future? There are two things, two important aspects of time, and you'll see that already, even at this very general level, we're really digging into the fundamental workings of this abstract notion that we call "time." The two aspects I want to highlight are 1) continuity—that is to say, from moment to moment, the world changes, but it only changes gradually. It does not completely rearrange itself from moment to moment. The universe happens over and over again; it's a succession of moments. In some ways it's like a movie reel; it's like a good old-fashioned non-digital filmstrip. There are frames that represent the different moments of time in the history of the universe, but those frames change only slightly from frame to frame, and that gives us the appearance of motion. The universe is like that except that the frames are an infinite number and there's no distance in between them. The universe is a continuous succession of moments that has gradual rearrangements, but not complete and utter rearrangements. If the universe through time were utterly different at every moment, then the universe would make no sense to us. We can be very thankful that continuity through time is a fundamental feature of how reality works.

The other feature is progress, or at least, change in a certain direction. Not only is this filmstrip of the universe a sequence of frames, it's an ordered sequence of frames. There is something called "the past" and something called "the future" and there's a difference between them. We have a feeling that one moment happens and then another moment and then another moment, and we all agree which moments come first and which moments go later. This is something we're going to have to think a lot about to reconcile our intuitive way of dealing with time with what the physicists have learned and what the mysteries really are. We talk all the time about moving through time. We either say that we move through time or time passes around us—we're a little bit ambiguous about that. When you dig right down into the details, it turns out that we don't move through time. That's sort of a metaphor that takes us a little bit of the way but doesn't get at what is really happening. We move through space. When you're talking about moving at a velocity—you're saying, I'm standing still, I'm moving in my car at 50 miles an hour, or whatever—what we're talking about is how our location in space changes as time passes. Time itself doesn't have a speed. We appear to experience time at different rates—there are psychological explanations for that—but time itself only moves at 1 second per second. Or, even better, it doesn't move at all; it just is. There is us in the past, there is us now, there will be us in the future.

Time happens, like it or not; we exist at every one of these moments. This is part of the mystery we would like to understand. Why do we have this feeling that we move through time if in a sense every moment exists independently of any other moment?

One clue is this division we have into past, present, and future. We treat the present moment obviously very special. That is when we are alive right now; that is when we are thinking and having these thoughts. But the past and future are completely different from each other. They're the two directions in which you could imagine extending your mental image of time, but we do so in very, very different ways. The past we think of as settled. The past happened; we can't change it; it is there. We might know it on the basis of evidence or we might have to reconstruct it somehow, but we don't think that we can affect it. Whereas the future, we think that we can make choices. The that puzzle, which we don't know the answer to right now, we will talk about relativity, cosmology, what we know about the Big Bang, and the possibility that the Big Bang was not only a unique event—the possibility that the Big Bang was one even among many and that our observable universe is imbedded in a much wider multiverse. Maybe the reason why our universe had such a low entropy at early times is because it is a side effect, it is an outgrowth of the much larger evolution of the multiverse. Or maybe not. This is a very, very speculative idea. It's an example of what we are pushed to when trying to understand the origin of the arrow of time.

As far as the mysteries of time are concerned, we're going to admit there are many mysteries, we're going to insist that many of them have answers that we know right now, but there are some—maybe the biggest of all—that we don't yet understand. That's why the future remains interesting.

My own interest in this subject actually started when I was in graduate school. I was learning the puzzles of modern cosmology, and in any scientific field there are certain questions that everyone is talking about, and other questions that you could ask but that aren't really that popular. So I read a paper by the very well-known physicist Roger Penrose, who did great work in gravity and black holes in the '60s and '70s, and Penrose was arguing that much of modern cosmology was missing the point, that the real puzzle that we should be trying to explain is why the early universe had low entropy. Furthermore, Penrose argued, none of the current ideas on the market did a good job in explaining that phenomenon. So I was very young, and I was sort of in the mainstream at the time. My own personal beliefs were that Penrose had done great work, but in this particular case he was missing the point. We had really good reasons to understand why the past was different from the future, blah, blah, blah.

It was only years later that I realized that Penrose had been right all along. In fact, the explanations that are typically given for why we think that the universe started with a low entropy really don't hold up to scrutiny. They always treat the past different from the future from the beginning. But the point is, the reason the past is different from the future is because the early universe had a low entropy. You can't use the fact that the past is different than the future to explain that feature of cosmology. Nowadays, I think this is the primary question that cosmologists are faced with, and others are beginning to catch on. It's becoming more and more popular for modern research cosmologists to think about the arrow of time.

What we're going to do to get there is go through a series of steps understanding how time works. We're going to talk about how to measure time. (What is a clock? What is a calendar?) We're going to talk about what entropy is. There is one definition and another definition—definitions at increasing levels of sophistication. We're going to talk about the concept of reversibility—running the clock backwards, making a movie and then playing it in the wrong direction. Then we're going to see what the connection is between the increase of entropy and all of the other arrows of time—the fact that we remember the past and not the future, that we can make choices now that affect the future but not the past. We'll be left with this puzzle: Why did the universe begin with such a low entropy? To address in this course is that that one fact about the difference between the past and the future, namely that entropy increases toward the future, underlies all of the different manifestations of the arrow of time. We talked about many of them. We talked about aging, we talked about evolution, we talked about cause and effect. We're going to claim that all of them can be traced to the increase of entropy at a fundamental level.

If that's true, we want to understand why entropy wants to go up. Well, we have a theory for that. This is one of the mysteries of time to which we know the answer. This was given to us by a set of tremendous advances in 19th-century physics under the rubric of thermodynamics and statistical mechanics. We invented the second law of thermodynamics which says that entropy always goes up. The first law of thermodynamics says that energy is conserved, so there's no arrow of time there. It's really the second law that gives us the arrow of time, that distinguishes the past from the future. When the giants of 19th-century thermodynamics figured out what entropy was and figured out that it tends to go up and figured out why it tends to go up, they realized there was one unanswered question, namely: We understand why the entropy tomorrow will be bigger than the entropy today; we don't understand why the entropy today is bigger than the entropy yesterday. If the fundamental laws of physics are completely symmetric, then given the universe today, if we can argue the entropy should be larger tomorrow, the same argument tells me it should have been larger yesterday, but that's not true. The entropy was lower in the past; that's another consequence of the second law. So in other words, we understand already why the second law says that entropy will increase toward the future. What we don't understand, what we don't truly have an explanation for, is why the entropy of the universe was lower in the past.

We will say that this mystery, even though it's not completely solved, can be traced back 13.7 billion years. The reason why the entropy was low in the past is because it started out low at the origin, at the beginning moment of our observable universe. That's the Big Bang, 13.7 billion years ago. To make the analogy perfectly clear, if the Earth is what gives us an arrow of space locally, it's the Big Bang that gives us an arrow of time in our observable universe.

the tendency of objects that are moving is to simply keep moving. They move in a way that if you played the movie backwards, it would look the same. In Aristotelian physics, if you made a movie of something and then played it backward it would look bizarre. But in Newtonian physics, you play a movie of an object moving in one direction, then play it backwards and it looks completely fine.

This is a deep feature of physics as we currently understand it. The fundamental laws of physics do not have an arrow of time. It's true for Newton, it's true for modern quantum mechanics or general relativity. If we study the object of individual small particles (atoms, elementary particles, small pieces of matter that are very, very simple), it is always true that they do something, we make a movie of it, we play it backwards, it looks fine. There is no fundamental difference at the level of microscopic particles between the past and the future.

Nevertheless, in our world, it's obvious there's a difference between the past and the future. Memory is one difference, but again, just mixing milk into coffee is another big difference. There's something involving macroscopic behavior that brings the arrow of time to life. When we make the transition from a small number of moving parts (like a pendulum or an atom) to a large number of moving parts (like all of the molecules in the cream and the coffee), somehow the arrow of time comes into play.

What's going on in one word can be summed up as entropy. Entropy is the scientific concept that we use to measure the disorderliness, the randomness, the disorganization of some collection of objects. The rule is that as time passes, that disorder increases. This is the second law of thermodynamics, which we will study in great detail. Entropy left to itself will always go up or remain constant. It will never spontaneously go down. The classic example is simply a messy room. If you have a room, your bedroom, and you make everything very, very neat, over time it will get messier. That is what happens. It is never true that you have a messy room and over time, without you cleaning it up, it spontaneously organizes itself. Systems in the real world left to their own devices only increase in disorder. That is a difference between the past and the future. It is a statement about what happens as time goes on. As we move toward the future, entropy goes up. What we will learn

All directions of space would be treated equally to us. Up, down, left, right, forward, backward—it wouldn't matter. We could do any operation we wanted to do—a physics experiment, read a book, whatever it is we want to do—we could be rotated to some different direction, do the same thing, get the same answer.

So even though space and time were purportedly united by Einstein, this arrow business puts space and time on a very different footing. The arrow of space to the extent that it exists at all seems very parochial and depends on our environment; the arrow of time seems to be universal. Everyone in the world is born young and then grows older. If we ever met an alien species we would think that they also remember the past and predict the future. It seems to be a deep feature of reality.

What we will learn is that, in fact, the origin of the arrow of time turns out to be very similar to the origin of the arrow of space. We have an arrow of space telling us the difference between up and down because the Earth is influencing how we think about space. We have an arrow of time telling us the difference between the past and the future because we live in the aftermath of the Big Bang. All of the differences between the past and the future are going to be ultimately traced to what happened at the origin of our universe. They're not traced to the laws of physics themselves. That's the great mystery that we need to confront. If you were Aristotle, for example, if you were an old-school person thinking about the world before modern physics came along, the arrow of time would not have been a puzzle that you were thinking about. It's obvious: There's an arrow of time past events come before future events. Causes precede effects, and so forth. The ways in which a Greek thinker like Aristotle would have thought about evolution through time would be totally embedded in the arrow. They would all have the feature that the past is different from the future. In Aristotelian physics, if there's an object and you want to keep it moving, you have to keep pushing it. If you push something for a little while and then let it go, it will come to a halt. That's how the world actually seems to work. But it says that the past is different from the future because if you find an object that is not moving, if you want ask what happened to it in the past, was it moving before or not, Aristotle has no way of telling you that. In contrast, when Galileo and Newton came along and laid the groundwork for modern physics, they realized that future we can change. By what we do right now, the future can be altered. That's a huge difference between the past and the future.

It's not the only difference. We remember the past. It is possible not only in our heads but in other forms of records—either in a computer record, or a photograph, or a fossil, or a historical document—to in some sense know something for sure about the past. We have artifacts that we think are reliable indicators of what transpired in the past. There are no artifacts in the world that are reliable indicators of what will transpire in the future. We can predict the future—sometimes we're very good at it, sometimes we're not so good—but predicting the future seems to us to be an utterly different process than remembering the past.

We age. We are born as young people, we grow older, and then we die. It's always in that direction, no matter what you might have thought by watching *Benjamin Button*. The direction in which we age from young to now to older is in the same direction for everybody.

This is the arrow of time—the fact that the past is different from the future. The arrow of time is simply the directionality from the past toward the future; our list of moments has an ordering on it. This arrow of time will be an absolutely essential preoccupation of us in this course.

Remember we said that Einstein taught us that time and space are actually quite similar; they're both part of spacetime. If the arrow of time is a very natural thing, you might ask, is there an arrow of space? The answer is yes and no. If you're sitting here on Earth, there is certainly an arrow of space in the sense that there is a difference between up and down. Everyone in the same room agrees on what the direction of down is. However, we all know that that's not a deep down feature of reality; it's sort of an accident of our environment. We happen to be standing on the Earth. If we were standing on the Earth in a different continent—we're comparing what happens in the United States to what happens in Australia—the so-called arrow of space would point in different directions. We would always be pointing toward the center of the Earth. If we were not standing on the Earth at all, if we were an astronaut out in space in our little spacesuit, zooming around, nowhere near a planet or anything like that, there wouldn't be any arrow of space.

What Is Time?
Lecture 2

Science and philosophy have a longstanding relationship, a sort of friendly rivalry. The two disciplines have different aims, but their subject matters often overlap, and the study of time is one area where the philosophical perspective is extremely helpful, even to physicists. Philosophers try to understand the logical inner workings of something, while physicists are often happy just to get a theory that works, not necessarily one that makes sense. We have an understanding of how time works in a physical way in certain well-defined circumstances, but philosophical questions remain. In this lecture, we'll look at some of those questions to help us understand the scientific aspects that we'll uncover in the rest of the course.

Time and Space in the Universe
- When we think of the universe, we generally think of space—not just outer space but the space around us, the location of things in the world. We also think of the universe as happening over and over again. Right from the start, we treat time and space differently.
  - Space seems to be somehow more important or relevant to what the universe is, whereas time is just a label that tells us which moment of the universe we're talking about.

  - In Lecture 1, we discussed the analogy of the universe as a movie reel. It's important to note that both each frame and the whole series of frames—the movie—define what we think of as the universe; it's a four-dimensional thing, with both space and time.

- Unlike space, the universe doesn't rearrange itself.
  - In space, what happens at one point seems more or less completely disconnected from what happens at another point. Space doesn't have any rules about what comes next to everything else.

- Time, on the other hand, has rules about what comes one moment after the other. That's how the laws of physics work: If you know everything that the world is doing at one moment in time, the laws of physics will tell you what happens next. And from that moment, the laws of physics will tell you what happens next, and so on.

- The laws of physics start from a moment—a state of the universe at one instance in time—and they tell you, using the equations that are the laws of physics, what happens at each subsequent moment.

The Difference between Time and Space

- Later in the course, we'll talk about relativity and the relationship between time and space, but for right now, let's examine the notion that time and space are completely different.

- We can choose to go to some other location in space, but we can't choose to go to some other location in time. Time is relentless, whereas how we move in space is up to us.

- This fact gives us a certain perspective on reality. We think of reality as one moment in time; however, we don't think of a distant location that may be inaccessible to us as not real.
  - Different locations in space are absolutely real, whether or not we're there, but what about the past and the future? Are they real?

  - We think of the universe right now as existing, but we think of the past as over with; we don't think of the past as real in the same way that the present is, and we certainly don't think of the future as just as real.

  - Why do we treat the past and future so differently? Why are we thinking of them in such a different way than we think of the different parts of space?

- To answer that question, let's consider how we go about describing space and time.
    - If you propose to meet a friend at a coffee shop at a certain time, what you're really doing is giving your friend coordinates in the universe—what a physicist would call an "event."

    - You need to specify space (where you're going to meet) and time (when you're going to meet). On Earth, to specify location, you need to give only two numbers—the street that you're on and the address on that street but in space, you would need to give three numbers, because space is three-dimensional.

    - Time is another dimension in the universe; we can marry the three dimensions of space to the one dimension of time to get four-dimensional spacetime.

Presentism and Eternalism

- Philosophers would call our everyday way of thinking about the world "presentism." This is the idea that what exists and what is real is the three-dimensional universe at some moment in time and everything in that universe. The past and the future are not real.

- But physics suggests a different point of view: If we know the universe exactly right now, we can predict what the future will be and can reconstruct what the past was. The laws of physics connect the present moment to the future moment and the past moment.

- From that perspective, we begin to think that the past, present, and future are perhaps all equally real. This point of view is called "eternalism."
    - As opposed to presentism—which says that the present is real, the past is a memory, and the future is a prediction—eternalism says that all the moments in the history of the universe are equally real. There's nothing special about the present moment except that you're experiencing it right now.

- o Eternalism is sometimes called the "block universe perspective" because it's like stepping outside the universe and seeing all four dimensions as one block of both space and time. Another term for it is the "view from nowhen," the view not from any one moment in time but outside the whole thing.

- The current laws of physics suggest that eternalism—this idea of treating the past, present, and future on an equal footing—is the correct view of the universe.

The Arrow of Time
- As we said, presentism views the present moment as real, but not the past or the future. A slight twist on presentism treats the present and the past as real, but not the future. The past is fixed—we might not be living there, but it happened—whereas the future is unsettled. This way of thinking seems natural to us as human beings, but it has no reflection in the laws of physics.

- A better way to understand the reason we treat the past and the future so differently is the arrow of time. It's not time itself that treats the past, present, and future differently; it's the arrow of time, which is ultimately dependent on the "stuff" in the universe, our macroscopic matter and the configurations that it is in.

- The arrow of time gives us the impression that time passes, that we progress through different moments. From that perspective, we understand that it's not that the past is more real than the future; it's that we know more about the past. We have different access to it than we have to the future.

Clocks
- One way of thinking about time is that it is what clocks measure. With a clock, we can say not only that time has passed but that a certain amount of time has passed. As we said earlier, a clock is a device that does the same thing over and over in a repeatable way.

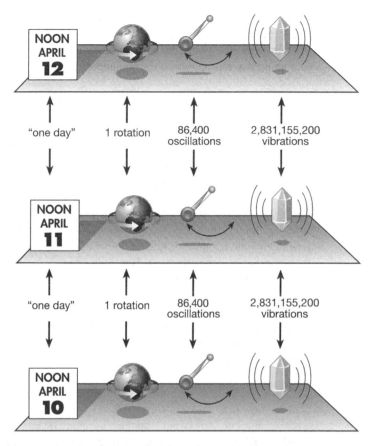

The "good clocks" in our universe include the rotation and revolution of the Earth, the rocking of a pendulum, and the vibration of a quartz crystal.

- Are we dealing in circular definitions here? We seem to be defining time as what clocks measure and defining clocks as devices that do the same thing over and over again as time passes.

- In fact, this is not a circular definition; there is some substance to the claim that time is what clocks measure. Further, the existence of things that do the same thing over and over again in a predictable way—like clocks—isn't something we can take for granted. We might have lived in a universe where everything that repeated itself did so unpredictably.

- An important feature of clocks is that there's more than one clock in our universe. Of course, "clocks" here means anything by which we can measure the passage of time.
    o The Earth rotates around its axis; it also revolves around the Sun. These are two different things that the Earth does, and it does them in a predictable way.

    o These two things are comparable to each other: Roughly speaking, the Earth rotates 365¼ times every time it revolves around the Sun. It's not a different number of days per year; it's the same number year after year. That's what makes the motion of the Earth give us reliable clocks.

- The rotation and revolution of the Earth make it an obvious choice for a good clock. These days, in wristwatches, the best clocks come in the form of quartz crystals, which can be made to vibrate at a precise rate. The motion of a pendulum is also a good clock.

- In a world in which there were no regularities, there would be no good clocks anywhere in the universe. Things would happen over and over again, but they would happen at unpredictable rates compared to each other. Time would still exist, but we could never say how much time had passed from one moment to another.

- Clocks give us an operational way of thinking about time; they refer to things that really happen, not just to abstract concepts.
    o What if time were to simply stop? Or what if time slowed down everywhere in the universe?

    o Actually, the answer is that it would mean absolutely nothing. If time stops everywhere for everything in the universe, there would be no way of knowing.

- What would happen if you could stop time for everything in the universe that was a distance of about 3 feet from you?
  - Suddenly, you wouldn't be able to see anything more than 3 feet away from you because there would be no light coming to you from anything in your time-stopping zone.
  - If you started to move through the air and the molecules 3 feet from you were absolutely stationary, they would be like a brick wall to you.
- The idea of stopping time or even that times moves at different rates for different people is a very slippery notion. When we talk about relativity, we'll see that there's a well-defined scientific sense in which different people can measure time moving at different speeds, but the only way they can do that is by being in different places in the universe or moving through the universe at different velocities.

Suggested Reading

Callender, *Introducing Time*.

Carroll, *From Eternity to Here*, chapter 1.

Klein, *Chronos*.

Questions to Consider

1. Do you think the past, present, and future are equally real? How would you try to convince someone who didn't agree with your viewpoint?

2. What processes around you would qualify as "good clocks"?

3. Can you imagine time speeding up, slowing down, or stopping altogether?

What Is Time?
Lecture 2—Transcript

As we ponder the mysteries of time, this is the one lecture in which we'll allow ourselves to get a little philosophical. By that I don't mean we'll just sort of sit back and ponder how awesome it all is and reflect upon the majesty of time and space. What I mean is we're going to think about the problem from a philosophical perspective, from the rigorous, academic discipline of philosophy.

There's, of course, a longstanding relationship between physics and philosophy, or between science and philosophy more generally; there's, let's say, a usually-friendly rivalry. The two disciplines are a little bit different—they have absolutely different aims—but their subject matters often overlap; and the study of time is one where the philosophical perspective is absolutely helpful, even to physicists like myself. What philosophers try to do is to really understand the logical inner-workings of something. Physicists like myself are often very happy just to get a theory that works; a theory that makes sense is a little bit too much to ask sometimes. You might think that this isn't possible, but we have it all the time: Quantum mechanics is a great example of a theory that doesn't completely make sense to us, but it works; so for a physicist that's fine, for a philosopher they'd like to do a little bit better. The same thing is true with time: We have an understanding of how time works in a physical way in certain well-defined circumstances, but there are still philosophical questions. In this lecture, we're going to sort of lay out what some of those questions are, and that will help us understand the scientific aspects that we'll uncover as we go through the rest of the lectures.

What do we mean when we say "time," especially as opposed to "space?" What do we mean when we say "the whole universe?" When we think of the universe, we generally think of space; we think of not just space as outer space—as planets, and stars, and galaxies—but the space around us; the location of things in the world. The world, the space that defines everything around us, and the stuff in the world, that's what we think of as the universe; but we think of that universe as happening over and over again, so we automatically, right from the start, treat time and space differently. We treat space as somehow more important or relevant to what the universe is, whereas time is just a label that tells us which moment of the universe we're talking about. Already in Lecture One we discussed this analogy of thinking about the universe as a movie reel or filmstrip, a series of frames, a series of moments; and what I'm trying to say here is that both each frame and the whole series of frames—the whole movie reel—define what we think of as the universe; it's this four-dimensional thing, both space and time. I have to caution, when we use that analogy it's very tempting to wonder whether or not, like a filmstrip, moments of time are discrete. As far as we know, nothing in physics tells us that time is discrete; as far as we can tell, time is perfectly smooth and continuous. People sometimes wonder whether or not a future reconciliation between quantum mechanics (the theory that happens at the very small scale) and relativity (Einstein's theory of gravity), maybe someday that will tell us that time really is discrete. After all, "quantum mechanics" comes from the word "quantum" meaning "coming in discrete packets," so maybe when you quantize space and time itself they will become indiscrete packets. But as far as we know, that's not true; our best understanding right now of time and space has time being absolutely smooth and continuous. That's not to say that we'll not get a better understanding of it in the future; but right now I'm not trying to use the filmstrip analogy to say that time comes in discrete packets, just that the universe is both each frame of the movie and the series of all the frames together.

Already the way that the filmstrip of the universe works involves a fairly subtle notion: the continuity of the universe from moment to moment in time. We said before the universe doesn't rearrange itself, unlike space; so let's sort of interrogate that concept a little bit more closely. In space, what happens at this point right here seems more or less completely disconnected from what's happening at this point right here. It's true that in the real world things don't completely change from point to point in space—here's the air in the room, here's the air in the room also—but I can put my hands very close to the ground and then you'd have air here and the ground right next to it; there's nothing to stop me in principle from having absolutely empty space at this point and some incredibly dense, or incredibly hot, or incredibly interesting object right next to it. Space, the stuff around us and the stuff inside space, doesn't have any rules about what comes next to everything else.

Time, on the other hand, has rules about what comes one moment after the other. That's how the laws of physics work; that's how we think of the laws of physics in the modern world. This isn't at all the only possible way the laws of physics could work; this is a really interesting, profound feature of the real-world laws of physics, namely that the way they work is the following: You tell me what the world is doing right now, or at some other moment in time, but you tell me everything in the world at one moment in time and the laws of physics will tell you what happens next; and then from that next moment, the laws of physics will tell you what happens the moment after that. The laws of physics start from a moment—start from a state of the universe at one instance in time—and they tell you, using the patterns, using the physics, the equations that are the laws of physics, what happens at each subsequent moment. That's completely different than space (obviously): In space, the laws of physics aren't able to tell you just because something is happening in this place what will happen in the place next to it. Even when it comes to time, it could've been different. Remember, Aristotle discovered laws of physics that worked in terms of the future: Things have goals; and to understand what will happen next, you need to contemplate what the future conditions could be. We could contemplate different laws of physics that involved the past in some important way; that knowing the universe right now wasn't enough, we needed to know how it got there. But as a matter of fact, the laws of physics that we use—whether they're from Galileo, Newton, Einstein, Schrödinger, or superstring theory that we talk about today—they all have this feature that what happens at one moment of time is enough to predict what will happen at the next moment of time, and after that you can just keep predicting all the way toward the future.

To really get our brains wrapped around this, think about the difference between time and space. Later on we'll talk about relativity and how they're related to each other, but let's for right now just accept the ordinary notion that time and space are completely different. We move through space as we like; we can choose to go to some other location in space. We can't choose to go to some other location in time; we inevitably move through time at the rate of one second per second. Time is relentless, whereas space is sort of up to us how to move in it. That gives us a certain perspective on what the world is. We think of reality as one moment in time; we think that location far away—20 feet away, or 20 miles away, or 20 million miles away—

might be inaccessible to us right now, but that doesn't mean it's not real. We think of different locations in space as absolutely real whether or not we're there; but now think about time, think about the past and the future: Are they real? This is a question where as soon as you start thinking about it you begin to suspect that the answer is just this is not a good question to ask. Nevertheless, it's important to understand why we think of different times differently than different spaces. We think of the universe right now as existing, but we think of the past as over with; we don't think of the past as real in the same way as the present is, and we certainly don't think of the future as just as real. The important thing is not to figure out what's real and what isn't real, the important thing is to wonder: Why do we treat the past and future so differently? Why are we thinking of them in such a different way as we think of the different parts of space?

To answer that question, we should think about how we use these concepts; how we actually go about describing space and time. Let's imagine you want to meet someone for coffee. You say, "Let's be at a certain café at 6:00 pm." What you're really doing is giving them coordinates in the universe; what a physicist would call an "event." You need to specify space (where you're going to meet) and also time (when you're going to meet them). Fortunately, we live in a world where we're basically stuck to the surface of the Earth, so giving someone a coordinate in space is giving them an address, or two numbers, let's say, what street you're on and which number you are on that street. But if you were to meet them anywhere in space, you'd need to give three numbers. Space is three-dimensional; you have length, breadth, and height. Time is another dimension on the universe; we can marry together the three dimensions of space to the one dimension of time to make something called a four-dimensional spacetime. That spacetime is the collection of all the different elements in the universe.

You might think that the notion of spacetime is something that Einstein gave us; but it's just as important to think that Newton, who preceded Einstein, could have talked about spacetime. He had space, he had time; Newton knew as well as Einstein did that to locate someone, to meet them somewhere in the universe, you needed to give four numbers: three locations in space and one number for what time you'd meet them. The difference is that to Newton space and time were just so different that there was no temptation to marry them together. You could've done it, it wouldn't have been against the rules, but there was no benefit to doing it either. Einstein, as we'll see, showed that what counts as space versus what counts as time will be different to different people, so you need to think of them as spacetime. Once you start thinking about the universe in that way, you hit upon a different possible way of thinking about the world. Our everyday way of thinking about the world, the philosophers would call "presentism." Presentism is the idea that what exists and what's real is the three-dimensional universe at some moment in time and everything in that universe. The past and the future, those are not real; the present moment is real, the past is memory, the future is prediction. But physics suggests a different point of view: Physics says that if I knew the universe exactly right now, I could predict what the future would be and I could also reconstruct what the past was. The laws of physics connect the present moment to the future moment, and to the past moment.

From that perspective, we begin to think that the past, the present, and the future maybe are all equally real. This is a point of view called "eternalism." As opposed to presentism, which says the present is real and the past is a memory and the future is a prediction, eternalism says all of the moments in the history of the universe are equally real. There's nothing special about the present moment except that you're experiencing it right now. You might think that this is a difficult way of thinking about the world, and it's difficult because it's not our usual way of going through life. Sometimes the point of view of eternalism is called the "block universe perspective," because you try to imagine stepping outside of the universe and seeing the whole four-dimension thing as one block of both space and time. This is sometimes called the "view from nowhen"; the view from not any one moment in time, but outside the whole shebang. Kurt Vonnegut once wrote a novel called *The Sirens of Titan* where he had characters, aliens, called *tral fam adorians*, and the *tral fam adorians* had a special feature that they lived outside time; they experienced the block universe just as you and I would experience three-dimensional space. They could choose to go to the year 1600 or go to the year 2500 after breakfast that day.

When you think about it carefully—which I know is not the job of a satirical science fiction novel—but if you were to think about this very carefully, you'd realize that *tral fam adorians* don't quite make sense. We can talk about the four-dimensional universe, but if you imagine an alien species that could decide to go and visit the year 1600, then clearly they have a personal notion of time. They weren't visiting the year 1600 yesterday; they will be visiting it tomorrow. Really, you're just adding a whole new dimension of time to the universe. It's something you can imagine doing, but it's not the universe in which we live. For us, the task is to take ourselves outside our everyday experience and imagine looking at this four-dimensional block universe; to treat the past, present, and future on an equal footing. This seems to be the right way of looking at things as suggested by the current laws of physics. I should be honest in saying that it's a controversial point of view; many philosophical stances are controversial. There are certainly philosophers out there who don't believe in eternalism. To me, it seems to be the logical consequence of the physicists' way of looking at the universe, but it should be kept in mind that there are other ways to do it.

What does it mean if it's true that the past, present, and future are equally real? We certainly think of them very differently; we treat the future very differently from the past. Another way of thinking about our sort of folk, traditional, everyday way of thinking about the universe is the growing block universe model. I said that presentism treats the present moment as real, the past and future are not real; so a slight twist on this idea is to treat the present and the past as real but the future not yet. This fits in very well with our notion that we get to make choices about the future. We don't get to make a choice about what's to have done yesterday, but we get to make a choice about what to do tomorrow. The past is fixed, it's in the books, it's real—we might not be living there, but it's happened; it's settled—whereas the future is up for grabs, so maybe we should treat them very, very differently. This seems natural to us as human beings, but it has no reflection in the ultimate laws of physics; it's not something that we get any warrant for from our best understanding of the universe.

The better way to say it, I believe, is to understand that the reason why we treat the past and the future so differently is because of the arrow of time. In other words, it's not time itself that treats the past and present and future differently, it's the arrow of time, which is ultimately—as we'll see—dependent on the stuff in the universe; on our macroscopic matter and the configurations that it's in; in the entropy of the stuff in the universe that started out low and is growing and will continue to grow toward the future. It's the arrow of time that gives us the impression that time passes, that time flows, that we progress through different moments; so from that perspective, we understand that it's not that the past is more real than the future, it's that we know more about the past. Our memories access moments in the past; and when I say "memories," I don't just mean the actual memories in our brains, I mean any record, any fossil, anything that we can look at that we believe gives us reliable knowledge about the past. We live in a world where memories exist, where there are history books. There are no history books written about the future; that's the difference between them, which can ultimately be traced to the arrow of time. It's not that the past and the future don't exist; it's that we have different access to them. We were in the past, and the memories of that past are still reflected in the present day; but that's ultimately because of entropy growing. We'll discuss that in great detail; but it bears on this question of what's real and what's not real. The answer is that all moments of time seem to be real, but some we understand better than others. That's something we're going to be developing when we get to the reflection of the arrow of time and how we think and about how our brain processes information.

All this is a little bit philosophical, as we promised, so let's get more concrete about how it plays out in the everyday world. We mentioned that one way of thinking about time is that time is what clocks measure. We live in a world full of clocks, and an important thing about clocks is that time doesn't simply pass—it's not just that there's earlier, and now, and later—but that we can measure it; we can say not only has time passed, but a certain amount of time has passed. One second is shorter than one minute, which is shorter than one day. There's an amount of time that we can quantify using clocks. So a clock is something that does the same thing over and over again in a repeatable way. Then you might ask, isn't this whole thing circular? Isn't it that we're defining time as what clocks measure, and we define clocks as things that do the same thing over and over again as time passes? I want to take the opportunity to say that it's not a circular definition; there really is some substance to the claim that time is what clocks measure. A clock is something that does the same thing over and over again, but the existence of things that do the same thing over and over again in a predictable way isn't taken for granted. We might've lived in a universe where everything that repeated itself did so unpredictably. The important feature of clocks is that there's more than one clock in our universe; we're lucky enough to live in a universe that's full of clocks.

I mean, of course, by "clocks" as something by which we can measure the passage of time. The earliest clocks weren't on our wrists or on our computers; they were in the sky. The Earth rotates around its axis; it also revolves around the Sun. These are two different things that the Earth does, and it does them in a predictable way. They're comparable to each other: Roughly speaking, the Earth rotates 365 ¼ times every time it revolves around the Sun. The nice thing about that is that it's predictable; it's not a different number of days per every year, it's the same number year after year after year. That's what makes the motion of the Earth give us reliable clocks. It's not the only one, of course: We have the Moon, we have the other planets, we have the stars in the sky; the sky gives us repetitive phenomena that we can use to measure the passage of time.

You might ask, what are the other good clocks in our universe other than those that we get in the sky? It's certainly not our memories. We experience the passing of time, but we all know that the experience isn't very reliable. Sometimes we think a lot of time has passed; other times, we're not so sure, we think that very little time has passed. What we're looking for in a good clock is something that can be relied on; that does the same thing over and over again. Obviously, the rotation and revolution of the Earth is an obvious choice. Another one is the rocking of a pendulum back and forth; that's why we put pendulums in our grandfather clocks. These days, on our wristwatches, the best clocks will be in the form of quartz crystals. You can actually make a certain quartz crystal in the shape of a tuning fork that's guaranteed to vibrate back and forth exactly 32,768 times per second. The reason why that number is chosen is because it's 2^{15}; so you can take 32,768 and divide it by 2, 15 times. You can divide the time periods up as finely as you want to. Quartz is both reliable and predictable. It won't matter what the humidity is, what the temperature is; that quartz crystal will tell you the time to very high accuracy.

This search for reliable clocks isn't always an easy one. Let's go back to the idea of the pendulum rocking back and forth. It turns out to be true that the rocking of the pendulum is a pretty good clock, but it's certainly not obvious. You might think that if the pendulum was just moving a little bit, maybe it goes faster than if it's moving a lot. This was actually shown not to be the case all the way back in 1583 by none other than Galileo. He was a young boy, he was going to church—like all young boys in Pisa in Italy did in that period of time—but occasionally he'd get bored sitting around in church, and because he was Galileo he did little mental science experiments. Galileo noticed there was a chandelier hanging from the ceiling of the cathedral in Pisa—which is still there, you can visit it today—and the chandelier rocked back and forth. Sometimes it was only rocking a little bit, sometimes some air had disturbed it and it was rocking a little bit more, and Galileo got the idea in his head that it seemed that the amount of time it took the chandelier to go back and forth was the same. No matter what the amplitude of the rocking was, the time was approximately the same. The question was, how would he check this idea? He's an empiricist; he wanted to do an experiment to figure out if this idea was on the right track; and again, because he was Galileo, he realized that he could compare the rocking of the chandelier to the beating of his heart. Galileo calculated that he could count his pulse while he was watching the chandelier go back and forth, and what he realized was that his conjecture was correct: that no matter what the amount of rocking was, the frequency—the time it took the pendulum to go back and forth; the chandelier hanging from the ceiling—it was going to be the same. This is the basis we have for pendulum clocks even today. It's not the absolute most reliable way of telling time we have, but for the 16th century it was a very, very good step forward. We still use that way of timekeeping today.

You can imagine, like we said, a different world: a world in which there were no regularities; that there were no such good clocks anywhere in the universe. There were things that happened over and over again, but they happened at unpredictable rates compared to each other. This would be a very strange world to live in. it would be completely crazy from our point of view; it's a world in which you couldn't measure the passage of time, there would be no reliable clocks. There'd still be time; there could still be the idea that there was a moment before some other moment; but you could never say how much time had passed from one moment to another. In a way, we're very fortunate—at least scientists are very fortunate—that we live in a world where time can be measured; a world in which clocks exist, and we can build them, and they will tell us how much time has passed.

The reasons why clocks are so important when we think about time is because they give us an operational way of thinking about time. You can easily get yourself mixed up thinking about the fundamental nature of time, and clocks help us because they refer to things that really happen, not just to abstract concepts. Questions like "Are the past and the future equally real?" may or may not be interesting questions, but you can ask, "What would a clock do?" That's an interesting question; that gives you an answer one way or another. Consider, for example, a favorite thought experiment: What if time were to simply stop? What if time ceased passing everywhere in the universe? Or, alternatively but just as well, what if time slowed down everywhere in the universe; what would that mean? I'll give you time to think about it: The answer is, it would mean absolutely nothing. If time stops everywhere for everything in the universe, there would be no way of knowing. You would stop, the processes in your brain would stop, your pulse would stop, your breathing would stop; so you wouldn't experience any time at all, because time has stopped. But at the same time, all of the clocks would stop, the Earth would stop rotating, the pendulum would stop rocking back and forth. So when you've compared the amount of time felt by any one object in the universe to the amount of time experienced by any other object in the universe, time stopping everywhere leaves no trace whatsoever. While I'm lecturing to you right now, you might imagine that time has stopped and you'd never know; but then you realize that if time has stopped, you can't even say, "Time has stopped for one century." That doesn't mean anything because time has stopped. When you get lost in these questions about the meaning of time—what's real and what's not real—it's always helpful to go back to "What do clocks do?"

There's a story that illustrates this very well. There's a novel written by Nicholson Baker called *The Fermata*. It's about a man who can stop time; he just has a magical ability, it's not explained, it's not hard science fiction or anything like that. Here's just a person who can decide to stop time whenever he wants to. But obviously he's not stopping time everywhere in the universe; he himself keeps time going. If you read the novel—which a good read; it's interesting and thought-provoking—the man himself isn't an admirable character, he mostly uses his ability to sort of make people look foolish in different ways; he's not saving the world. But we as scientists can ask—or philosophers—what would it mean for someone to be able to stop time? How would that really work? Forgetting about could you do it, forgetting about the technology problem of how you'd build a time-stopping machine, what would actually happen if you could stop time for the rest of the universe while you yourself kept going? You'd quickly realize that even if you grant the possibility of such a thing, it doesn't actually work like it says it would work in the novel. For one thing, if you stopped time for the rest of the world—let's say everything more than three feet away from you time stopped—suddenly you can't see anything more than three feet away from you; there's no light coming to you from any of those things that are out of your time-stopping zone because time has stopped. You can't breathe anymore if the air in the room has stopped moving. If you started to move through the air and these other air molecules were absolutely stationary, they'd be a brick wall to you; you couldn't even walk in such a room. When you actually sit down to think about what it means to stop time, or even for time to move at different rates for different people, it's actually a very, very slippery notion.

We'll later see when we talk about relativity that there's a well-defined scientific sense in which different people can measure time moving at different speeds; but the only way they can do that is by being in different places in the universe, or moving through the universe at different rates of good old-fashioned velocity. If you're stuck near a black hole, or if you're in a spaceship moving very, very close to the speed of light, then the amount of time that passes on your wristwatch will be different than the amount of time experienced by someone who stayed back at home. But if you're looking at your wristwatch, it'll seem to move at the absolute conventional rate; because you, and your watch, and any clocks you bring along with you all are traveling together, you all experience the same passage of time.

I think that in some ways in the modern world time has a bad reputation. We feel that we're slaves to the passage of time; that there's a clock, there's a calendar, we have to get things done, there are deadlines, and they get in our way. But if you think about it, all of these aspects of time are social. The reason why time presses on us is because we're trying to coordinate with the outside world. We have a clock, someone else has a clock, we agree to meet for coffee at the same place at the same time. Time is only pressing us when we want to match our actions in the world with the actions of somebody else. I'd argue that it's not fundamentally a bad thing, we should not blame time for this; we should be grateful we live in a world that lets us coordinate and be social creatures.

Keeping Time
Lecture 3

In our last lecture, we got a little philosophical, but in this lecture, we want to be practical. If time passes and we experience the passage of time, the obvious practical question is: How do we measure the passage of time? Clearly, this is a question about building clocks. But as we become more technologically advanced—as society moves through the arrow of time—the demands for measuring time precisely become greater. We want to know: What are the best clocks? To answer that question, we'll look at how the notion of keeping time has developed through history.

Units and Measurement of Time
- A solar day is the amount of time it takes the Sun to go from one position, such as noon, to the same position the next day. A solar year is the amount of time it takes for the Earth to go around the Sun. The lunar month is the amount of time it takes the Moon to go from one phase—new moon or full moon—back to the same phase again.

- Our task is to build a device to measure those time periods—the day, year, and month—as precisely as possible.
    o The simplest such device would be a sundial. We could put a stick in the ground and watch the progress of the shadow of the stick over time as the Sun moves through the sky.

    o Even this, however, is not as simple as it seems. If we put the stick exactly vertically, the time of day it measures would be different in the spring, summer, fall, and winter. The key is to align the stick with the axis of the Earth's rotation.

    o The sundial represents a small version of a bigger problem we have: What is the best way to tell time that is not subject to the vagaries of astronomical whims?

Stonehenge performs the most basic function of an astronomical observatory; it tells us the time of year.

- We can tell what time of year it is using an observatory. We know—and ancient people knew, as well—that the Sun is more northerly in summer and more southerly in winter. Stonehenge is a classic example of a simple astronomical observatory. The stones at Stonehenge are aligned so that at sunrise on the summer solstice—the longest day of the year—the Sun bisects the design.

Building a Calendar

- Ancient people would have found it useful to know when summer begins, but more generally, we would like to build a calendar, a system of units that lets us tell, in a uniform way, where we are in the year. That means taking a long period of time, such as a year or a month, and dividing it into shorter periods of time.

- A year, a lunar month, and a day are all specific time periods that we can compare to each other, but the problem is that they don't match up in easy ways.

- o A year is 365¼ days, but even that is not precise. At a more accurate level, a year is 365.2422 days, a very awkward number.

- o Even worse is the lunar month. It is 29.5306 days, another completely awkward number, and a year is 12.3683 lunar months.

- o The universe has given us no relationship between a month, a year, and a day. For thousands of years, humans have struggled to find clever ways to reconcile these astronomical timekeepers.

- This struggle resulted in different notions of a calendar in different cultures. The Islamic calendar, for example, based on a lunar month (29.5 days), alternates months of 29 days and 30 days. The problem is that when we add all the lunar months, we get 354 days, not quite a year.

- The desire to reconcile months with years resulted in the clever notion of the leap month, which can be traced back to ancient Babylon.
  - o The Babylonians hit on the curious fact that 19 solar years is very close to 235 lunar months.

  - o They further realized that if they let 12 months per year be the rule for 12 years in a row, and then 13 months per year be true for the next 7 years, they would have a 19-year cycle that synchronized the lunar calendar with the solar calendar.

The Gregorian Calendar

- The Gregorian calendar that we use today has given up on the notion of the lunar month, which is why some months have 30 days; some, 31; and February, 28 or 29.

- The idea of the months having different numbers of days unsynchronized with the Moon comes from the Julian calendar. The Julian calendar basically had the same months that we have except it had a leap year every 4 years. This seems close to our current system, but if we had a leap year every 4 years, we would go out of synchrony with the Sun by about 11 minutes per year.

- The system embedded in the Gregorian calendar is as follows: Every 4 years, we have a leap year, but then we skip leap year every 100 years.
  - The last time we had a year divisible by 100 was the year 2000, and the fine print in the Gregorian calendar says that we do have a leap year every 400 years.
  - Thus, 1700, 1800, and 1900 were not leap years; 2000 was; 2100, 2200, and 2300 will not be leap years; 2400 will be; and so on.

- The Gregorian calendar was developed and implemented in 1582, based on the realization that the Julian calendar was getting out of sync with the solar cycle.

- The Julian calendar was wrong by 11 minutes every year, but the Gregorian calendar is accurate to 2.6 seconds per year. In the year 4000, we might need another leap day to realign the Gregorian calendar with the Sun.

The Leap Second

- The leap second is completely different conceptually than the leap month or the leap year. The leap day and the leap month exist to fix the fact that one cycle—the rotation of the Earth or the orbit of the Moon—gets out of sync with another cycle—the Earth going around the Sun. Leap seconds exist to fix the fact that the Earth does not rotate at a completely constant rate.

- As we've said, the whole point of an accurate clock is to get something that repeats itself over and over again in a predictable way, but now that we have much more accurate clocks, we realize that the rotation of the Earth is not perfect, and it is not perfect in an unpredictable way. Things like earthquakes and hurricanes change the rate at which the Earth rotates.

- Rather than fixing that unpredictability once and for all, we have to wait for the Earth to get out of sync and then add a leap second to the calendar. You do not even notice it, but occasionally, New Year's Eve is 1 second longer than it was the day before. This is a controversial issue because it presents problems for computers.

More Accurate Timekeeping

- As technology evolves, it is not good enough to look at the sky to measure what time it is. In crossing the ocean, for example, it's necessary to know the time to know where you are. You can figure out your longitude by looking at the stars but only if you know what time it is.

- This puzzle—how do you know what time it is when you are on a boat far away from any land-based clocks—was so important that in the 18th century, the British Parliament offered a prize of £20,000 for figuring out how to measure time accurately, even at sea. The motion of the boat meant that a pendulum couldn't be used, and changes in temperature and humidity meant that metal springs couldn't be used.

- The problem was ultimately solved in 1761 by a man named John Harrison, who invented the marine chronometer.

Personal Timepieces

- These days, of course, we live in a world where keeping time is something that everyone wants to do, whether or not you are a navigator on a ship.

- In the 19th century, what that meant was a pocket watch, but the early pocket watches were maybe accurate to 15 minutes a day.

- In World War I, military personnel started wearing wristwatches to keep their hands free to carry weapons, and the fad eventually spread to people outside the military. Building mechanical wristwatches became a major industry, especially in Switzerland.

- In the 1960s, Japanese manufacturers realized that they could base a wristwatch on a vibrating quartz crystal. A watch based on quartz is much more accurate than any mechanical timepiece and much cheaper.

Atomic Clocks
- Today, we use atomic clocks, not quartz crystals, for maximum accuracy. A quartz crystal can be very accurate, but we want increasingly more accurate clocks because we are putting more and more demands on them.

- We can increase accuracy by looking at the transitions of electrons in individual atoms. In practice, we use cesium or rubidium, specific atomic elements that give up very precisely defined frequencies.

- The best atomic clocks are more accurate than 1 billionth of a second per day. Such accuracy is needed for a global positioning system (GPS).
  - By comparing how much time it takes for the signals from at least four satellites to reach your GPS, the system can determine your exact location on Earth.

  - A GPS can tell you exactly where you are to within a few feet, but that depends on the fact that the clocks onboard the satellites are incredibly accurate.

  - If the satellite clock were off by 1 microsecond, your location would be off by 1000 feet. Given that the speed of light is so fast, to triangulate correctly, it's necessary to know exactly when that signal left the satellite. That's what a good atomic clock can do.

Clocks and the Arrow of Time
- As we've said a good, accurate timepiece always does exactly the same thing, which means that it doesn't have an arrow of time in the sense that many things in the universe do.

- If we took the hands off a clock and made a movie of it, it would be impossible to know whether the movie was playing backward or forward.

- The arrow of time is a feature of things that do not always do exactly the same thing. We can turn an egg into scrambled eggs, but that process goes only in one direction.

- That is the difference between pristine accurate timepieces and the universe. Understanding where that difference comes from will be our goal for the rest of the course.

Suggested Reading

Barnett, *Time's Pendulum*.

Falk, *In Search of Time*.

Questions to Consider

1. If you didn't have any clocks or calendars, how would you personally keep time? Would you be able to fashion (or at least design) a sundial or astronomical calendar?

2. Do you wear a watch? Have your watch-wearing habits changed over the years? What kind of clocks do you usually use to tell the time?

3. Do you usually know what time it is?

Keeping Time
Lecture 3—Transcript

In our last lecture, we got a little philosophical. We were thinking about the reality, for example, of the past versus the present versus the future. The goal there was to set up things that we will be talking about in future lectures where we talk about why the past, present, and future are very different.

In this lecture, we want to do the opposite thing. We want to get down and dirty. We want to be as practical, as utilitarian as we can be. If time passes and we experience the passage of time, the obvious practical question is, how do we measure the passage of time? How much time is it that has passed?

Clearly, this is a question about building clocks. Time is what clocks measure, and we defined what clocks are. We said that clocks are things that do the same thing over and over again in a predictable way. We are lucky enough, fortunate enough, to live in a world that is full of clocks, but some clocks are better than others.

Once we get more and more technology advanced, as society moves through the arrow of time, the demands on measuring time very precisely become stronger and stronger. We want to know, what are the best clocks? What is the best way to build accurate timekeepers?

The best way to answer that question is actually historically, to see how we have developed the notion of keeping time through the ages. The first thing you might want to do is say what day is it. To give a calendar rather than a wristwatch or something like that, ancient timekeeping came down to what was happening in the sky. Where was the Sun? Where was the moon? Where were the stars?

The problem here is that, already, you are faced with choices. Are we measuring the days? Are we measuring the years? Are we measuring the months? Remember, the month originally was the time that the moon took to go from one new moon to the next moon. That is why the word month sounds very similar to moon.

What we are doing here is choosing units for measuring time. Is it days? Is it years or month? You might not think that choosing units is a very important thing. It is trivial, you do it once and for all.

When I was in college, one of my professors, to stir up trouble, once asked the class, what is the one opinion you have that you think your fellow students will find very controversial. I could have brought up a lot of things that I would have been able to say to that my fellow students would not have agreed with, but I decided to play it safe. I chose the fact that I thought we should switch to the metric system. You would have thought that I said we should all be chewing on babies for breakfast or something that. The entire class was outraged that I would think that the metric system was a better system than miles and pounds and inches.

Units matter. We have an emotional attachment to the units we use to measure time and space. When it comes to time, when it comes to calendars, the Sun gives us a day, the amount of time it takes the Sun to go from one position, like noon, to the next position, noon the next day is a solar day.

The solar year is the amount of time it takes for the Earth to go around the Sun. From our position here on Earth, the way that we see that is that the Sun, at noon, is at its highest point every day, but from day to day, the amount of height the Sun attains is different. It goes up and down, and its highest point, as the year passes, and allows us to measure the solar year.

Finally, we have the lunar month, the amount of time it takes the moon to go from one phase, new moon or full moon, back to the same phase again. It is about what we currently call a month.

What we want to do is build a device to take those time periods—the day, the year, the month—and to measure them as precisely as possible. If you want to know what time of day it is, as long as it is daylight outside, you can look up into the sky. You can see how high the Sun is.

You should not be looking at the Sun too much, and in any case, that is a very crude measurement. We want to build something that will tell us what time it is more carefully, more accurately, and more precisely. The simplest thing to do, if you do not know how to build machines, is to build a sundial. Put a stick in the ground and watch the progress of the shadow of the stick over time as the Sun moves through the sky.

Even that is not that simple. It requires a little bit of technological know-how to figure out exactly how tilted that stick should be. You can always put a stick exactly vertically, but then what you would realize is that if that stick stayed there all year long, the time of day that it was measuring would be different in springtime, in summer, in fall, and winter.

Eventually they realized that if you aligned the stick with the axis of the Earth's rotation, that is to say if you were at the North Pole, the stick would be vertical. If you were at the equator, the stick would be almost horizontal, and in between, it will be at some angle. If you did that, the Sundial would tell accurate time. It would tell you the same hour of the day whether you were in winter, spring, summer, or fall.

That is a little nice version of this bigger problem we have. What is the best way of telling time in a way that is not subject to the vagaries of astronomical whims? We want to be able to tell time in such a way that it is the same every time we want to tell it.

Sundials, of course, have other problems. They don't work at night. They do not even work on cloudy days, and even when they work, you cannot move them from place to place. Certainly, you want to do better than a sundial when it comes to telling time.

Something you might think is easier is telling what time of year it is, what is the day of the year? You can do that using an observatory. We know that the Sun is more northerly in summer and more southerly in winter, and we see that ancient people understood this very well. Stonehenge is a classic example of what we think of as a simple astronomical observatory.

These days, when we think of observatories, we think of giant telescopes and peering far out into the heavens, but before the telescope was invented, there were still observatories. What they were measuring was not trying to get very close up pictures of astronomical phenomena. What they were trying to do was to measure the positions of different astronomical phenomena.

The simplest thing you can do is measure the position of the Sun as it rises, and that moves as the year changes. Stonehenge is a collection of many different stones, and to be honest, we don't know what all the stones are for.

This is a British monument, an archeological find, that has been around since at least 2300 B.C. We don't have written records of what it was for. All we can do is look at where the stones are and try to figure out what good they might be.

It is easy to check that the stones in Stonehenge are aligned so that at sunrise, on the summer solstice, at the longest day of the year, the Sun bisects the design. That is to say by looking in a certain direction through the keystones at Stonehenge, you can tell what day the longest day of the year is. You can pinpoint the summer solstice. That is the most basic thing you might want an astronomical observatory to do. It lets you tell what time of the year it is. You cannot be more accurate than that.

Telling whether it is the beginning of summer or the end is something that was very crucial for ancient people, so it was a very useful thing to know. More generally, we would like to build a calendar. We would like to build a system of units that lets us tell, in a uniform way, where in the year we are.

What that means is taking a long period of time, like a year or a month, and dividing it into shorter periods of time. A year we might want to divide into months; a month, we might want to divide into days. Sometimes this is very easy. We say a week is seven days, but a week is a completely made up unit of time. We define it as seven days.

A year, a lunar month, and a day are all very specific time periods that we can compare to each other. The problem is they it is match up in easy ways. It is not as if a week is made of seven days.

A year is made of 365¼ days, and even that is not precise. At a more accurate level, a year turns out to be 365.2422 days. That is a very awkward number, and that is eventually going to give rise to the notion of a leap year. Even worse is the lunar month. It is 29.5306 days, another completely awkward number, and finally a year is 12.3683 lunar months.

The point is that all of these are a mess. There is no relationship that the universe has given us between a month, a year, and a day. For literally thousands of years, human beings have struggled to come up with clever ways to reconcile these different astronomical timekeepers.

That is what leads different cultures to have different notions of a calendar. For example, the lunar month was very, very important to a lot of ancient people. A lunar month is 29.5 days, so it is close to 29½ days. The simplest thing you might imagine doing is having a month be 29 days half the time and 30 days the other half of the time. You would have one month 29, the next month 30, the next month 29, and keep that cycle going on forever.

In fact, some calendars do this. The Islamic calendar, still used today, alternates months of 29 days and 30 days. The problem is when you add all lunar months, you do not quite get to a year; 29½ times 12 is 354 days in what we call the lunar year.

The 354 days is something like 11 days short of the actual year, so the Islamic calendar slides with respect to the true solar year every year. It is nice for the moon. You know exactly what the phase of the moon is if you tell time using the Islamic calendar.

You do not know what the season of the year that you are in is. You talk about a certain month 1,000 years ago, you have no idea what time of the year that corresponds to if you use a lunar calendar.

You might say you do not care. You might say they are for purposes of planting or whatever. Knowing how bright is going to be outside, whether the moon is in the sky is more important to you. As time goes on and society becomes more advanced, the importance of knowing where the moon is becomes less and less until most of us today who live in cities could not even tell you whether it was full moon or new moon right now.

Nevertheless, the desire to reconcile months with years is very strong, and so people came up with a clever notion called the leap month. This goes all the way back to ancient Babylon over 2,000 years ago.

They did a little bit of arithmetic. The Babylonians had invented a lot of very advanced mathematics, and they were also great astronomers. They figured out what was going on in the sky. One of the realizations the Babylonians hit on was the following curious fact that 19 years, solar years, is very, very close to 235 lunar months.

You might ask yourself why should I care that 19 years is 235 lunar months? I am not going to make one notion of time that lasts 19 years rather than the good old ordinary year. That would be impractical for various reasons, but 19 can be divided up.

What the Babylonians realized is that if they let 12 months per year be the rule for 12 years in a row, and then 13 months per year be true for the next 7 years, they would have a 19-year cycle that synchronized the lunar calendar with the solar calendar. That is what they did.

The Babylonians had leap months 7 out of every 19 years. They would go 12 months a year for 12, 13 months a year for the next 7, and then repeat the cycle in perpetuity.

This is very close to the modern day Hebrew calendar. That is why Jewish Holidays keep moving around compared to the dates that the Gregorian calendar has.

Back in the 6th century B.C., the Hebrews underwent what is called a Babylonian captivity. They borrowed their calendar at the time, and it still survives with us to this day. It is one of the last remaining remnants of a lunar calendar and the use of the cycle that 19 years is 365 days.

We have an elaborate system called the Gregorian calendar. The Gregorian calendar took a long time to invent. It is not at all obvious. We have given up on the notion of the lunar month. Our months do not try to be 29.5days. This is why our current calendar has some months that are 30 days, some months that are 31, and February is 28, except, sometimes it is 29. This is completely irrational.

It is a complete mess. There is no system underlying it. That is why every time that I want to know how many days April has, I have to go through the jingle, 30 days hath September, etc. There is no rhyme or reason to it.

You have to memorize the rule. That is the Gregorian calendar. This idea of the months being different numbers of days unsynchronized with the moon is older than the Gregorian calendar. It comes from something called the Julian calendar. The Julian calendar basically had the same months that we have except that they simply had leap year every four years. Every four years, February would get an extra day.

You say to yourself that is pretty close to the system we have right now. Is it not true that every four years we have a leap year and we get an extra day? That is not exactly true. If you did that, if you had a leap year every four years, then you would go out of synchrony with the Sun by about 11 minutes per year.

The system embedded in the Gregorian calendar says the following thing. You have a leap year every four years. Every four years, there is 366 days in the year, but then you skip leap year every 100 years. In the year 1700, the year 1800, the year 1900, those were not leap years even though they were years divisible by four.

We got lucky because the last time that we had a year divisible by 100 it was the year 2000, and the little fine print in the Gregorian calendar says do have a leap year every 400 years. So 1800 and 1900 were not leap years; 2000 was. But 2100, 2200, and 2300 will not be leap years; 2400 will be, and so on.

That is the Gregorian calendar. It was actually developed and implemented all the way back in 1582. That is a long time ago, but the astronomical measurements were good enough for people to realize that the Julian calendar that just had leap years every four years was getting out of sync with the solar cycle.

In October 4, 1582, you went to sleep, you woke up, and it was October 15. When they got the calendar back in synchrony in the year 1582, they did it by eliminating 11 days from the calendar. This was a very controversial thing at the time. People worried whether they were losing 11 days out of their lives, but it was just a label. It did not really matter.

In fact, different countries did this at different times. In the United Kingdom and in the United States we adopted the Gregorian calendar much after most of the European world, which is why dates during our revolutionary war do not match up with dates that were going on in Europe unless you make the switch for the calendars.

With the Gregorian calendar, even though it is a little bit complicated to keep everything in mind, it is much more accurate than the Julian calendar was. The Julian calendar was wrong by 11 minutes every year. That seems like a small amount, but over the course of centuries, it will add up to quite a bit. The Gregorian calendar is accurate to 2.6 seconds per year, so it takes a long time for the solar system, for example, the summer solstice, to move out of sync with where the Sun is in the sky.

Of course, people have sat down and calculated how long will it take for the Gregorian calendar to get out of sync with the Sun. The answer is we might need another leap day in the year 4000. It is not part of the official system, but it is an obvious thing to do. Just have a leap day in the year 4000. We will be fine. Someday in the far future, that is going to be an important issue. Right now, it is not.

There is an important issue which is called the leap second. The leap second is actually completely different conceptually than the leap month or the leap year. The leap day and the leap month are trying to fix the fact that one cycle, the rotation of the Earth or the orbit of the moon, gets out of sync with another cycle, namely the Earth going around the Sun. Leap seconds are there to fix the fact that the Earth rotates at a not completely constant rate.

Remember the whole point of an accurate clock is to get something that repeats itself over and over again in a predictable way, but now that we have much more accurate clocks, we realize that the rotation of the Earth is not perfect, and it is not perfect in an unpredictable way. There are different things like earthquakes and hurricanes that change the rate at which the Earth rotates.

Rather than fixing that once and for all, we have to wait for the Earth to get out of sync and then add a leap second to the calendar. You do not even notice it, but occasionally, New Year's Eve is one second longer than it was the day before. This is a hugely controversial issue these days because it is a tremendous pain for computers. The computers would like to be built with clocks that work once and for all, but our current system of timekeeping has randomly inserted leap seconds depending on what the Earth is doing.

There is a movement afoot to get rid of the leap second, to let the day slowly move out of sync with what we use to keep time. This depends on what you consider to be convenient. It makes the computers a lot more convenient, but ultimately, some day down the road, you are going to need a leap minute or a leap hour. That would seem like to be a mess. Right now, the consensus is we will keep the leap seconds, keep putting them in, but as computers become more and more important, we might change our minds.

Eventually, you want to move beyond astronomy. As technology evolves, it is not good enough to look to the sky to measure what time it is. Then it becomes even harder. Remember, Galileo figured out a mechanical device that told time, namely, a rocking pendulum back and forth. However, as time went on, it was more and more important to know what time it was when you are on a boat. This is the age of exploration. People were crossing the seas, and it became very important to know what time it was.

A pendulum does not work when you are on a boat. The boat itself is rocking back and forth. That completely messes up the nice regularity that the pendulum has. You have to do better if you want to keep time at sea.

At sea is a particularly important place to keep time because you want to know not only when you are, but also where you are. If you are on the land, you can ask somebody where you are, but if you are at sea in the middle of the ocean, you cannot look around and know where you are.

You could figure out where you are. You could figure out your longitude, how far east or west you were, by looking at the stars in the sky but only if you know what time it is. If you do not know what time it is, then maybe you are farther east than you think, you are farther west, or it is earlier or later.

The stars move through the sky. That is because the Earth rotates, and because of that, it is not enough to look to the sky if you want to figure out your longitude. You need to look at the sky and know exactly what time it is.

This puzzle, how do you know what time it is when you are on a boat far away from any land-based clocks, was so important that the British parliament offered a prize of 20,000 British pounds if someone could figure out how to measure time accurately even at sea.

I have to note, of course, that 20,000 British pounds at the time corresponds to $4.5 million today. That is because of inflation which is one aspect of the arrow of time. The value of the British pound changes over time. It is not a particularly uniform or predictable change, but it is a reminder that the past is very different than the future.

The challenge was, back in the 18th century, build a mechanical device that could tell time on a boat, and these days, we have more than enough options. We have mechanical clocks. We have atomic clocks. We have quartz watches or whatever, but back in those days, in the 18th century, that was the challenge, build something that would be reliable, not only against the rocking of the boat, but also against the fact that the temperature would be different from moment to moment. Maybe you are in the South Seas. Maybe you are in the North Sea. The humidity would vary by a lot. You need a very reliable mechanical device.

The general idea, if you cannot use a pendulum, is to use a spring. You wind up a clock, and then the spring releases its energy into a ticking device that tells the time. An obvious problem with that is that springs are made of metal. Metal responds differently depending on what temperature it is, what pressure it is, etc.

The problem was ultimately solved in the year 1761 by an inventor named John Harrison who invented the Marine Chronometer. It was not one little breakthrough. It was that Harrison worked very, very carefully to get every little piece of his chronometer working well. He first built huge machines that were bigger than a person, and they worked okay. He eventually hit upon the idea that it was easier to make a small clock accurate than a big one.

His ultimate prize-winning chronometer in the year 1761 was only about five inches across. Harrison did things like building his springs out of strips of metal that were different kinds of metal on different sides, so as one expanded in the temperature or the pressure, the other one would contract and they would cancel out.

Harrison's chronometer turned out to be accurate to better than $1/10^{th}$ of a second per day, more than enough to win the prize from the British parliament. Nevertheless, they did not actually want to pay. It took him some years of legal maneuvering, but eventually he got the prize of 20,000 pounds.

These days, of course, we want to do more than travel throughout the world. We want to have personal timepieces. We live in a world where keeping time is something that everyone wants to do whether or not you are a navigator on a ship.

In the 19th century, what that meant was a pocket watch, similar ideas to Harrison's chronometer, but the early primitive pocket watches were maybe accurate to 15 minutes a day. Essentially, you could use them to tell what hour it was with approximately the same precision as you would get by looking at the sky.

They were things that men could carry that were status symbols that said I can afford a mechanical device that tells me what time it is. At the time, in the 19th century, the idea of a wristwatch, something you wore on your wrist rather than a pocket watch was thought to be disdainful. That was a low-class thing, and real men did not do that.

That changed only during World War I. It was the military personnel in World War I that realized they could not use one of their hands to tell what time it was. They needed both of their hands free to carry their weapons, so the military started wearing wristwatches, and the fad spread to people outside. It was the early 20th century that first got people wearing watches on their wrists as a regular thing.

This became, over time, a major industry, building accurate wristwatches for, especially, the country of Switzerland. It turns out that building a mechanical wristwatch, even if you know how to do it, even if you have the plans, is incredibly difficult. Starting a watch making manufactory from scratch is an enormously expensive, time-consuming operation. The Swiss became very good at it, and essentially developed almost a monopoly. It was recognized that all of the best watches came from Switzerland. This changed a little bit in the 1960s.

In the 1960s, something happened that, in Japan, is called the quartz revolution. In Switzerland, it is called the quartz crisis. What happened was that Japanese manufacturers realized how to take a vibrating quartz crystal and base a wristwatch on it.

Quartz, as we said before, is much more accurate than any mechanical timepiece, and not only that, it is also cheaper. The first quartz wristwatches went on sale in the 1960s, and during the 1970s, the way that we keep time on our wrists totally changed.

The idea that everyone had an elaborate hand-tooled mechanical timepiece was completely replaced by the idea that you would have a quartz wristwatch which was both cheaper and more accurate. Why would you ever want to spend more money for the Swiss mechanical timepiece that did not tell time as accurately?

It was in the 1980s that the Swiss figured out they could do the same things that were done in the 19th century. They could market themselves as a luxury item. Today the Swiss watch making industry is thriving, but only as an upscale market kind of phenomenon. If you are willing to pay for a very fine Swiss watch, that tells something about who you are, even though, even today, that fine Swiss watch does not tell time as well as the $25 cheap quartz watch that you can get on the street.

These days, most kids, most younger people do not wear watches at all, and I mean by kids everyone under the age of 25. You carry your cell phone with you. You have all sorts of devices all around you that tell you what time it is. The idea of wearing a wristwatch is becoming a little bit passé.

For the role of a status symbol, that is exactly what you want. You want it to be a little exclusive, something that says I want to wear this fine mechanical piece of equipment.

Today we use atomic clocks, not quartz crystals, for maximum accuracy. A quartz crystal can be very accurate, but we want higher and higher accuracy clocks because we have more and more demands that we are putting on them.

You can do that by looking at the transitions of electrons in individual atoms. In practice, we use cesium or rubidium, specific atomic elements that give up very precisely defined frequencies. The best atomic clocks are more accurate than one billionth of a second per day.

That is not just showing off. You might think, why do I need to be accurate to a billionth of a second per day? The answer is the global positioning system. If any of you have ever used the device that lets you tell where you are in your car or on your phone, what that is doing is the GPS system is getting signals from satellites flying overhead. At any one time, there are at least four GPS satellites sending signals to your car or your phone, and what you do is you simply triangulate.

You use the fact that the speed of light is finite. It is not infinitely fast. It is light; it is radio waves that are traveling the signal from those satellites to your GPS system, and by comparing how much time it takes to get the signals from different satellites, you can figure out exactly your location here on Earth.

Of course, as you know, if you use a GPS system, they can be very, very accurate. You can know exactly where you are to within a few feet, but that depends on the fact that the clocks on board the satellites are incredibly accurate themselves.

If they were off by one microsecond, your location would be off by 1,000 feet. That is because the speed of light is so fast, to triangulate correctly, you need to know exactly when that signal left the satellite. That is what a good atomic clock lets you do.

The final thing to say here is that a good timepiece does not have an arrow of time in the sense that many things in the universe do. Our primary consideration, our real goal here in these lectures is to understand why the past is different from the future.

Obviously, a clock has an arrow in the sense that it goes from 12 o'clock to 1 o'clock to 2 o'clock, but that is an arbitrary direction. If I took all of the hands off a clock, made a movie of it, and played it backward, you wouldn't know whether I was playing it backward or whether I looked at it from the opposite side.

The nice thing about a good, accurate timepiece is that it always does exactly the same thing. The arrow of time is a feature of things that do not always do exactly the same thing. You take an egg and you scramble it. You turn it into scrambled eggs. It only goes on one direction.

That is the difference between the pristine accurate timepieces and the universe. Understanding where that difference comes from will be our goal for the rest of the course.

Time's Arrow
Lecture 4

In the past few lectures, we have talked about the philosophy of time; the reality of the past, present, and future; and the down-to-earth aspects of how we measure time. Now, we are free to concentrate on the important issue that will motivate the rest of the course: why there is an arrow of time—why the past is different from the future. In this lecture, we'll try to understand what is meant by the arrow of time and why its existence in our world is so startling.

How Is the Past Different from the Future?

- All aspects of memory (meaning any form of historical knowledge) refer to things that already happened. Memory is a correlation between some artifact here and something that happened in the past.
  - There are no memories of the future; there are predictions, but they are different than memories. They are not artifacts; they are not connected in some recordkeeping way with what happens in the future.

  - Our difference of "epistemic access"—what we know about the past versus what we know about the future—is probably the most obvious and the most important difference that the arrow of time gives to our everyday lives.

- Another obvious difference between the past and the future is, of course, aging. We are all born young and grow older with time. Aging applies not only to people and to other animals but to physical objects, as well.

- Of course, we all know that art, politics, literature, and music are all things that evolve in time. Although cultural changes are not as closely tied to the arrow of time as physical changes are, they still represent one-way evolutions. There is a past, a present, and a future, and all of them are different from each other.

- The same thing is even true for society, in what we might call social progress. Societies change through time in interesting ways.
    - We would like to believe that social changes through time represent true progress.
    - That statement is arguable from a historical point of view, but it is unarguable that there is an evolution in society that points in a certain direction.

- Such evolution is also reflected in the physical universe. The past of the universe is very different from what we think the future will be like.
    - The universe at early times was hot, dense, and smooth. Currently, it is more or less empty, and the matter that is present is lumpy. It is very different in one place, where there is a galaxy, than in another place, where there is interstellar space. In the future, the universe will continue to empty out.
    - The universe has a very strong arrow of time, which is reflected in the life cycles of stars and planets.

- That physical change in the universe is reflected in biological change here on Earth. The Earth is only about 4.5 billion years old, so it is a substantial fraction of the age of the universe.
    - Life formed fairly quickly after the Earth formed, but there is a clear evolution. This is not just a Darwinian evolution, but evolution in the sense of change over time in a certain direction.
    - The first life forms were simple single-celled organisms. The fact that individual species became more complicated as time progressed is not a requirement of evolution.
    - Because there is room for differentiation and change, our biosphere supports increasingly complicated systems of species.

- Besides physical and biological change, there are some features of the arrow of time that seem ingrained or logically necessary.

- o For example, we have the idea that a cause will always precede an effect. If the laws of physics do not account for what happens first and what happens after, we will clearly have to think harder about what causality means.

- o We also have to think harder about free will. We do not have the ability to make choices about the past.

- All of these different areas—from biology and sociology, to the universe and physics, to logical issues of causality and free will—reflect the arrow of time. A universe without the arrow of time would not have progress or differentiation from the past to the future.

Storytelling

- The arrow of time plays a crucial role in storytelling. As you know, most novels are chronological narratives, but novels can be made more interesting by playing with the conventional way time works—incorporating flashbacks or reverse narration. The idea behind these literary devices is that the information we have changes the way we perceive events.

- If we are given a certain scene happening right now, the meaningfulness of that scene depends on things that happened in the past; those past occurrences may change our perception in the future. The notion of information and what it means to us is absolutely central to the arrow of time.

- Many attempts to use time in clever ways in storytelling actually miss the impact of the arrow of time. It is almost impossible for us to imagine two characters talking to each other from two different directions of time. One feature of the arrow of time is it must be the same arrow for everyone.

- As scientists or philosophers, the imaginative storytelling aspects of the arrow of time help us think of different ways that it could work and compare them to our world.

The Universe without the Arrow of Time

- We have basically two choices for what the universe would look like without an arrow of time. One is a world in which nothing changes over time. In this world, there would be no arrow of time because time would not do anything.

- Another possibility is a random world. Things would happen one way sometimes and another way at other times. In this world, you could not make plans. You would remember tomorrow just as much as you remembered yesterday, which means you probably wouldn't remember anything at all.

- The need for an arrow of time is something about being human. We cannot imagine a life in which the past and future were truly built on symmetric principles.

Irreversibility

- Breaking an egg and scrambling it (increasing its entropy) is an example of the arrow of time in action. Other examples include putting an ice cube in a glass of water and waiting for it to melt, releasing the scent of perfume in a room, or breaking a glass.

- The common thread in these examples is irreversibility: Something happens in one direction, and it is easy to make it happen, but it does not happen in the other direction, or if it does, it is because we put effort into it. It does not spontaneously happen. Things go in one direction of time. They do not go back all by themselves.

- That difference between going from the past to the future is consistent throughout the universe as far as we know. This is clearly not a feature of our biology but a feature of the way the universe works.

- We actually define the past versus the future using the arrow of time. The fact that we remember the past and not the future is so obvious to us, so inherent in how we think about the world, that we automatically assign an intrinsic difference to the past versus the future.

- Time could exist in a universe without an arrow, and time is not the arrow itself. The arrow is a feature of the stuff in the universe—the eggs, the glasses, the scent of perfume in a bottle. It is these things that evolve in certain ways always in the same direction, from the past to the future.

- The arrow of time is the arrow of stuff evolving in time. Thus, it is not time that we need to understand but matter. It is the motion of particles and objects in the universe.

The egg is a classic example of a physical system in which the arrow of time is obvious; we can scramble an egg, but we cannot unscramble it.

The Second Law of Thermodynamics

- As we have said, the feature of matter that changes with time is called entropy. The feature of the universe that we are trying to understand is that the increase of entropy is associated with the passage of time.

- The second law of thermodynamics—entropy of the universe increases—underlies all the ways in which the past is different from the future. That is a very surprising claim.
    - The claim makes sense when you say that the reason you can scramble an egg but not unscramble it is ultimately that entropy increases.

    - But we're saying that the fact that entropy increases is the reason you remember the past and not the future, the reason you are born young and grow older, the reason you can make a choice about what to have for dinner tomorrow but not about what to have for dinner yesterday.

- It's important to note that the second law of thermodynamics does not imply that every single object in the universe needs to experience increased entropy.
  - Here on Earth, the biosphere is more organized now than it used to be. The existence of complicated multicellular organisms is a low-entropy phenomenon. Through the course of evolution, however, the Sun shining on the Earth increased the entropy of the universe enormously.
  - We usually find that when entropy decreases in one small system, it is because the universe was increasing its entropy greatly in the wider system.
- The more we understand nature at a fundamental level, the more mysterious the arrow of time becomes. Deep down, time is just like space; it is a label on events in the universe. There is no arrow of space—no preferred directionality—but there is an arrow of time.
- The same kind of reasoning that tells us why entropy will be greater tomorrow would seem to tell us why it should be greater yesterday.
  - That is a reflection of the fact that the fundamental laws of physics do not distinguish between the past and the future, and if we think we can prove that entropy will be greater tomorrow, then we should be able to prove that it was greater yesterday.
  - Nobody believes that the entropy of the universe was greater yesterday than it is today, but we cannot prove that using just the laws of physics.

Suggested Reading

Carroll, *From Eternity to Here*, chapter 2.

Greene, *The Fabric of the Cosmos*.

Price, *Time's Arrow and Archimedes' Point*.

Questions to Consider

1. How many different aspects of the arrow of time can you think of?

2. What is your favorite story involving flashbacks, reverse chronology, or time travel? Is the passage of time portrayed realistically?

3. Can you imagine a world in which time existed, but there was no arrow to it?

Time's Arrow
Lecture 4—Transcript

We have gotten some of the groundwork out of the way. We have talked about the philosophy of time, the reality of the past, present, and future, and also about the down-to-earth aspects of how we measure time using calendars and clocks. Now, we are free to concentrate on the important issue that will motivate the rest of the course, why there is an arrow of time, why the past is different from the future.

Mostly we will be focusing on the physics of why there is an arrow of time. This will show up in our understanding of why there is sociology or psychology, or cultural aspects to the arrow of time.

What we will be arguing is that the physics underlies it all. If we can understand why physics gives time an arrow, we will understand why the past is different from the future in all aspects of our lives.

The difficult thing about this task is that the arrow of time is so important to how we think about the world that it is impossible to imagine the world without it. We are going to try to step outside of our four-dimensional universe to take that view from no end that we talked about to look at what the universe is like at all times all at once so that we can ask why, in that four-dimensional progression, the past of one moment is different from the future.

Because we cannot think about the world without an arrow of time, we need to be reminded that there is a question here to be addressed, that there could have been a universe without an arrow at all, that it is actually very startling, very surprising that we live in a world where the past and future seem so different to us. If we take the objective you standing outside, there is absolutely a challenge, and that challenge has to be met by physics.

The goal of this lecture is to understand what the challenge is. What do we mean when we say the arrow of time? Why is it unusual at all? We talked about clocks, for example. A clock has an arrow in some primitive sense, but it is not the kind of arrow of time that we are trying to explain in this course.

A clock goes from 12 o'clock to one o'clock to two o'clock, but it could go the other way. There is nothing wrong, there is nothing to be challenging if you asked a clockmaker to build a clock that went from two to one to 12. If you looked at a clock, and you saw that it was one o'clock, you would know that in the past, it was 12 o'clock. In the past of that, it was 11 o'clock. You can reconstruct the past from the present, but the whole world is not like that.

There is information exchange and loss and gain as time goes on. That is what the arrow of time is all about. Think about it this way. Compare the arrow of time to what would be an arrow of space. A clock goes from noon to one o'clock to two o'clock in exactly the same way that a ruler goes from one inch to two inches to three inches, but if you took 100 people separated from each other, gave them a ruler, and said put the ruler down on the table, they would all align their rulers slightly differently.

There is no best way to point that ruler going from one inch to two inches to three inches, but if you told these hundreds of people to build clocks, they would all move their clocks in the same direction. Everyone would agree what time was earlier, what time was later.

That is a feature of the arrow of time, that it is universal within the universe we have so far perceived that everyone agrees on which is the past, which is the present, and which is the future. One thing to emphasize is that this difference between the past, present, and future has many, many different aspects, and yet all of those aspects come down to the same thing.

This is what makes it difficult to imagine a world in which, for some reason, time moves backwards. It is very difficult for us to wrap our heads around the idea that there would be such a thing as time moving backwards. When we try to imagine it, we often do not do a very good job.

You might know the movie *Benjamin Button* about a man who was born as an old man. His first days on Earth were as a wrinkled old person, and as time goes on, his body gets younger and younger. In some sense, this is an example of an imaginative, fantastical take on the idea of time running backwards, but when you think about it—and maybe it is not a good idea to think about the movie too hard—but, if you think about that movie, you see how they cheated in the storytelling.

The body of Benjamin Button was wrinkled and old when he was born, and it got younger and younger, but his brain, his consciousness, his thinking, moved in the absolute ordinary direction. Benjamin Button did not live backward in time. He remembered yesterday and predicted tomorrow just like you and I would. When was a wrinkly old-looking person, he was still a young person at heart. His personality was that of a child.

It would be very, very difficult to tell a story about a character who literally lived backwards in time. The reason why is because the many different ways in which the past differs from the future are all tied together. What we call the arrow of time, this singular thing, has many, many different aspects and in this warm-up lecture let's list a bunch of ways in which the past is different from the future and recognize that they all line up, that they all are similar ways of telling yesterday from tomorrow.

The most obvious feature is memory, and again, by memory we mean any form of recordkeeping, any form of historical knowledge. It might be memory in your brain, or it might be memory in a computer memory. It could be a photograph or a historical document.

The point is that all aspects of memory refer to things that already happened. They tell us about the past. What memory really is, is a correlation, a connection between some artifact here and some thing that happened in the past.

If you find a photograph of your birthday party and you see that you were wearing a red sweater, you go, ah-ha, this is evidence. This photograph that exists now tells me that I was wearing a red sweater. That is a connection between this moment now, this photograph in front of you during the present time and a past event. That is a simple model of memory, something that exists now that tells us about the past.

The point is that there are no memories of the future. There are predictions. There are attempts to guess at what will happen in the future, but they are very different than memories. They are not artifacts. They are not things that exist that are tangible, you can touch them, and that are connected in some recordkeeping way with what happens in the future.

Our difference of epistemic access, a philosopher would say, between the past and the future, what we know about the past, versus what we know about the future, is probably the most obvious and maybe the most important difference that the arrow of time gives to our everyday lives.

Another obvious difference is, of course, aging. Benjamin Button put aside, we are all born young. Young people look a certain way. We all grow older with time. If you show people pictures of the same person at different moments in their lives it is usually easy to tell when they were young and when they were old. Our common aging all moves in the same direction as time passes.

Aging applies not only to people and to other animals, but to physical objects as well. We can tell the difference between a newly built building and an older one. Buildings and objects tend to wear down, tend to decay over time. That is a fairly uniform progression, although, to be careful, we have to admit that we can clean things up, so it is not absolute.

That is a very common feature of evolution through time, that there is a general trend and you can interrupt the trend in one way or another. Changes in style, in art, in culture, in politics, in literature and music, these are all things that evolve in time. Very often if we hear, for example, a piece of music, you can tell which piece of music was from today versus which was from the 60s, certainly which is from today versus which was from the 17th century.

There are fads. There are retro styles that go in art, politics, and literature, so you have to be careful. The underlying point is that even if social changes are not as closely tied to the arrow of time as physical changes are, they still represent one-way evolutions. There is a past, a present and a future. All of them are different from each other.

The same thing is even true for society, for what you would call social progress. Societies change through time in interesting ways. We would like to believe that social changes through time represent true progress, that things get better. We become more free, more prosperous. This is arguable from a historical point of view, but it is unarguable that there is an evolution that points in a certain direction and we try to make plans about the future.

This is reflected in the physical world as well as the human world. The universe evolves. The past of the universe is very different from what we think the future will be like. The universe at early times was hot, dense, and smooth.

Currently, it is more or less empty, and what matter there is, is lumpy. It is very different in one place where there is a galaxy, than in another place where there is interstellar space, and in the future, it will continue to empty out. The universe has a very strong arrow of time, and that arrow of time is reflected in the lifecycles of stars and planets. Young stars are very different than old ones.

That physical change in the universe is reflected in the biological change here on Earth. The Earth is only about 4.5 billion years old, so it is a substantial fraction of the age of the universe. Life formed fairly quickly after the Earth was formed, but there is a clear evolution, not just in the Darwinian sense, but simply in the sense of change over time in a certain direction.

The first life forms were simple. They were single-celled organisms. They became more complicated as time goes on, and this is a very subtle issue because evolution does not say that individual species must grow more complicated. It is that because there is room for differentiation and change, our biosphere supports more and more complicated systems of species. The interplay between complicated, complex life, and entropy and the arrow of time is a fascinating one that we will talk about later.

Finally, besides physical change and biological change, there are features of the arrow of time that seem more ingrained, that seem logically necessary, but as you will see, that is again, a more subtle point. For example, we have causality. We have the idea that a cause will always precede an effect. If I knock over a glass of wine and it spills on the floor, it makes sense to say the reason why the glass spilled is that my hand moved and then knocked it. No one would say the reason why my hand moved was because the wine was going to spill on the floor and the hand was necessary to make it happen.

We live in a world where we think that causes happen, then their effects. If the laws of physics do not have any such difference between what happens first and what happens after, we are clearly going to have to think harder about what causality means. What that means is that we are going to have to think harder about free will. We think that we have the ability to make choices. What does that mean?

We do not have ability to make choices about the past. Nobody thinks that I can decide right now what to have had for dinner last night. That is already in the books. That is not something I can make a choice about. I can make a choice about what to have for dinner tomorrow night, but not about yesterday.

Our ability to make choices seems to reflect the arrow of time, and of course, our individual metabolism reflects the arrow of time. We eat, we get fuel, and then we can run around.

All of these different aspects, from biology and sociology to the universe and physics to logical notions like making choices and causes and effects, they all reflect the arrow of time which makes the universe interesting. We can imagine a fake universe, a toy model universe, in which there is no arrow of time. It would be boring. It would not have this progress, this differentiation, from the past to the future.

This feature that the arrow of time is what makes time interesting is one reason why the arrow of time plays a crucial role in storytelling. One of the ways to make your story more interesting is to play with the conventional way the time works.

In most novels, the first page is about what happened first and the next page is about what happened next, and there are chronological narratives. You can change that a little bit in the context of the story because you have the freedom to say things about what is happening now before you say about what happened in the past.

As far back as Virgil's *Aeneid*, an ancient roman poem, Virgil told individual scenes using reverse chronology, starting with what happened last and then telling what happened earlier.

In the modern day, we have TV shows, novels, and movies. If you have ever seen *Lost*, the TV show *Lost* often used flashbacks to fill in the stories of their characters. If you have read the novel *Remains of the Day*, the narrator told things in a reverse narration and that actually filled in the meaningfulness of what happened in the present day.

It can get as complicated as the movie *Memento* where there are two timelines both telling the same story, one moving forward and one moving backwards. The idea behind all these literary tricks is that the information we have changes the way we perceive events.

If we are given a certain scene happening right now, the meaningfulness of that scene depends on things that happened in the past, so the author, the storyteller, can give us information about things that already happened but that in the future, will change the way we perceive it. The notion of information and what it means to us is absolutely central to the arrow of time.

I would be remiss not to say that many attempts to use time in clever ways in storytelling actually miss the impact of the arrow of time. We have already mentioned Benjamin Button who just aged physically in one direction, but everything else was absolutely typical.

There are other examples of the same thing. In T. H. White's *The Once and Future King*, he tells the story of King Arthur and Merlyn, and White's character, Merlyn, we are told, lives life backward in time. When we first meet him, King Arthur is very young, and Merlyn is old and tired and has been through it all before. Later on in the book, after Arthur has grown up, Merlyn thinks that he has just met him.

It does not work that way because, even though we are told that Merlyn lives backward in time, it would make no sense to have Merlyn talk in scene-by-scene language as if he did not remember what happened five minutes ago, but he did remember what happened five minutes from now.

It is almost impossible for us to imagine two characters talking to each other, one of which lives in one direction of time, and the other one of which lives in the other direction. One feature of the arrow of time is it must be the same arrow for everyone or else all hell breaks loose.

An example of where it works is in Martin Amis's novel, *Time's Arrow,* but the only reason it works is because the one character that experiences the arrow of time backwards never interacts with the outside world. In *Time's Arrow*, the narrator is what we call a homunculus, a conscious being that lives inside of another person, but does not talk, does not have any control over what the other person does, and simply experiences what happens around him. He experiences the life of this other person from the moment of death backward in time to the moment when he was a young boy and what you learn is, of course, this literary trick of revealing past events can color our impression of present events. It is probably the best example of where a backwards pointing arrow of time made a story work better.

As scientists or philosophers, as thinkers trying to understand the world, the imaginative storytelling aspects of the arrow of time help us think of different ways that it could work. Our challenge will be to step outside of the universe, to imagine worlds without an arrow to wonder what they would be like and to compare them to our world.

If you did not have an arrow of time, what would that universe look like? There are basically two choices. One is that there is no time at all, or at least nothing changes over time. In a world where everything was exactly the same, everything in the world, every place of every object, was the same from moment to moment in time, then there would be no arrow of time because time would not do anything. It would be impotent. The world is the same. It is static. That is not an interesting universe to contemplate.

Another possibility is just a random world. Things happen one way sometimes, the other way the other time. There is no intrinsic arrow of time. That is, in some sense, not an interesting world either. You could not make plans. You would remember tomorrow just as much as you remembered yesterday which means you probably would not remember anything at all.

The need for an arrow of time is something about being human. We cannot imagine a life in which the past and future were truly built on symmetric principles, and therefore, understanding the physics underlying that helps understand how we do live our lives.

Let's go back to that physics to try to understand the essence of why the past is different from the future. We have simple physical systems in which it is obvious. The egg is probably the classic example. We have an unbroken egg. It is orderly. It has low entropy, we would say.

We can break the egg, and the entropy goes up. We can scramble the egg, and the entropy goes up even a little bit more. Those are easy things to have happen. It will be very difficult to take the scrambled eggs and to reconstruct Humpty Dumpty to make an unbroken egg out of them.

That is an example of the arrow of time. It is hard to go backwards. There are many other examples. Put an ice cube in a glass of water. Over the course of time, that ice cube will melt. It would be very, very surprising to put down a glass of water and watch, over the course of time, an ice cube form inside of it. That is not the kind of thing that we experience in our everyday lives.

If you have a cup of coffee, you can mix cream into it. That is easy to do. It is hard to unmix it and separate the coffee from the cream. This is the kind of thing that we experience over and over again in the world. If you open a bottle of perfume, the scent fills the room. This is an example that was used in the 19th century by Ludwig Boltzmann, one of the giants of thermodynamics, trying to understand the growth of entropy.

If you knock over a glass and it breaks, that is easy. Unbreaking the glass is not so easy. If you shuffle cards, they tend to disorganize. They tend to become more random. It is never the case that you shuffle cards and they spontaneously order themselves, ace, king, queen, jack, ten and so forth.

If you are playing pool, you start with all the pool balls arranged in a nice orderly fashion. You hit them with the cue ball, and they scatter across the table. This happens all the time. Start with pool balls scattered, hit them with a cue ball, they never rack themselves. They never come into an orderly arrangement.

We can do it. We can put them into an orderly arrangement. It does not happen by itself. What is the common thread? If these are all different examples of the arrow of time in action, what is going on that makes all of these examples similar.

The common thread is irreversibility that something happens in one direction, and it is easy to make it happen. It does not happen in the other direction, or if it does, it is because we put effort into it. It does not spontaneously happen. It is easy to break the egg. It is hard to unbreak the egg. Things go in one direction of time. They do not go back all by themselves.

That difference between going from the past to the future, is consistent throughout the universe as far as we know. It is easy to go into your kitchen break an egg, and say I can break the egg. I cannot unbreak it, but it is a remarkable fact that if we ever meet an alien species that lives on another world, they will not be able to unbreak eggs either, or at least it will not be any easier for them than it is for us. The aliens will remember yesterday. They will not remember tomorrow. Even if we are completely different species, we experience the same arrow of time. It is clearly not about our biology, it is about the way the universe works.

We actually define the past versus the future using the arrow of time, the fact that you can break an egg but not unbreak it, the fact that you remember the past and not the future, this is so obvious to us. This is so inherent in how we think about the world, that we automatically assign an intrinsic difference to the past versus the future.

What we have learned by thinking about these processes is that the arrow is not belonging to time itself. Time could exist in a universe without an arrow. Time is not the arrow itself. The arrow is a feature of the stuff in the universe, the billiard balls, the eggs, the glass of wine, the scent of perfume in the bottle, it is these things that evolve in certain ways always in the same direction from the past to the future.

The arrow of time is the arrow of stuff evolving in time, and that is a clue that we are going to try to use to understand the arrow of time better. It is not time that we need to understand. It is matter. It is the motion of particles and objects in the universe.

As we have said, the feature of matter that changes with time is called entropy. It is not that everything goes up in entropy, but that the universe, the whole shebang, increases in entropy as time goes on. Entropy is a way of thinking about the messiness, the disorderliness, the randomness of a configuration of stuff.

That is only a simple gloss on the notion of entropy. We get to be much more careful over the course of 24 lectures. We will give you a much better definition of entropy, but you are not wrong if you think of the fact that entropy increases as things going from being organized to being disorganized.

The unbroken egg is more organized. There is a shell. There is an egg white. There is an egg yolk. They are all separate from each other. When we break the egg and scramble it, everything is mixed together. The disorderliness has gone up, and the feature of the universe that we are trying to understand and explain is that that increase of entropy is associated with the passage of time. As we go from the past to the future, the entropy always goes up in the universe as whole. That is the law of nature, the second law of thermodynamics.

As we have said, it is this law, the second law of thermodynamics, entropy of the universe increases, that underlies all the different ways that the past is different from the future. That is, and should be, a very surprising claim. It makes sense when you say that the reason you can scramble an egg but not unscramble it is ultimately because entropy goes up.

It makes sense that the melting of an ice cube or the spreading of perfume through a room is because entropy is going up, but what we are saying is that the reason why you remember the past and not the future is because entropy is going up. The reason you are born young and then grow older is because entropy is going up.

The reason why you can make a choice about what to have for dinner tomorrow, but not a choice about what to have for dinner yesterday is ultimately because entropy is going up. That is a surprising claim. It requires a lot of work to build us up from the increase of entropy to all of these different manifestation of the arrow of time.

That is especially true because you have to be very, very careful about what the second law of thermodynamics actually implies. It does not imply that every single object in the universe needs to experience increased entropy.

After all, you can clean your room. The claim is that while you clean your room, the entropy of the universe is going up. You are doing work. You have eaten food; you have fuel in you. You are going to burn that fuel. You are going to sweat, and you are going to make noise.

You are going to do things that if you really counted carefully increase the entropy of the universe even as you were cleaning your room, so it is not true that the second law says everything gets more and more disorganized. Here on Earth, the biosphere is more organized now than it used to be. The existence of complicated multi-cellular organisms is a low entropy phenomenon.

Along the way, the Sun shining on the Earth, as we will see, increased the entropy of the universe by an enormous amount. What usually happens is what we will find when entropy goes down in one little system it is because the universe was increasing its entropy greatly in the wider system.

That is a mystery. That is something we would like to understand better. It did not used to be a mystery. Back before we understood the laws of physics very well, there was a difference between the past and the future. That is how the world was. That was not something that required explanation. It was just a fact.

The more we understand nature at a fundamental level, the more mysterious the arrow of time becomes. Deep down, time is just like space. It is a label on events in the universe. There is no arrow of space. There is no preferred directionality, but there is the arrow of time.

It is that difference that we need to explain. Why does entropy increase if it is not found in the laws of nature? That is what we are going to try to understand.

What we will find over the course of these lectures is that it is very easy to explain why entropy goes up. Starting today, explaining why the entropy is larger tomorrow is the simplest thing in the world. That is no longer a difficult question. We understood that over 100 years ago back in the 19th century.

The hard part will be understanding why the entropy was lower yesterday, just that it is more logical, easy to understand, why entropy will be higher tomorrow. The same kind of reasoning would tell us that the entropy should be higher yesterday.

That is a reflection of the fact that the fundamental laws of physics do not distinguish between the past and the future, and if we think we can prove that entropy will be higher tomorrow, then we should be able to prove that it was higher yesterday.

What we will find out, of course, is that that is not true. Nobody believes that the entropy of the universe was higher yesterday than it is today, but we cannot prove that using just the laws of physics.

We are going to need something more. We are going to need an external assumption, which we will call the past hypothesis, which basically comes down to the fact that the early universe 13.7 billion years ago, started with very low entropy. The universe started in a very organized state. It is like a toy that was all wound up and has been ticking along ever since.

Why was the early universe like that? I do not know. Nobody else knows. We have some theories. We are going to investigate those theories, and you can judge for yourself whether these theories are promising or not.

The Second Law of Thermodynamics
Lecture 5

The central theme of this course is that entropy is the reason for the existence of the arrow of time, the fact that the past is different from the future. In the past, because of initial conditions near the Big Bang, the entropy of the universe was lower. In this lecture, we begin to put together the pieces of our picture of time with an exploration of the second law of thermodynamics: In an isolated system, entropy either increases or stays constant; it never decreases. Things happen in one direction of time—not the other.

The Laws of Thermodynamics
- There are actually four laws of thermodynamics, but we are primarily concerned only with the first and second ones.
  - According to the first law, energy is conserved; according to the second, as we know, entropy increases or remains constant in isolated systems.

  - The second law is crucial to these lectures. It says that there is an irreversibility—a direction—of time. Entropy increases in one direction and decreases as we go to the past.

- Although it is one of the most rock-solid laws of physics, when we look at the second law carefully, we realize that the statement of it is actually only an approximation. It is not absolutely impossible for entropy to decrease spontaneously; it is, however, extremely unlikely.

- The reason the second law seems to be immutable is that it is not a specific model of some fundamental interaction, like Maxwell's theory of electricity and magnetism. The second law is a metalaw; it refers to how different kinds of laws of physics can possibly work.

Development of the Second Law

- The second law of thermodynamics was first formulated in 1824 by a French military engineer named Sadi Carnot. It was Carnot who came up with the idea that entropy increases, although he didn't know about the idea of entropy itself.

- Motivated by a desire to catch up to the British in the science and technology of steam engines, Carnot set himself the task of formulating a concept for the most efficient possible steam engine. In the process, he discovered that there is a maximum efficiency that a steam engine can achieve.
    - Carnot further realized that this maximally efficient steam engine had an interesting property, namely, that it could be run backward.

    - We have already hinted at the idea that reversibility is at the heart of the arrow of time. A process that goes forward and backward equally well is a reversible process, and it does not have a direction of time.

- Most real-world steam engines are very inefficient and are not reversible. Today, we would say that they generate entropy. But Carnot's formulation of the second law is that it's possible to have either a perfectly reversible engine, which means that entropy stays constant, or any other engine, which means that entropy is increasing.

- About 40 years later, a German physicist, Rudolf Clausius, realized that Carnot's idea was a law of nature. Clausius was interested in thermodynamics—the science of heat; at the time, heat was considered a substance, called caloric. Clausius's discovery was that heat flows in only one direction.
    - If we put a box of hot gas next to a box of cold gas, the hot box heats up the cold one, and they equilibrate. We go from one hot box and one cold box to two warm boxes. Once we have two boxes at the same temperature, the heat can no longer travel.

- Notice that this is a one-way process; it is an arrow of time. We go from a difference in temperature to a sameness in temperature.

- Clausius's version of the second law can be thought of as the statement that heat gradients tend to even themselves out. Boxes of gas at different temperatures or other things, solids or liquids, tend toward equilibrium.

- Clausius invented the word "entropy" to quantify exactly what was happening, and his formula for the second law says that entropy is the change in heat divided by the temperature. The result will always be a positive number.

Our Intuition about the Second Law

- It might seem that Clausius's version of the second law is not compatible with how we know the world works. It seems to be saying that temperature gradients always smooth out, that different objects always come to the same temperature, no matter what we do.

- Such a statement seems to be at odds with our experience that we can heat things up. We can start with a room-temperature situation, such as a stove and some eggs we want to scramble, and we can turn on the stove and heat up the eggs. How is that possible if temperatures only even out? How is it possible to cool something down from room temperature by putting it in a refrigerator?

- The first law of thermodynamics says that energy is conserved, but we can change energy between different forms.
  - When you turn on the gas burner of a stove, you are not creating new energy. You are taking energy that was stored in natural gas, and you are combusting that natural gas to release its energy in the form of heat. The energy was always there; you are changing its form from unconsumed natural gas into heat.

- That is easy to do, and it does not violate the second law. Clausius's version of the second law tells us how heat moves around. It doesn't say that we cannot create heat by burning something.

- Decreasing the amount of heat in an object is more difficult than increasing it because the process does not involve just releasing energy. If we release energy, we increase heat, so how can we ever cool things down?
  - Again, the answer can be found in thermodynamics. Instead of applying heat to a box of gas, we expand the size of the box. The energy inside the box is the same, but because we are moving things apart, the molecules slow down and become colder.

  - That is the principle behind a refrigerator. We take Freon or some other appropriate gas, expand it, and it cools down.

  - Both freezers and refrigerators work in this way, but note that you cannot cool off your living room by opening the door of your refrigerator. Why not? The answer is that you are not actually destroying the heat that was in the Freon in your refrigerator; you are moving it around.

  - When we expand Freon, we cool it off, but at some point, we have to recompress it to make it go through the cycle again and again. That is why the backside of your refrigerator is hot.

The Uselessness of Energy
- This way of thinking about heat suggests another way of thinking about entropy. We said that entropy measures disorderliness, and we also said that Clausius gave us a formula for it, the change in heat divided by temperature. Yet another way to think about entropy is as a measure of the uselessness of a certain amount of energy.

- Energy is conserved, but it can change forms. If you have energy in a low-entropy form, you can do useful work with it. You can lift something up, drive a car, or fly an airplane. If you convert that energy into a high-entropy form, it becomes useless.

- A low-entropy concentration of energy is called fuel. We have fossil fuels sitting in the ground with energy in them in a concentrated form. We can extract the energy because the entropy of the fuel is low. Once we burn the fuel, we cannot go back. You can heat a room in your house by burning wood, but you cannot cool off a room in your house by unburning fuel and turning it into wood.

Does Clausius's Law Always Work?
- Clausius's version of the second law says that if we have two boxes at different temperatures and we put them next to each other, the temperature smoothes out; the gradient decreases.
    - If the system we are looking at is two fixed-sized boxes of gasses, that is a perfectly good formulation of the second law, but it would be a mistake to think that everything always smoothes out under all circumstances.

    - Think, for example, of a cloud of gas—not a box. If you went out into empty space and released a small amount of gas, what would happen? It would fly out in different directions. That is what you might expect according to the second law. It would become smoother and smoother, spreading throughout the universe.

    - Now imagine that you have a giant cloud of gas, an amount equivalent to billions of solar masses. Then, gravity kicks in. Of course, there is gravity even for a small box of gas, but it is not that important because the mass is very tiny. Once you have enough gas to make a galaxy, gravity becomes very important. The galaxy contracts, stars and planets form, and so on. That's how our galaxy started.

- It's important to realize that the process of forming a galaxy increases the entropy of the universe. The formation of a galaxy makes the universe lumpier, not smoother, so increasing entropy does not mean increasing smoothness.

* A more widely applicable definition of entropy was given to us by Ludwig Boltzmann in the 1870s, based on his understanding of the existence of atoms.
  - In this definition, heat is actually thermal energy—the random motions of atoms. This understanding based on atoms led from thermodynamics to statistical mechanics, the study of the probabilities of atoms being in different arrangements.

  - Boltzmann realized that arrangements of atoms are macroscopically indistinguishable, and his insight was that entropy is simply a way of counting the number of arrangements of atoms inside a certain system.

  - In other words, the reason entropy increases, according to Boltzmann, is simply that there are more ways to be high entropy than to be low entropy. That is a rigorous definition that corresponds to our intuitive feeling that entropy measures disorderliness.

  - When entropy is low, the macroscopic configuration is very precisely arranged. There are only a few such configurations that look the same. When entropy is high, the configuration is spread out. There are many different ways to arrange the atoms, and all of them look alike.

  - Boltzmann's definition of entropy is the one that makes the arrow of time go. Once we understand it, we can ask why entropy was so low in the early universe.

Suggested Reading

Albert, *Time and Chance*.

Carroll, *From Eternity to Here*, chapter 2.

Von Baeyer, *Warmth Disperses and Time Passes*.

Questions to Consider

1. How does the second law play out in common household appliances, such as an oven or a refrigerator?

2. How does the second law play out in the evolution of the Earth? The universe? The biosphere?

The Second Law of Thermodynamics
Lecture 5—Transcript

The central theme of this course is that the mysterious thing about time is the arrow of time, the fact that the past is different from the future, and the single reason why there is a pronounced arrow of time in the real world is because of entropy, the disorderliness or the randomness of the universe is increasing. The entropy was lower, and it will be higher.

The reason why the entropy was lower is because of the initial conditions near the Big Bang. That is the picture we want to put together. We have not identified these various steps yet, and in this lecture, we are going to begin that justification.

We want to start with the second law of thermodynamics. That is the law of nature we have already mentioned that says that in an isolated system, in a system that you close off from any influences from the rest of the universe, or for that matter, in the universe itself, which we think is an isolated system, entropy either goes up or it stays constant. The entropy never goes down all by itself if you leave a system isolated from the rest of the world.

If you put a glass of water in a room temperature environment, and you put an ice cube into the glass of water, the ice cube will melt, and the water will cool down. But if you put a cool glass of water in a room temperature environment, it will never spontaneously form an ice cube. Things happen in one direction of time, but not the other. The way that things happen is the way that entropy goes up. That is the second law of thermodynamics.

This law was put together in the 19th century, in the 1800s, and originally it came from people who were not worried about the arrow of time or anything anywhere near that abstract. They were engineers. They were practical people worried about steam engines and machines and how to make them as efficient as they could possibly get. These people developed an entire science of thermodynamics. The engineers' ideas were taken by the physicists and they were generalized and made into a whole beautiful picture that contained laws of nature. The second law is obviously a law of thermodynamics.

The fact that we call it the second law should suggest to you there is probably a first law, and that is true. In fact, there are four laws of thermodynamics, and they are slightly oddly numbered. There is the zeroth law, the first, the second, and the third. Only the first and second ones are the ones you need to know. The second law says that entropy increases or remains constant in insolated systems.

There is a first law, but it is a little bit boring. The first law of thermodynamics says energy is conserved. It is a little bit more explicit than that. It says there is this kind of energy, there is that kind of energy, there is another kind of energy, and if you add up the change in one kind of energy plus the change in another kind, etc., the total change of energy is always exactly zero. That is the first law of thermodynamics.

What about the zeroth law of thermodynamics? That refers to the concept we call thermal equilibrium. We will talk about that a little bit more explicitly later, but thermo equilibrium says the temperature of two objects has come to a common value.

When we talk about the laws of thermodynamics, the thought experiment laboratory has boxes of gas. We take boxes of gas, and we put them next to other boxes of gas. We mix them together, we heat them up, and we do things like that. That is what we do over and over again when we are discussing thermodynamics.

If you take two boxes of gas and put them together, they come to the same temperature, and once they are there, that is thermo equilibrium. The zeroth law of thermodynamics says that if two boxes of gas are in thermo equilibrium with a third box, they are also in thermo equilibrium with each other. It sounds perfectly obvious, if a equals b and b equals c, then a equals c. It is that simple. That is why it is called the zeroth law and we do not worry about it that much.

The third law of thermodynamics is also one we do not worry about that much. It refers to the idea of the temperature absolute zero. We did not realize at first, when we measured temperature way back when, that there was a lowest possible temperature we could reach, but the science of thermodynamics implied that we could decrease the temperature of something only so far.

Today, this does not surprise us at all. Today, we know that stuff is made of atoms. The atoms move around, they jiggle back and forth, and the temperature has to do with how fast the atoms are moving, so clearly once the atoms are not moving at all, that is the lowest temperature we can have. That is what we call absolute zero.

The third law of thermodynamics says that the entropy goes to zero as the temperature of something goes to absolute zero. There is fine print there. There is a little subtext because in some special circumstances, you can have more than one kind of absolute zero, and then the entropy does not need to be zero. What you need to remember about the third law is nothing at all, but if you are going to remember it, it says there is a lowest possible temperature and the entropy is as low as it can get at absolute zero.

It is the second law of thermodynamics that is interesting, puzzling, and crucial to our dialog here in these lectures. The second law says that there is an irreversibility, that there is a direction of time. Entropy goes up in one direction and decreases as we go to the past.

We mentioned that C.P. Snow used the second law of thermodynamics as an example of a scientific concept everybody should be aware of. Snow was a Cambridge professor. He was being a little bit curmudgeonly. He was trying to tweak his colleagues in the humanities because Snow felt that people in his era were felt to be uneducated if they did not know the works of Shakespeare, but they were proud of the fact that they did not know anything about science.

He was trying to make the argument that knowing the second law of thermodynamics was as important to being a cultured person as knowing the works of Shakespeare was. Amusingly, later in life, Snow went and said that was not a good example to use because it is true that even though the second law is very important, and everyone should know it, we do not understand the second law at a fundamental level. There are still mysteries. That is why we have these lectures here. People should at least know what the second law of thermodynamics says, but asking that people understand it is a little bit too much.

Nevertheless, the second law is very, very important. It has been said more than once that if you wanted to choose a law of physics that we believe is true today and we will still believe is true in the future 1,000 years from now, the second law of thermodynamics is probably your best bet.

The interesting thing about that is that even though we do believe the second law is rock solid and true, when you look at it very, very carefully, as we will eventually do, you realize that the statement of it is actually only an approximation. It is not absolutely impossible for entropy to decrease spontaneously. It is just really, really unlikely. It is very, very improbable that it would ever happen in the lifetime of the universe from a macroscopically sized object, but it is so improbable that we treat it like a law.

The reason why we think the second law is going to stick around, that it is not going to be replaced, is because the second law is not a specific model of some fundamental interaction. It is not like the standard model of particle physics, general relativity as a theory of gravity, or Maxwell's theory of electricity and magnetism, the second law is a meta law. It refers to how different kinds of laws of physics can possibly work, and let's explore that to see exactly how it came about.

It is very interesting to note that the second law has an unusual history. It is almost as if the arrow of time went backwards when it comes to inventing the second law of thermodynamics. You might think that if we have a law that has to do with some concept called entropy, you would first define the concept, then you would figure out how the concept plays out in the laws of physics, and finally, you would formulate a rigorous law.

For entropy, we went exactly backwards. If we play the movie in reverse, we can understand the development of the second law by starting in 1877. That is when Austrian physicist, Ludwig Boltzmann, finally gave us the definition of entropy that we still use today. The difference between Boltzmann and his predecessors is that he believed in atoms, and we will discuss how, once you believe in atoms, that gives you a clue to how to understand entropy in a perfectly general context, far away from the idea of steam engines and machines.

It was before Boltzmann, it was in 1865, that the second law was written in terms of entropy and the word entropy was introduced. This was by a German physicist, Rudolf Clausius. He invented the word entropy from a Greek word. He invented the word entropy in German, but we borrowed it in English. It was Clausius who wrote down the statement of the second law that says the entropy of a closed system either increases or remains constant.

The second law itself predated the concept of entropy. The second law was formulated in 1824 by a French military engineer, not a physicist, named Sadi Carnot. It was Carnot who invented the idea that entropy increases, but he did not know about the idea of entropy itself.

You are going to ask how you can formulate a law that says that entropy always goes up if you do not know what entropy is and you have not invented the concept. It is because there are different equivalent formulations of the same law.

As we will see, Carnot invented a law of physics which we now know to be equivalent to the second law of thermodynamics, even though he did not know what entropy was. In fact, it is going to happen over and over again that there are different definitions of the word entropy, and people get into arguments about which definition is right or wrong. Those arguments are utterly a waste of time. There is no right or wrong definition of entropy. All that is there is the appropriate definition for the circumstances that we are talking about.

Let's look a little bit more carefully about how this notion of the second law came about. It was a combination of practical concerns with machinery and also nationalism. Sadi Carnot was a military engineer. He served in the French army, and the French army lost under Napoleon to the British. There was a rivalry at the time in the early days of the 1800s between the French and the British, as there has been often throughout history. In particular, the French were upset that the English were way ahead of them in the science and technology of steam engines.

It was James Watt who in 1769 invented the good, working steam engine, and the French were struggling to catch up. This bugged Carnot that his rivals were so far ahead of the French, so he set out to study the science of steam engines. What Carnot did was not to write down some particular design for an individual steam engine. What he wanted to do was to say, what is the most efficient possible steam engine we can imagine?

Forget about individual details of materials and what we are burning and so forth. What is the ideal steam engine? He wrote a wonderful book called *Reflections on the Motive Power of Fire* where he set out his thoughts about how efficient an engine could possibly be.

This sounds pretty dry and down-to-earth stuff, but Carnot made a wonderful discovery that would reverberate through the centuries, namely that there is a maximum efficiency that a steam engine can possibly have. It does not matter what you make it out of. It does not matter what materials you use or how fine you cut your pistons and so forth. There is a maximum efficiency you can possibly get by burning a certain amount of fuel and using it to power a steam engine. That was a surprising result in and of itself, and what Carnot realized was that his maximally efficient steam engine has a very interesting property, namely that when you run it, the cycle is reversible. You can run it back.

Remember, we have already hinted at the idea that reversibility is at the heart of the arrow of time. If you have processes which go forward and backward equally well, that is a reversible process, and it does not give you a direction to time.

We will talk about how the ultimate laws of physics are reversible in the next lecture, but what Carnot realized is that is a maximally efficient steam engine would be reversible, but any other steam engine would not be. Most steam engines, most real world engines, are very inefficient. They are not reversible. They are irreversible. Today, we would say they generate entropy.

Carnot's formulation of what we now call the second law is that you either have a perfectly reversible engine, which is like entropy staying constant, or you have any other engine which is inefficient. That means you are increasing the entropy of the world. That is why it is related to the second law. We know this sort of at an intuitive level in our everyday engine experience. If you have a gasoline engine in your car, what does it do? You burn the fuel, you make your car go, and that is it. You can only burn that fuel once. If you have gasoline or whatever it is, diesel in your car, you burn it, and you are done.

There are better kinds of engines as far as efficiency is concerned. That is why, these days, it is becoming increasingly popular to look at electric cars or hybrid engines that have both an electric engine and a gasoline engine.

The point is that for gasoline, you burn it, and it is done. You never get it back, but electricity can be regenerated. If you have an electric engine in your car and you go up the hill, you are doing work. You are taking energy out of the batteries. You are putting it into the car, and you are lifting the car up the hill.

In the electric engine, once you go down the hill, what can happen is you put on the brakes to slow yourself from going too fast, and in an ordinary car, brakes are entropy-generating machines. The brakes in a regular car smash two metallic plates together. The friction creates heat and noise and that is what slows down your car, dissipating energy into the environment.

In an electric car, you can have what is called regenerative braking. Instead of creating entropy, you can use your brakes to put energy back into your batteries. You are never perfectly efficient. Even the most efficient electrical car still generates entropy. The second law is still true, but you can be more efficient than if you burn the fuel once and for all.

This discovery of Carnot's was certainly interesting, but it seemed like an engineer's toy. It seemed like something you would keep in mind when you try to build a steam engine.

It was Clausius, Rudolf Clausius the German physicist, who promoted this idea of Carnot's and realized it was a law of nature. Carnot was interested in thermodynamics, the science of heat, and back in those days, in the mid 1800s, we thought that heat was a thing. Today we know that the heat is actually a measure of how fast atoms are moving, but back in the mid 19th century, we thought that heat was a substance. It even had a name, they called it caloric, and they would talk about the caloric moving from one box to another as one box heated up the box next to it.

What Clausius realized is that heat only flows in one way. If you take two boxes of gas, one that is hot and one that is cold, and put them next to each other, the hotbox heats up the cold one and they equilibrate. You go from two boxes at very different temperatures to two boxes at the same temperature. You go from one cold and one hot to two warm boxes.

There was heat in the hot box. There was not as much in the cold one, and the heat got shared. It moved from the hot to the cold. That can only happen so long. Once you have two boxes at the same temperature, the heat cannot travel anymore.

Notice this is a one-way process. It is an arrow of time. You can go from a difference in temperature to having the temperature be the same thing. You see where the irreversibility is coming in.

Clausius's version of the second law can be thought of as saying that heat gradients tend to even themselves out. Boxes of gas at different temperatures or other things, solids or liquids, they tend toward equilibrium, a state where everything has the same temperature and everything is uniform.

The brilliance of Clausius was that he did not simply state that this was true, he quantified it. He invented this concept that we now call entropy. He even invented the word entropy to quantify exactly what was happening.

Clausius formulated the second law. He said that things tend toward maximum entropy, thermodynamic equilibrium where everything is smoothed out. In any isolated system, the entropy will always go up.

Clausius did not know about Atoms. That would have to wait for Boltzmann, later, so Clausius invented a formula for entropy that only referred to the energy and the heat of the boxes of gas. Clausius's formula said that the entropy you create is the change in heat divided by the temperature which means if you have two boxes at two different temperatures, as the heat flows from one to the other, you can calculate the change in entropy.

You figure out how much heat left the hot box to move into the cold box. You divide by the temperature, and you get a number. The second law of thermodynamics says that number, the entropy created, is going to always be a positive number. It is never going to go down.

We can even do the math for this particular case. If you are a math expert, you know that I am going to cheat a little bit, because we said that the entropy generated is the change in heat divided by the temperature, but the temperature is changing. As this thing is happening, as the box is losing heat, it is decreasing in temperature.

To do this correctly, we need calculus. We need a way to integrate and get the total change in the heat in a box even as the temperature is also changing. We do not have that now. We are not going to need that. All I want to demonstrate to you is that Clausius's formulation matches our intuition that entropy goes up.

Think about what Clausius is saying in the context of these two boxes, different heat contents, different temperatures. We say that the hot box has a lot of heat and the high temperature. The cold box has less heat and a low temperature, so the entropy we create is the change of heat in the cold box divided by its temperature minus the change in heat in the hot box divided by its temperature. The reason why there is a minus sign is because the hot box is losing heat.

The amount of heat that is lost by the hot one is equal to the amount of heat that is gained by the other one, so the heat change is the same, just with an opposite sign. The entropy change in the cold box is the change in heat divided by a small number, its temperature. The negative change in entropy, in the hot box, is the change of heat divided by a large number, its temperature.

When we calculate the total entropy created, we get a big number, the change in heat of the cold box divided by the temperature, minus a small number, the change in heat in the hot box divided by its temperature, and a big number minus a small number is a positive number. It is greater than zero, so if you did not exactly follow that, that is fine. I wanted to let you know that these words have math backing them up.

Clausius did not simply write a poem; he wrote down equations that tell you how to calculate the entropy, what it is, and match this statement that entropy always increases with this picture of boxes of gas being put next to each other and equilibrating, coming to the same temperature.

You might worry. You might worry that this statement of the second law is not compatible with how we know the world works. It seems to be saying, Clausius's version of the second law, that temperature gradients always smooth out. That is to say that different objects always come to the same temperature no matter what we do.

That would seem to be at odds with our experience that we can heat things up. We can start with a room temperature situation like an oven and some eggs that we want to scramble, and we are able to turn on the oven and heat up the eggs. How is that possible if temperatures only even out? Likewise, we have refrigerators. We are able to take something at room temperature and cool it down by putting it in the refrigerator. How is that compatible with Clausius's formulation of the second law?

There are various different things going on. One is that heating things up is much easier than cooling things down. Creating heat is very easy to do. It is very difficult to destroy heat, and the reason why is because heat is a form of energy.

The first law of thermodynamics says that energy is conserved, but we can change energy between different forms. When you turn on the gas burner in your stove, you are not creating new energy. You are taking energy that was stored in the fuel and the natural gas that you are burning, and you are combusting that natural gas to release its energy in the form of heat. The energy was always there; we are changing its form from unconsumed natural gas into heat.

That is easy to do, and it does not violate the second law. The second law in Clausius's version tells us how heat moves around. It does not say that we cannot create heat by burning something.

Cooling something off, building a refrigerator, is harder, and this is something you should be familiar with. A refrigerator is more expensive than an oven or a stove. Air conditioning your house is more difficult and more expensive than heating your house. That is because decreasing the amount of heat in an object is more difficult than increasing it because you cannot just release energy. If you release energy, you increase the amount of heat, so how can we ever cool things down?

The answer is again in thermodynamics. You can take a box of gas, and you do not put any heat into it. You expand the size of the box. What that means is that the energy inside the box is the same, but because you are moving things apart, the molecules slow down and become colder.

That is the principle behind a refrigerator. You take some gas, Freon or some other appropriate gas, you expand it, and it cools down. This can be a very big effect. Freezers as well as refrigerators work that way, but note that you cannot cool off your living room by opening the door of your refrigerator. Why not? Because you are not actually destroying the heat that was ever in the Freon in your refrigerator.

What you are doing is you are moving it around. You expand the Freon. You cool it off, but at some point, you are going to have to recompress it to make it go through the cycle again and again. That is why the backside of your refrigerator is hot. It gives off as much heat as you cool things off inside the refrigerator box.

The total effect of your refrigerator is not to cool off the room it is in. If anything, it is to heat up the room you are in a little bit because nothing is ever perfectly efficient. That is the second law of thermodynamics.

You see how these laws that were created in the 19th century apply to the situations of our everyday life. In fact, this way of thinking about heat suggests another way of thinking about entropy. We said that entropy measures the disorderliness of stuff, and we also said that Clausius gave us a formula for it, the change in heat divided by the temperature.

Yet another way to think about entropy is as a measure of the uselessness of a certain amount of energy. Energy is conserved, but it can change forms. If you have energy in a low entropy form, you can do useful work with it. You can lift something up. You can drive your car. You can fly your airplane.

If you convert that energy into a high entropy form, it becomes useless. When you have a low entropy concentration of energy, that is called fuel. You have fossil fuels sitting in the ground. We pull them up. There is energy in them in a concentrated form. We can extract the energy because the entropy of the fuel is low.

Once you burn it, once you create that heat, the combustion, all of the fumes that come out of your car, you cannot go back. You can heat up a room in your house by taking wood and burning it. You cannot cool off a room in your house by unburning fuel and turning it into wood. That is the irreversibility of thermodynamics.

Along the way, we have to notice that some of the intuitive ideas we get about entropy do not always work in broader circumstances. Clausius's version of the second law says that if you have two boxes at different temperatures, put them next to each other and the temperature smoothes out. The gradient decreases.

If the system you are looking at is two fixed-sized boxes of gasses, that is a perfectly good formulation of the second law, but it would be a mistake to think that, therefore, the second law says that everything always smoothes out under all circumstances. It depends on what is going on, what the environment is of the box of gas or whatever it is you are looking at.

Think, for example, of a cloud of gas, not a box that is confined in some container, but gas particles all by themselves in the universe. If you went out into empty space, and you released a small amount of gas that you could hold in your hands, what would happen? It would expand. It would fly out in different directions. It would dilute away. That is what you might expect according to the second law. It would become smoother and smoother, spreading throughout the universe.

Imagine that you have a giant cloud of gas. Imagine you have an amount of gas equivalent to billions of solar masses, billions of times the mass of the Sun. Then something new kicks in called gravity. Of course, there is gravity even for a small box of gas, but it is not that important because the weight, the mass, is very tiny. Once you have enough gas to make a galaxy, gravity becomes very important and then what happens is the galaxy contracts. The gas pulls together, and you start lighting up. You form stars and planets and so forth. That is how our actual galaxy started.

It is important to realize that the process of forming a galaxy increases the entropy of the universe. It does not decrease. There are more ways to be looking like a galaxy than to be thinly spread out gas throughout the universe, but it does not make the universe smoother. If we wait long enough, eventually the universe would smooth out. You make that galaxy. The stars burn their fuel, and that fuel eventually gets spit out into the rest of the universe.

Along the way, the universe becomes lumpier, not smoother, so increasing entropy does not mean increasing smoothness. It often means that in certain circumstances, but in the more general case, we need a better definition.

That better definition was given to us by Ludwig Boltzmann in the 1870s. Remember, it was Clausius who formulated the second law and gave us the word entropy. Boltzmann came up with a better definition of entropy that is more widely applicable, and what Boltzmann had that Clausius did not, is a belief in the existence of atoms. These days, of course, we would say molecules, particles, and so forth. In the mid 19th century, you could not see atoms. Chemists believed in atoms because they helped them understand the rates of different reactions, but physicists were more skeptical. They said show me a picture of these atoms. If I do not see them, then I think that you are making things up.

People like Boltzmann and his supporters who included Maxwell and Gibbs and other of the giants of thermodynamics of the time, said that if you believe in atoms, we can understand the laws of thermodynamics. We do not need to simply postulate that they are true. We can derive them under certain very simple assumptions. What you call heat is actually the thermal energy, the random motions of the atoms, and by this kind of reasoning, they went from thermodynamics, the study of heat and its motion, to statistical mechanics, the study of the probabilities of atoms being in different arrangements.

That was a huge step forward in our understanding of entropy. What Boltzmann realized is that when you look at a glass of water or a box of gas, you do not see every atom. You see certain gross macroscopic features. That means that different arrangements of atoms will look the same to us. They will be macroscopically indistinguishable, and Boltzmann's brilliant insight is that entropy is simply a way of counting how many arrangements there are of the atoms inside a certain system.

In other words, the reason why entropy increases, according to Boltzmann, is simply that there are more ways to be high entropy than to be low entropy. That is the rigorous definition that corresponds to our intuitive feeling that entropy measures the disorderliness, the randomness. When entropy is low, you are in a macroscopic configuration that is very precisely arranged. There are only a few such configurations that look the same. When entropy is high, things are all spread out. There are many different ways to arrange the atoms. All of them look alike.

We are going to make Boltzmann's definition very precise. We are going to go into the details about its assumptions and its implications, but it is the right definition. There are other ideas about entropy in information theory, computer science, but Boltzmann's definition of entropy is the one that makes the arrow of time go. Once we understand it, we can ask why it started so low in the early universe, and that will help us appreciate the mystery of time.

Reversibility and the Laws of Physics
Lecture 6

We have seen that the arrow of time is fundamentally connected to the notion of irreversibility, but our best understanding of physics does not include this notion. For this reason, it's useful to think about the idea of reversibility and where it came from. In this lecture, we'll trace the development of the reversible laws of nature from Aristotle to Avincenna and Galileo and to Newton and Laplace. This exploration leads us to the question of why reversibility on the microscopic level is not reflected in our macroscopic world.

Newton's Insight

- Isaac Newton put the earlier ideas of Avincenna, Galileo, and others into a mathematical framework that allowed for the development of physics. His laws of motion and gravity suddenly made the world make sense.
    - Perhaps Newton's most famous law is the law of universal gravitation: the idea that what makes an apple fall from a tree and what explains the motion of the planets in the solar system is the same.

    - Newton realized that the force pulling the apple down from the tree could be the same force that explained the motion of the Moon around the Earth or the planets around the Sun. It was the universality of gravity that was new for Newton, and the reason that was a breakthrough is that the motion of the planets does not look like the fall of an apple.

- The motion of the planets is reversible, whereas the motion of the falling apple seems irreversible. It required the genius of Newton to realize that the laws governing the motion of the apple are reversible.
    - If you filmed the Moon going around the Earth with a movie camera and then you played that movie to an audience, no one

in the audience would be able to tell whether the movie was playing backward or forward.

- o A solution to the equations of motion that is played backward in time is still a solution. That means that if we know the state of a system right now, we can figure out what the state was in the past. There is a match between where we are today and where we were.

- Since Newton's time, every attempt we have made to understand the laws of physics works in that way. In other words, we think that there is a state of the system right now, and if we know that state, we can evolve it forward or backward in time. It is perfectly reversible. How does that accord with the apple falling from the tree?
  - o The motion of the apple does not look reversible. It starts at the top and ends at the bottom, never the other way around.

  - o The key is that the initial state—when the apple was tied to the tree by its stem—and the final state—when it lands on the ground—are irreversible processes because entropy is involved. The trajectory of the apple from the tree to the ground is one where air resistance is not that important—the trajectory is caused by the influence of gravity on the apple, and that trajectory is perfectly reversible.

  - o If you try to throw an apple into the air and catch it using the same motion for both actions, you will find that the trajectory of the apple is the same played forward and backward.

  - o This was part of Newton's great insight: that the fundamental laws of nature do not pick out a direction of time. It is only the interaction of the apple with the tree or the ground that gives us the impression of irreversibility.

- The same thing is true for many other simple systems in nature. Newton applied it to planets and moons. Physicists apply it to billiard balls on a frictionless surface.

- The lesson we derive from this is that when we have simple systems, the Newtonian laws of physics apply, and the rules are reversible.
  - You may guess that the difference between Newton's laws of physics and the complex, irreversible processes we see around us is simply a matter of how many moving parts there are.
  - Note, however, that complicated systems are made out of individual simple systems. If the rules of physics govern what simple systems do, then complicated systems should obey similar rules. The challenge of reconciling those two facts will be the challenge of understanding how entropy works in the real world.

Determinism
- Newtonian physics is self-contained. If we know the state of a system at any one moment in time, Newton's laws of physics let us figure out what the state of the system will be at any moment in the future and what it was at any moment in the past.

- We can build up the entire history of the system starting from the information at any one moment. The present completely determines every other picture of the universe in the past and the future.

- This is a particular way of thinking about physics that is not at all obvious. It was not, for example, a property of Aristotelian physics.
  - According to Aristotle, if you see something that is not moving, you do not know whether a little while ago it was moving—someone was pushing it—or whether it had been sitting still forever. There would be no way, in Aristotle's physics, to figure out the past given the present.
  - Newton said that if an object was not moving, the cause was friction; the object was interacting with the environment. If we could recover all the information about the object and the environment and work backward, we could figure out exactly the state of the system arbitrarily far into the past.

- The French scientist Pierre-Simon Laplace realized that Newton's laws contained within them a feature called determinism. Knowing the state of the universe now, we can determine its state with 100% accuracy arbitrarily far into the future or the past.
  - Laplace further posited a vast intelligence—now called Laplace's demon—that knows everything possible about the world at one moment of time, knows all the laws of physics to arbitrary accuracy, and has infinite computational capacity.

  - To this demon, there is no difference between the present, past, and future. If the demon knows the state of the universe at one moment in time, it knows it at all moments of time.

- We can imagine other kinds of laws of physics that are not deterministic. In an Aristotelian world, for example, the laws of matter and motion are teleological, that is, oriented toward a goal.
  - We can also imagine laws that are past-dependent. In fact, we generally think of the laws of nature as past-dependent; we think that what will happen to us in the future depends on where we are now and where we came from.

  - However, if we have *complete* information, we do not need to also remember the past in order to predict the future. The past is embedded in the current situation.

Conservation of Information

- If we have an object moving in some direction, what do we need to know to predict what it will do next? According to Newtonian mechanics, what we need to know to figure out the future and past of a moving object is its position and momentum (velocity).
  - But we need to know the position and velocity of every single part of that object.

  - With a human being, we would need to know the position and velocity of every single atom in the body to be able to predict what the human will do next.

- One implication of Newton's laws is that information is conserved, meaning that the total amount of knowledge that we could possibly have about a system is the same at every different moment of time. The position and velocity of every particle in some collection of particles is the total amount of information we can possibly have, and that amount of information is the same at every moment. Information is neither created nor destroyed.
    - Of course, particles can be created or destroyed. An electron and a positron can come together and be destroyed, but they do not disappear into nothingness. They create photons that carry off both their energy and their information.

    - Not only energy but information is conserved in the universe. That is why we can predict the future and retrodict the past. That is why the laws of physics are reversible—because the total amount of information does not change.

    - That is also why time has continuity, why the universe does not rearrange itself fully from moment to moment in time. The information contained in the universe is present at any one moment of time and is preserved from one moment to the next.

The Irreversibility of the Real World

- The tiny particles of which the world is made—quarks, electrons, neutrinos, and so on—basically obey Newtonian mechanics; their motion seems to be reversible. But when we get large collections of many, many particles, the real world is not reversible. Why not?

- Another way of stating this is to say that macroscopically, information does not seem to be conserved. If you believe that a glass of water with an ice cube in it is made of atoms, and those atoms have positions and velocities, then you believe that if you were Laplace's demon, you could tell the future and the past of that glass of water.

- But you are not Laplace's demon. You do not know the position and velocity of every molecule of water inside the glass. Instead,

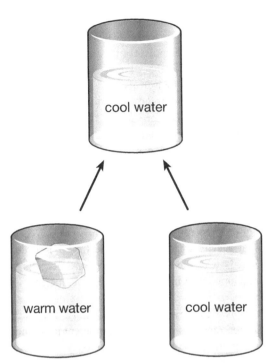

Consider the apparent loss of information with a glass of water: A glass of warm water with an ice cube in it and a glass of cool water can evolve into what appears to be the same state—a glass of cool water.

you have macroscopic information, and from that, you cannot reconstruct the past. In the evolution from the glass of water with or without an ice cube to the glass of water in the future, information is lost.

- Clearly, there is something going on when we change our perspective from the microscopic to the macroscopic laws of physics. This is a complex topic, but in the next lecture, we will sketch out a few simple assumptions that will enable us to match the irreversibility of the real world to the reversibility of Newton's laws of physics.

Suggested Reading

Albert, *Time and Chance*.

Carroll, *From Eternity to Here*, chapter 7.

Price, *Time's Arrow and Archimedes' Point*.

Questions to Consider

1. Think about everyday events, such as writing, watching TV, or reading. How is information conserved? How is it lost at a macroscopic level?

2. What are the implications of Laplace's demon for human history? Do you think we could ever have enough information about human beings to accurately predict the future?

Reversibility and the Laws of Physics
Lecture 6—Transcript

We have talked about how the arrow of time is fundamentally connected to the notion of irreversibility. Certain things in the world happen in one direction of time but not the other. You can take an unbroken egg and turn it into a scrambled egg. You cannot take scrambled eggs and turn them back into unbroken eggs.

What we have not talked as much about is why you should be surprised by any of this. After all, ice cubes melt. Causes precede effects. You burn fuel in your car and make it go. You do not recreate fuel, unmelt ice cubes, or have effects preceding causes. These are obvious features of the world. What is it that demands that we explain it?

The answer is that our best understanding of physics does not include the notion of irreversibility at a very deep level. That is a feature of the specific ideas of physics we believe are true. It is not a necessary feature. It is useful to think about the notion of reversibility and where it came from.

The reversible laws of nature that we know and love are based on those of Isaac Newton from the 1600s, but he did not spring out of nature fully formed. Newton built on what went before. To understand how he came up with these laws, it is useful to go back into history and think about where they came from.

Some of the first laws of physics, as we would call them, were actually developed by Aristotle and his fellow ancient Greeks. Aristotle gets a little bit of a bad reputation among modern scientists. He said a lot of things about physics that do not accord with the way that we think about things now, but we should be a little bit more charitable than that.

Aristotle was one of the great geniuses of human history. If he said something that we now think is untrue, it is not because he was being a dummy. It is because the context in which he was working was a little bit different, and we can learn a lot by thinking about what Aristotle was saying.

For example, when Aristotle talks about motion, he says that if you want something to keep moving, you have to keep pushing it. In modern physics, we have the idea that motion just continues, that momentum, the amount of impetus that something has, is conserved over time, that when something slows down it is because of friction, air resistance, and all that stuff.

The point is that Aristotle's statement at face value is simply true. In the real world, there is friction, there is air resistance, and if you do want something to keep moving, you do have to keep pushing it. The idea that you can ignore things like friction and air resistance when you talk about the motion of objects in the world is not an obvious idea. It is one that needs to be developed.

As far as we can tell, this idea goes back to about the year 1000. There was a Persian philosopher named Avincenna who was a philosopher in the Islamic golden age, and he was one of those polymaths who wrote about astronomy, medicine, and also physics. It was Avincenna who said that if it were not for air, if it were not for stuff in the world that things keep bumping into, if you were in a vacuum, a moving object would simply keep moving forever. The natural state of motion is to have continual motion, to move something away from continual motion requires a force of the kind we would now call air resistance or friction.

Avincenna's ideas were not quite developed into a mathematical theory, but Galileo, later on in the 1500s developed a very similar point of view. What Galileo did was develop experiments, and to show that you could decrease the effects of air resistance more and more, and you would find that for example, two objects of different masses would fall at the same rate. If you could have the air resistance on them being approximately equal.

What this story tells you is that a lot of the phenomenological features of our everyday world are not fundamental. They are features because we live in a world of friction, noise, and air resistance, what we would now call a world full of entropy. If you could get rid of all that entropy and dissipation, and look at the pure motion of particles, you might find something very different.

That is exactly what Isaac Newton found. What Newton did was to finally put these ideas from Avincenna and Galileo and others into a mathematical framework in which you could do physics. He developed laws of motion and laws of gravity that made the world suddenly make sense.

Perhaps his most famous law is the law of universal gravitation, the idea that what makes an apple fall from a tree and what explains the motion of the planets in the solar system, is the same law. There is the famous story of Newton sitting under a tree and seeing the falling apple and understanding gravity for the first time.

Probably, this is a famous story because Newton loved to tell it in his later life. Newton was definitely one who likes to burnish his own legend as a genius. It did not need a lot of burnishing because everyone recognizes that Newton was arguably the greatest physicist of all time, maybe the greatest scientist.

Newton did not invent gravity when he looked at that apple. What he realized was that the force pulling the apple down could be the same force as what explained the motion of the moon around the Earth or the planets around the Sun. It was the universality of gravity that was new for Newton, and the reason why that was a breakthrough is because the motion of the planets in the sky does not look like the fall of an apple.

The motion of the planets is reversible. It is smoothly running. It is not running down. It is not falling into disrepair or anything like that whereas the motion of the apple does not seem reversible to us. It starts in the tree. It falls down, and then it lands. That is a directionality to time. It requires the genius of Newton to realize that the laws governing the motion of the apple are reversible.

What we mean by reversibility here is that if you took a movie camera, which Newton did not have, obviously, but you made a picture of let's say the moon going around the Earth, and then you played that movie backwards for someone, they would not be able to tell which direction the movie was being played in. A solution to the equations of motion, we would say, that is played backwards in time is still a solution.

What that means is that if you know where you are right now, if you know the state of your system, you can figure out what the state was in the past. There is a match between where you are today and where you were. You can literally reverse the trajectory to find out where you came from if you knew everything about the system.

Since Newton's time, it has been the case that every attempt we have to understand the laws of physics works in that way. In other words, we think that there is a state of the system right now, and if you know the state, you can evolve it forward in time or backward in time. It is perfectly reversible. You might say, how does that accord with the apple falling from the tree? It does not look reversible. It starts at the top and ends at the bottom, never the other way around.

The trick is, of course, that the initial state when the apple was tied to the tree by its stem, and the final state when it lands on the ground and rolls in the dirt, are irreversible processes because entropy is involved. The flight of the apple, the trajectory of the apple from the tree to the ground is one where air resistance is not that important. It is the influence of gravity on the apple, and that trajectory is perfectly reversible.

If you try very hard to throw an apple into the air and catch it using the same motion for throwing as you do for catching, you will find that the trajectory of the apple is the same played forwards and backwards. This was part of Newton's great insight, that the fundamental laws of nature do not pick out a direction of time. It is only the interaction of the apple with the tree or the ground that gives you the impression of irreversibility.

The same thing is true, of course, for many other simple systems in nature. Newton applied it to planets and moons, the celestial bodies. We could also apply it to billiard balls on a frictionless surface. This is why, if you have taken a physics class, they like you to imagine things moving on frictionless surfaces. This was the great insight of Galileo and Newton, that you can make progress by ignoring the effects of friction, air resistance, and so forth.

There is the fundamental way that things move, and then there is the annoying little features that are added to that by friction. If you had billiard balls or hockey pucks that are moving on absolutely frictionless surfaces, if you have a pendulum that is rocking back and forth so much that the air resistance is not important, the laws of physics that describe those motions are completely reversible.

The lesson that we tend to derive from that is that when we have simple systems, systems with very few moving parts, the Newtonian laws of physics apply and the rules are reversible. You may guess that the difference between Newton's laws of physics and the complex irreversible processes we see around us is simply a matter of how many moving parts there are.

If you have a lot of moving parts, you get irreversibility. If you get very few moving parts, you get reversibility. That is a good idea. It is more or less on the right track, but there is a little bit of a tension there you should be aware of. After all, the pieces of many moving parts, complicated systems are made out of individual simple systems. If the rules of physics govern what the simple systems do, then the complicated systems that are made out of them should also obey similar rules. The challenge of reconciling those two facts is going to be the challenge of understanding how entropy works in the real world.

In the 1800s, along came a guy named Pierre-Simon Laplace. Laplace was a very interesting character in his own right. He made contributions both at the end of the 18th century and the beginning of the 19th century. He was a bit of a political climber, and a very successful one. He managed to be a success under the king during the French Revolution, under Napoleon, and also after Napoleon, so he clearly knew what he was doing.

Nevertheless, he was also famous for not always being completely politically correct. There is the famous story that Laplace dedicated a book he had written on celestial mechanics to Napoleon when Napoleon was the emperor of France, and so he gave a copy to Napoleon. It was very clear that Napoleon did not read this thick tome about celestial mechanics, but one of his advisors did, and to stir up trouble, Napoleon's advisor told the emperor that nowhere in his book did Laplace mention the idea of God.

Napoleon quizzes Laplace about this. He says, professor, why did not you mention the idea of God in your book about the working of the heavens, and Laplace could not resist coming back by saying I had no need of that hypothesis.

It is true. To understand the motions of the planets, you do not need God's help. Newtonian physics works. You might need God for other things, but the great thing about Newtonian physics is that it is self-contained. If you tell me the state of a system at any one moment in time, Newton's laws of physics let you figure out what the state of the system will be at any moment in the future, and what it was at any moment in the past, as long as you are giving me enough information about the entire system.

The important thing for our present purposes is that the information you need is instantaneous. It is about what the system is doing right now. You do not need information about what it will be doing in the future or about what it was doing in the past.

You can figure that out from the laws of physics as long as you have enough information, the laws of physics tell you, starting now, what will the system be doing a moment later and then a moment after that or a moment before and a moment before that. You can build up the entire history of the system starting from the information at any one moment. The now, the present time, completely determines every other picture of the universe toward the past and the future.

That is a particular way of thinking about physics that is not at all obvious. It was not, for example, a property of Aristotelian physics. If you see something that is not moving, according to Aristotle, you do not know whether a little while ago it was moving, someone was pushing it, then they stopped, and so it stopped or whether it was sitting still there forever. There would be no way, in Aristotle's physics, to figure out the past given the present.

Newton said that if you are pushing something and it comes to a stop, that is because of friction. That is because the object you are pushing was interacting with the atoms in the environment it was stuck in. If you could recover all of the information in those atoms, all the noise it made, all the heat that it generated while being pushed along a surface, and you could work backwards, you could figure out exactly the state of that system arbitrarily far into the past.

This is what Laplace figured out was an implication of Newton's laws. Laplace was not changing Newton's laws. He realized that they contain within them this feature called determinism. If you know the state now, you can determine what the universe will be with 100 percent accuracy arbitrarily far in the future or the past.

The way that Laplace talked about this idea of determinism was in terms of a hypothetical intelligence. He almost wrote a little science fiction story. What Laplace said was imagine a vast intelligence which we now call Laplace's demon.

He did not call it a demon. He had no need for that hypothesis either, but we imagine some creature which has supernatural abilities in the sense of supernaturally precise knowledge of the state of the world. If Laplace were writing today, he probably would have talked about an infinitely good computer with exact data about what the universe was doing, but he did not have computers in the 1800s.

Laplace says imagine what we now call Laplace's demon, which knows everything there is to know about the world at one moment of time. The demon also knows all the laws of physics to arbitrary accuracy and the demon has infinite computational capacity. The demon can do any calculation you can imagine doing.

What Laplace said was if you believe there could be such a demon and you believe that Newton's laws are right, then to the demon there is no difference between the present, the past, and the future. To Laplace's demon if you know the state of the universe at one moment in time, you know it at all moments of time.

We cannot stress enough, there is no Laplace's demon. Sometimes people say the introduction of chaos theory has made Laplace's demon impractical. Chaos theory says that if you have imperfect information about a physical system, so if you know its positions and its velocities, but not precisely, with a little bit of uncertainty, chaos theory says that as the time evolves, as you go into the future, your uncertainty will grow dramatically. A tiny bit of imprecision at one moment of time grows to a huge precision at some other moment of time. It is, therefore, impractical to imagine you can realistically predict the future even if you are Laplace's demon.

I have never been that impressed with this argument to be honest because the point is that Laplace's demon was never practical, has nothing to do with chaos theory. There is no way to have enough information about the universe, enough information about the laws of physics, and enough computing power to actually be Laplace's demon. The computer that we would need to have, the computer brain, to reproduce the Laplace's demon experiment would be the size of the universe and have the computational capacity of the universe. It would be another copy of the universe.

The universe knows what it is going to do but nothing else knows. Once you have imprecise information, then you are not Laplace's demon anymore, so of course, chaos theory makes it impossible to accurately predict the future, but it never was possible to accurately predict the future in a realistic case where we had imperfect information.

It did not have to be that way. The laws of physics work instantaneously. We know what happens at one moment of time. From that, we can predict the future and the past. That is what Laplace says Newton's laws imply, but we can imagine other kinds of laws of physics.

If Aristotle had been right, you would not be able to reproduce the past and future from our present state. In an Aristotelian world, the laws of matter and motion are teleological. That means they are oriented toward a goal. There is a certain place that matter wants to be, and that is where it goes. You discuss what will happen in terms of a future boundary condition.

We could also imagine laws of physics that were past dependent, laws of physics that it depended on where you were in the past not only what you are going to be doing in the future, but also if you want to predict further into the future than that. In fact, we generally think of the laws of nature as past dependent. We think of what will happen to us in the future as depending not only on where we are now, but also where we came from.

As one very down to earth example, I often find myself in Baker, California. This is a small town, about 700 people on I-15 going east from Los Angeles. There are not many reasons to find yourself in Baker, California. I like to go to the Mad Greek Café that has some great falafels. There is also the world's tallest thermometer in Baker, California, but you are passing through. You might say, given that I find myself in Baker, what will I be doing in the future?

There are two choices. You are going to get on I-15. You are going to go west to Los Angeles or you are going to go east to Las Vegas. That is what you are going to be doing. You might say, how do I predict that future? I can only know where I am going to go if I know where I came from. If I left L.A., and went to Baker, probably I am going to keep going east to Vegas. If I left Vegas and went west to Baker, probably I am going to keep going west to L.A. You think that knowing your current state is not enough to predict your future. You also need to know something about the past.

What Newton and Laplace are trying to tell you is that you are not being careful enough about your current information. The current information you have sitting there in the Mad Greek Café is not just, here I am in Baker. You also have the memory of where you were. You also have the mental image of the plans that you have of the future. You might have things written down in an itinerary. All of those count toward your current instantaneous state.

All of that information is required to predict what you will do next. If you are good enough to include all that information, then you do not need to also remember the past. The past is embedded in your current situation in the form of memories and records. Once you know all of the information, that is when you can predict the future.

The question becomes what is all of the information that you need. It is simplest to think about one falling apple or one baseball moving through space. If you have one object moving in some direction, what do you need to know to predict what it will do next?

Let's say I have a baseball, and I tell you its position. It is in the air. It is sitting at some place in the universe. Do you have enough information to predict what it will do next? The answer is clearly no because I have not told you the direction it is moving in.

Let's say that I give you the position of the baseball and its velocity, I tell you its speed and the direction it is moving in, can you then predict where it will be? The answer is yes. Do I need to tell you its acceleration? The answer is no.

Using the laws of physics, if you give me the position and the velocity, I can figure out the acceleration. The way the velocity is changing through time is something the laws of physics predict. What you need to know to figure out the future and the past of an object according to Newtonian mechanics, is the position and the momentum of the object, but what we mean is the position and the momentum of every single part of the object.

Momentum, by the way, is the velocity combined with the mass, so either the velocity or the momentum is equally good. You need the position and the momentum. You predict the future of the object if you know the position and the momentum of every single particle. Inside the baseball that is not that important because all of the atoms of the baseball move along with each other, but if you are a human being that can make decisions, you cannot predict what you will do next knowing your overall position and momentum. You need to know every single thing inside you, every position of every atom, every velocity or momentum of every atom, then in principle, you could predict what that human being will do next.

In practice, of course, no, you cannot predict what any human being is going to do next. This is a matter of principle versus practice. Newton's laws work in a certain way. They have certain implications, and those implications are important. It has nothing whatsoever to do with what we can actually realistically achieve in the real world. We cannot write down the position and velocity of every atom in any one human being. We do not have enough paper in the entire universe to pull that off.

Nevertheless, the implications are important. One way of stating the implications of Newton's laws is to say that information is conserved. When I talk about information being conserved, we have to be careful because the word information is thrown around in different contexts. Sometimes we refer to the known information, the information we actually possess, as the information.

When I say information is conserved, what I mean is the total amount of knowledge that you could possibly have about a system is the same at every different moment of time. If I know the position and the velocity of every particle in some collection of particles that is the total amount of information I can possibly have and that amount of information is the same at every moment. Information is not created. New things do not come into existence nor is it destroyed.

Of course, particles can be created or destroyed. An electron and a positron, which is the antiparticle of an electron, can come together and be destroyed, but they do not disappear into nothingness. That is crucially important. They create photons that carry off their energy and also their information.

Not only is energy conserved in the universe, but information is conserved. That is why we can predict the future and retrodict the past. That is why the laws of physics are reversible because the total amount of information does not change.

That is also why the laws of physics have this feature we have already mentioned, that time has continuity, that the universe does not rearrange itself fully from moment to moment in time. They are the same basic kinds of stuff. That is because the information contained in the universe is there at any one moment of time, and it is preserved from one moment to the next.

There is no comparable law when it comes to space. There is no preservation of information from point to point in space. From point to point in space, things can be wildly different. There can be a lot of stuff here and very little stuff over there, but time is special. Time has the property that if you know one moment of time, in principle, you know all other moments. That is the lesson of Laplace's demon.

It is philosophically evidence for externalisms, for the idea that all moments are equally real. If you are Laplace's demon, you would practically be one of Kurt Vonnegut's Tralfmadorians. To you, all moments of time would be obvious. You would have them all figured out, but it is not a practical concern for us because we live in a world of very, very incomplete information.

Even after Newton and Laplace had figured out that the fundamental laws of physics were reversible, we still thought that the macroscopic world was intrinsically irreversible. We knew that ice cubes melted, but did not unmelt, so there was some tension there already built in from the start.

Newton seemed to describe a world that worked equally well if you played the movie forward or backward, but the real world in which we live is not like that. Of course, when Newton formulated his laws, it was always possible to imagine that they only applied to certain special cases. Maybe Newton's laws tell you about apples falling from trees and planets moving around the Sun. Maybe they do not tell you about ice cubes changing.

In the 19th century, once we understood the second law of thermodynamics, the idea that entropy increases with time and does not decrease, we had what seemed to be a law of nature that is manifestly irreversible, one that works one way in time, but not the other. Maybe the second law of thermodynamics sits next to Newton's laws as a completely independent feature of the world.

The alternative, of course, and one that might be more attractive from the point of view of simplicity, is to imagine that somehow we can derive the second law of thermodynamics from Newton's laws of motion. This became a little bit more plausible in the 19th century once physicists figured out that the world is made of atoms, that macroscopic objects are not, in and of themselves, fluids or solids or something like that. They are combinations of little particles fit together in different ways, and you can then imagine that those little particles that we call atoms or molecules obey Newton's laws.

The tiny particles of which we are made, themselves, do not have any moving parts. Today we would really think about quarks, electrons, neutrinos, and so forth, but the Greeks called them atoms, so we still call them atoms.

If the atoms obey Newtonian mechanics, then we ask the question, is it possible to understand how the macroscopic world is irreversible? It is true. The atoms actually kind of do obey Newtonian mechanics. We have updated that to quantum mechanics, which we will talk about very quickly, but the basic rule is that the laws that govern the motion of atoms seem to be reversible. But, when you get large collections of many, many atoms, the real world is not reversible. What is going on?

Another way of putting this is to say that macroscopically, information does not seem to be conserved. If you believe that a glass of water with an ice cube in it is made of atoms, and those atoms have positions and velocities, then you believe that if you were Laplace's demon, you could tell what the future and the past of that glass of water was.

We are not Laplace's demon. We do not have the knowledge of the position and velocity of every molecule of water inside the glass. We have macroscopic information. We can see things like, oh, that is a warm glass of water. That is a cold glass of water. That is a warm glass of water with an ice cube in it.

If you think about it, let's say you see on a table a cool glass of water, and you want to know, there is no ice in it. You are asking yourself what was the state of this glass of water 10 minutes ago? It is possible that 10 minutes ago, this was a warm glass of water with an ice cube in it, and the ice cube melted and cooled it off.

It is also possible that 10 minutes ago, it was just a cool glass of water, so just with the macroscopic information we have, you cannot reconstruct the past. In the evolution from the glass of water with or without an ice cube to the glass of water in the future, information is lost. There is some answer to the question was there an ice cube in the glass of water or not, but once the ice cube melts, that information is lost.

Clearly, there is something going on when we change our perspective from the microscopic laws of physics to the macroscopic laws of physics. It is not, by itself, answering our question. We are raising a very difficult issue here. This is something that throughout the 19th century, physicists and philosophers fought over: How can it be true that the fundamental laws of physics are reversible and the macroscopic world is not if the macroscopic world is made of the microscopic world?

It took a lot of geniuses, including Ludwig Boltzmann and others, and even today, it remains a controversial topic, but we will sketch out how, with a few simple assumptions you can match the irreversibility of the real world to the reversibility of Newton's laws of physics.

Time Reversal in Particle Physics
Lecture 7

The reversibility of the fundamental laws of physics raises a puzzle for our understanding of the arrow of time: How can it be true that the deep laws of physics are perfectly reversible, yet the macroscopic laws strongly pick out a direction of time? The fundamental laws of physics on which that tension is based came from Newton's classical mechanics, but today, quantum mechanics and particle physics have affected our notion of reversibility. In this lecture, we'll begin to understand exactly why quantum mechanics and particle physics are not fully reversible yet still do not help us explain the arrow of time.

The Four Forces of Nature
- Particle physicists tell us that there are four forces of nature, two that are long range and two that are short range. To a particle physicist, a force is not like friction or centrifugal force; it is one of the fundamental interactions that allows different particles to bump into each other and go their own separate ways or pull together.

- The long-range forces are the ones we know about in our everyday lives: gravity and electromagnetism. Of course, gravity pulls us to the Earth and keeps the planets moving around the Sun. Light, X-rays, and radio waves are waves in the electromagnetic field.

- The short-range forces, or nuclear forces, are known as the strong force and the weak force. The strong force is what holds the nuclei of atoms together. An atom is an electron or a set of electrons orbiting a nucleus. The nucleus is a set of protons and neutrons, which are, in turn, made of quarks. That collection of quarks that makes up protons and neutrons in the atomic nucleus is held together by the strong nuclear force.

- The weak nuclear force has almost no effect on our everyday lives. What we need to know about it is that weak interactions are not invariant under time reversal.

Varying Degrees of the Arrow of Time
- We need to recognize that there are different degrees of having an arrow of time. There are different ways in which you could be different from the past going to the future.
    - The weakest, most trivial kind of arrow of time is represented by something in the world that goes one way and not the other—a car going one way down a road. In some sense, that's an arrow of time, but it's not a deep, fundamental insight about nature.

    - The next level of an arrow of time is represented by fundamental processes going one way and another way but perhaps at a different rate or in a different way. Note that this is not the same as entropy.

    - The third level of an arrow of time is that things can only go one way. That is true irreversibility, and that is the arrow of time that we care about.

- Even Newtonian mechanics includes the weak first-level arrow of time; things happen in one direction but not the other.
    - For example, the Sun always rises in the east and sets in the west. That is, in some sense, an arrow of time. We can tell which was morning and which was evening from where the Sun is, but it's clearly an accident of nature.

    - If we made a movie of the Sun and played it backward, an audience would be able to tell that we were playing the movie backward because the Sun would be setting in the east, but we could easily imagine a universe in which the Sun really did set in the east still obeying the same laws of nature.

- o Likewise, the hands of a clock move in the direction we call clockwise, but there's no difficulty in building a clock that goes counterclockwise. Even the expansion of the universe is something that just happens to be the case in our actual world. We can invent hypothetical universes that obey the same laws in which the universe is contracting.

- o All these are just accidents of nature in our world. They are different from the past to the future, but they are not fundamental arrows of time.

- The weak interactions of particle physics are in the second level. They are not irreversible, but one way they happen in time is different or at a different rate than when they happen another way in time.
    - o In the language of Laplace, if you know the state of the universe at one time, you can evolve it forward using the laws of physics, including the weak interactions, and from wherever it gets to, you can know where you came from.

    - o You can unevolve the universe—you can go backward—but you need to know that you are going backward in time because the weak interactions work differently going in one direction than the other direction.

Weak Interactions

- In particle physics, a paradigmatic experiment is to smash two particles together and watch what comes out. For a weak interaction, we could smash a proton and an electron together, and they could interact through the weak nuclear force and turn into a neutron and a neutrino.

- What about the time reverse of that interaction? Can we smash a neutron and a neutrino together and create a proton and an electron? The answers is yes; that is allowed by the weak nuclear force, but remember the force is weak, so the probability of that happening is very small.

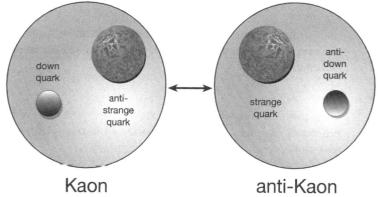

Nature gives us a perfect candidate for experiments in particle oscillation: the neutral kaon.

- It makes sense to us that the interaction of two particles can generate two different particles, but the phenomenon of particle oscillation involves only one particle turning into another kind of particle and then turning back.

- Particle oscillation illustrates a distinction in the concept of reversibility.
  o In previous lectures, we've said that reversibility means you can play the movie backward or you can recover where you evolved from. In the weak interactions, these are not quite the same.

  o You can recover where you evolved from, but if you really understood the weak interactions and you saw one particle oscillating into another one, you could figure out whether you were playing the movie backward or forward.

- A particle called the neutral kaon exhibits particle oscillation. The neutral kaon has an antiparticle that has the same mass and is also neutral. The fact that the kaon has no charge means that it has no conserved quantities. It can oscillate into an antikaon and back into a kaon.

- We know this oscillation occurs because neither the kaon nor the antikaon is stable. They both decay into different particles. A kaon decays into a negatively charged pion, a positron, and a neutrino. An antikaon decays into a positively charged pion, a regular electron, and an antineutrino.

- The important point for us is that if time reversal is violated in weak interactions—if the weak interactions treat going forward in time differently than they treat coming backward in time—then kaons and antikaons will decay at slightly different rates.

The CPLEAR Experiment

- In 1998, an experiment was conducted at CERN to test whether time reversal invariants were a good symmetry of nature.

- The researchers started with a 50/50 beam of kaons and antikaons. If the weak interactions respected going forward in time and going backward in time equally, then an equal amount of decay products from kaons and from antikaons would be expected.

- But the results showed more kaons decaying than antikaons. In other words, it takes more time for a kaon to turn into an antikaon than for an antikaon to go back to a kaon. The time it takes to go in one direction is different than the time it takes to go in the other direction.

- Time reversal is not a symmetry of the weak interactions. Unlike the other forces, the weak force does not respect time reversal invariant.
  o Laplace seems to give us an implication of reversibility, and because the world is reversible, there must be something symmetric about the past and the future. What is the symmetry between past and future that is a reflection of the underlying reversibility?

  o The answer is found in a symmetry called CPT; the different letters here stand for different hoped-for symmetries of nature that are all violated, but the combination of all three yields a good symmetry of nature.

CPT

- The C in CPT stands for charge reversal. Basically, the C operation changes a particle for its antiparticle. You might think that if we had a collection of particles in a box and we changed all of them for antiparticles and did not let the box interact with the outside world, it would be the same—it would weigh the same, act the same, and so on. It turns out that is not true.

- P stands for parity. This literally means that we flip the orientation—we flip clockwise for counterclockwise. In other words, we hold the experiment up to a mirror. Again, you might think that if we did an experiment and then a mirror-image of that experiment, we would get the same answer, but that's not true either. Nature violates parity.

- T stands for time reversal. Here, you would expect that if we did an experiment forward in time and then did it backward in time, we would get the same answer, but we have just seen from the CPLEAR experiment that this is violated.

- All of these transformations, C, P, and T, are things that you might guess are symmetries, but all three are individually violated by the laws of physics.

- The violation of parity in weak interactions was a tremendous discovery, made in the 1950s by three Chinese-American physicists. In 1964, researchers showed that the combination of parity and charge conjugation (CP) also violated symmetry. However, CPT seems to be a perfectly good symmetry of nature.

- The fact that the weak interactions violate time reversal invariance still does not explain the arrow of time.

Suggested Reading

Carroll, *From Eternity to Here*, chapter 7.

Greene, *The Fabric of the Cosmos*.

Questions to Consider

1. How are violation of time-reversal invariance and the arrow of time different, as defined by the second law?

2. Can you think of other experiments that would test time-reversal invariance in particle physics? Can you think of another test of parity invariance? What about CPT?

Time Reversal in Particle Physics
Lecture 7—Transcript

In the last lecture, we talked about the reversibility of the fundamental laws of physics and how that raises a puzzle for our understanding of the arrow of time. How can it be true that the deep down laws are perfectly reversible? There's no difference between moving toward the future and moving toward the past and yet the macroscopic laws, the world that we phenomenologically see around us, so strongly picks out a direction of time. However, our notion of the fundamental laws of physics on which we based that tension came from Isaac Newton, what we call classical mechanics, the whole system of Newtonian physics that says that there are positions, there are momenta, and if you knew the positions and the momentum of every individual atom you could predict things arbitrarily far into the future and retrodict them into the past.

Now, we know a little bit better than that. We know that the atoms of which real matter is made do not, strictly speaking, obey Newton's Laws of Physics. We have changed the laws of physics since Newton's time and you may ask does that change have an important impact. We now know that there is quantum mechanics. We now know that there is particle physics. Has that affected our notion of reversibility and the arrow of time? The bad news is that it has, but in a very subtle way. I do not like subtlety; I like things that are very, very simple to understand. The point is that quantum mechanics and particle physics change the story of reversibility, but not in a way that helps us explain the arrow of time.

To be honest, to be completely correct, we need to understand exactly why quantum mechanics and particle physics are not fully reversible and yet they do not help us explain why the past is different from the future. First let's look at particle physics; in the next lecture we will look at quantum mechanics, but right now let's take that for granted. Let's talk about the different interactions that govern our physical world.

Particle physicists say that we have four interactions, four forces of nature. There are two of them that are long range and two of them that are short range. To a particle physicist, a force is not like friction or centrifugal force or something like that. It is one of the fundamental basic interactions that lets different particles talk to each other, bump into each other, and go their own separate ways or pull each other together.

The long range forces are the ones we know about in our everyday lives. There is gravity; gravity pulls us to the Earth, it keeps the moon moving around the Earth, and the planets going around the Sun. The other long range forces, electromagnetism, sounds like two forces. There are two different phenomena, electricity and magnetism, but in the 19th century Maxwell and Faraday and others realized that the phenomenon of electricity and the phenomenon of magnetism are actually just two different manifestations of one underlying force, electromagnetism.

These long range forces are absolutely obvious to us in our everyday lives. We have magnets that tell us which direction we are walking in. We have the electromagnetic field; the thing that tells the electromagnetic force how to stretch across a space, and electromagnetic radiation is just a wave in the electromagnetic field. That means that light, X-rays, radio waves, basically everything that makes the modern world go is all just electromagnetism and the waves in the electromagnetic field. It is these long range forces that are very obvious to us in our everyday lives.

It is the short range forces, the nuclear forces they are sometimes called, that are of most interest to particle physicists. The short range forces have the boring names, the strong force and the weak force, or sometimes the strong nuclear force and the weak nuclear force. The strong nuclear force is what holds the nuclei of atoms together. An atom is an electron or a set of electrons orbiting a nucleus. The nucleus is a set of protons and neutrons which are in turn made of quarks, so that bag of quarks that make up protons and neutrons in the atomic nucleus are held together by the strong nuclear force.

There is also the weak nuclear force that has almost no effect on our everyday lives. It is, in fact, very, very weak. It is very hard to detect and it is also the weirdest of the fundamental forces. They all seem difficult to understand if they are all just forces of nature, but as a physicist you look more carefully at the four fundamental forces and you realize that three of them kind of make sense; gravity, electromagnetism, and the strong force are theories that we have that fit the data very well.

For the weak nuclear force, we also have the theory that it is a mess. It is much more complicated and it is a story that has to do with spontaneous symmetry breaking and the Higgs Boson, and I'm sure there are other Teaching Company lectures that talk about this. For our purposes, the weird thing about the weak interactions is that they are not invariant under time reversal.

We have talked a lot about the arrow of time, but if we are going to get subtle and careful about it, we should recognize that there're different degrees of having an arrow of time. There are different ways in which you could be different from the past going to the future. Let's distinguish between three different kinds of arrow of time from sort of the weakest to the strongest. The weakest, the least interesting, most trivial kind of arrow of time you could have is if you just have something in the world that is going one way and not the other. If you have a car going one way down a road, in some sense that's an arrow of time, but it's not a deep fundamental thing about nature.

The next level of an arrow of time is if fundamental processes go one way and they go the other way, but maybe at a different rate or in a different way. That's not what we talk about when we talk about entropy. The point about entropy is that you can take an egg and you can scramble it, but you simply cannot unscramble it. Let's imagine there was some fundamental feature of reality that could go back and forth, but when it went backwards and forwards it did it in different ways. That would be the second level of an arrow of time.

The third level of an arrow of time is that things can only go one way. That is true irreversibility and that is the arrow of time that we care about; the arrow of time that lets us remember the past and not the future, make choices about the future and not the past, that is a true irreversibility. That is the strong third level arrow of time.

Now even Newtonian mechanics, includes the weak first level arrow of time; things happen in one direction, but not the other. For example, the Sun always rises in the east and then it moves over and it sets in the west. That is in some sense an arrow of time. You can tell which was the morning and which was the evening from where the Sun is, but it's clearly an accident of nature. If you made a movie of it and played it backwards you would be able to tell that you were playing the movie backwards because you see the Sun setting in the east, but we could easily imagine a universe in which the Sun really did set in the east still obeying the same laws of nature.

Likewise the hands of a clock move what we call clockwise, but there's no difficulty in building a clock that went counterclockwise. Even the expansion of the universe is something that is just something that happens to be the case in our actual world. We can invent hypothetical universes that obey the same laws in which the universe is contracting. All these are just accidents of nature in our world. They are different from the past to the future, but they are not fundamental arrows of time. The increase of entropy in the universe is a fundamental arrow of time. That's a real irreversibility. You can mix milk into your coffee; you cannot unmix the milk out of your coffee. That's the kind of arrow of time that we care about.

The weak interactions of particle physics are in that no man's land of number two. Things are reversible, the weak interactions happen one way in time, and you can imagine undoing them. They are not irreversible, but when they happen one way in time it's different, it's at a different rate than when they happen the other way of time. That's the kind of arrow of time that you get in the weak interactions. It's not an irreversibility.

In the language of Laplace, if you know the state of the universe at one time, you can evolve it forward using the laws of physics including the weak interactions, and from wherever it gets to you can know where you came from. You can unevolve it, you can go backwards, but you need to know that you are going backwards in time because the weak interactions work differently going in one direction than the other direction.

You might ask how can this possibly be the case and if it is the case how could we know? What we are talking about here are interactions within particle physics. The stage in which particle physics acts is that of quantum mechanics, so we have the whole next lecture where we will talk about quantum mechanics in great detail. All we need to know right now is that quantum mechanics says that when you do an experiment you can't predict the outcome of that experiment with 100 percent certainty. Instead what you predict are the probabilities that one thing is going to happen versus some other thing.

In particle physics, a paradigmatic experiment is to take two particles, smash them together, let them interact, and watch what comes out. For the weak interactions for example, you can take a proton and an electron, smash them together and they can interact through the weak nuclear force and turn into a neutron and a neutrino. The neutron is the slightly heavier version of the proton that does not have an electric charge. The neutrino is a very, very light particle that just zips away. That is an allowed interaction according to the weak nuclear force; a proton hits an electron, turns into a neutron, and a neutrino.

Then you could ask the question, what about the time reverse of that interaction? Take a neutron and a neutrino smash them together and you can create a proton and an electron. That can happen. That is allowed by the weak nuclear force. If one thing can happen then the time reverse can happen. You could ask the question, do they happen at the same rate? Do they happen with the same probabilities, the way in which they happen precisely the same moving forward in time as moving backward in time.

The bad news is that as a practical fact, as a matter of actual procedure, those experiments would be extremely hard to do. Remember about the weak nuclear force; it is weak. It very rarely happens. When you take a neutrino, for example, and smash it into a neutron, you can hope that the weak interactions kick in and you create a proton and an electron out of it, but the chances, the probability of that happening is very, very small. As a practical matter, if you actually want to build the experiment and ask what happens, smashing two particles like that into each other is not the way to go.

Fortunately, there is a way to go. There is a very unusual and amazing phenomenon within particle physics called particle oscillations. It kind of makes sense to us that if you take two particles and interact them with each other, they can turn into two other particles. There's another thing that only involves one particle. You can have a single particle according to the rules of particle physics simply turn into another kind of particle and then turn back. This isn't something that happens in our macroscopic everyday world. It is very rare and the conditions need to be just right, but this does happen and it's known as a particle oscillation.

Neutrinos oscillate, and this is a whole industry within particle physics measuring the oscillations of neutrinos back and forth. The problem is that neutrinos are hard to capture and do experiments with because they're so weakly interacted. What we want is one kind of particle that oscillates into another and then oscillates back into the first kind and we can ask how fast does it happen? How fast does particle number 1 oscillate into particle number 2, versus particle number 2 oscillating back into particle number 1?

What you see here is we're distinguishing between the concept of reversibility meaning I can figure out where you came from and the related concept of if I play the movie backwards does it look the same to me. In previous lectures, we have been sloppy about drawing the distinction between these two things. We've said reversibility means you can play the movie backwards or you can recover where you evolved from. In the weak interactions it's are not quite the same. You can recover where you came from, but if you really understood the weak interactions and you saw one particle oscillating into another one, you could figure out whether you were playing the movie backward or forward.

Now it is not that often that these particle oscillations happen. If one particle oscillates into another one, it better be the case that none of the laws of physics are violated when that happens. For example, the mass of one particle better be the same as the mass of the other particle because mass is not coming into existence. The electrical charge of one particle better be the same as the one that it oscillates into because electrical charge is also conserved. When we want to do experiments in particle oscillation we need two different kinds of particles that look more or less identical, but truly are different at some fundamental level.

Fortunately, nature gives us the perfect candidate, a particle called the neutral kaon. This is a meson and if you are into particle physics you know that what that means is it's a combination of one quark and one antiquark. Quarks come in different flavors. There are up quarks, down quarks, top quarks, bottom quarks, charm quarks, and strange quarks, six different flavors of quarks. The kaon, the neutral kaon, is a combination of a down quark with an anti-strange quark. It has an antiparticle, the neutral antikaon, which is an anti-down quark and a strange quark.

None of these details are at all relevant. What you need to know is that there's a kaon that is neutral, it's a particle, it has an antiparticle that has the same mass and is also neutral. All particles have antiparticles with the same mass. The best thing about the kaon is because it is neutral it has no charge. It has no conserved quantities whatsoever. A kaon can oscillate into an antikaon and then they can oscillate back.

How do you know when it happens? This is the tricky experimental question. I say I have a kaon here, look it oscillated into an antikaon. The problem is they have exactly the same mass, the same charge, they interact in the same way, how do you know whether you have a kaon or an antikaon? The answer is that neither the kaon, nor its antiparticle are stable. They both decay into other particles and the good news is that a kaon decays into different particles than an antikaon decays into.

If you must know, a kaon, an ordinary regular particle kaon decays into a negatively charged pion plus a positron, which is the antiparticle of an electron; that's a particle with a positive charge, plus a neutrino. Kaon goes to negative pion, positron, and neutrino. The antikaon goes to the antiparticles of all those things. The antikaon decays into a positively charged pion, a regular electron, and an antineutrino.

The point is that we can wait and see. If you have something that is either a kaon or an antikaon, just wait for it to decay. If it's really a kaon it will decay into a negative pion, a positron, and a neutrino; the neutrino you don't notice so who cares, but if it's an antikaon it will decay into a positively charged pion and an electron. The point then is if time reversal is violated, if the weak interactions treat going forward in time differently than they treat coming backward in time, then kaons and antikaons will decay at slightly different rates. The rate for one will be a little bit different than the other.

What you do is you start with a box of kaons. In the real world you don't have kaons in a box, but you can create a beam of kaons. You smash particles into each other, get rid of everything that is not a kaon and just let the beam of kaons travel. As the kaons are travelling they're oscillating so you start with a 50/50 mixture of kaons and antikaons, they will oscillate back and forth into each other and eventually they will decay. If time reversal invariant is violated the rate at which a kaon turns into an antikaon will be different than the rate at which an antikaon turns back into a kaon. What that means is that even if you started with a perfect 50/50 mixture of kaons and antikaons, particles and antiparticles, you will not decay, ultimately, into the same 50/50 proportion of positrons and electrons or negatively charged pions and positively charged pions.

This is an experiment you can do and even though it is quite a subtle experiment to do, it requires a great deal of heroic experimental effort. It does not violate any of the laws of physics, it's not implausible in terms of very, very rare interactions, you just need to be very, very careful while you're doing it. This experiment has been done. It was done by an experiment at CERN called the CPLEAR experiment. CERN is the giant particle accelerator laboratory outside Geneva in Switzerland. Right now the CERN is most famous for the large Hadron collider, which is smashing protons together to look for new particles of nature, but the nice thing about a huge particle colliders that you get little spinoff experiments. You can take the particles that they don't need and use them to do other experiments.

The CPLEAR experiment was one of these little side experiments and it went in 1998 and they built a beam of kaons and antikaons. They started with 50/50, kaons and antiparticles, and you would expect if time reversal invariants were a good symmetry of nature, if the weak interactions respected going forward in time and going backward in time equally well, you would expect to see an equal amount of decay products from kaons and decayed products from antikaons. What you actually saw was different. There were more kaons decaying in that beam than antikaons by 2/3 of 1 percent, some tiny little fraction of a difference that says that the weak interactions like being kaons more than they like being antikaons. In other words, it takes more time for a kaon to turn into an antikaon than for an antikaon to go back to a kaon. The time it takes to go in one direction of time is different than the time it takes to go backwards.

Time reversal is not a symmetry of the weak interactions. That is an amazing fact. This is a deep feature of reality that, unlike all the regular forces that we know and love in our everyday lives, electromagnetism, gravity, even the strong nuclear force that keeps nuclei together, all of those forces respect time reversal invariant. They do things with the same speed forward in time or backward in time, but the weak interactions don't, but the crucially important feature of the world. You might wonder, why haven't they won the Nobel Prize for this? They discovered something crucial about the world, the world does not treat going forward in time the same as going backward in time.

The answer is that even though the CPLEAR experiment did verify that the weak interactions violate time reversal, it was not a surprise to us. Everyone in the world expected that that is exactly what they would find. You should think about why this is the case. There's sort of the vague reason and the more precise reason. The vague reason is that if you believe in reversibility, if you believe that Laplace was right that if you have the information about the current state of the universe to perfect precision that lets you predict both the future and the past then that predictability seems to treat the past and future on an equal footing and maybe there's an implication of reversibility that says well because the world is reversible there is something symmetric about the past and future. After all information is conserved as you go from the past to the future so shouldn't other things be conserved as well. That's kind of a vague hope that you might aspire to and you would like to make that more specific so what is the symmetry between past and future that is a reflection of the underlying reversibility.

There is an answer to that. There's a symmetry called CPT where the different letters stand for different hoped for symmetries of nature that are all actually violated. C and P and T are things that you might expect to be symmetries, but are not. The combination of all three of them gets you a good symmetry of nature.

C stands for charge reversal. Basically the C operation takes a particle and changes it for its antiparticle. You might think that if you had a collection of particles in a box and you changed all of them for antiparticles and you did not let the box interact with the outside world it would be the same; it is just particles change for antiparticles. It would weigh the same, it would act the same and so forth. It turns out that is not true.

P stands for parity. This literally means that you flip the orientation, you flip clockwise for counterclockwise. In other words, you hold the experiment up in a mirror and you might think that if I did an experiment and then did the mirror image of that experiment I would get the same answer, but the answer is also no. Nature violates parity.

T, of course, stands for time reversal and you would expect maybe that if you did an experiment forward in time and then did it backward in time, it would get you the same answer, but we have just seen from the CPLEAR experiment that that is violated. All of these transformations C, P, and T, are things that you might guess are symmetries, the things that if you do it once and then do it again you come back to where you started, if you change particles for antiparticles and then again you get back to the original particles, likewise for taking the mirror image or taking the time reverse, but all of thee transformations are individually violated by the actual laws of physics.

This by itself was a tremendous discovery back in the 1950s, the old wisdom before 1956 was that C, P, and T are all individually good symmetries and it was three Chinese-American physicists who proved to us that that was not the case, T. D. Lee, C. N. Yang, and C. S. Wu. Lee and Yang were theorists and they went through the literature because they had a sneaking suspicion there was something we did not understand about the weak interactions. The physics literature, the experiments, that they were able to discover that had already been done, showed us that parity, taking the mirror image of a physics experiment, seems to be experimentally a very good symmetry of the strong interactions, the electromagnetic interactions, and as far as anyone could tell the gravitational interactions as well, but what Lee and Yang realized was that no one had checked the weak interactions.

They went to their colleague C. S. Wu, who was an experimentalist at Columbia University, and they encouraged her to do an experiment to check whether or not the weak interactions respected parity. It turns out that this is one of my favorite stories about experimental physics. C. S. Wu was initially sort of skeptical. This was an interesting thing to do, it would take a lot of time, but her colleagues Lee and Yang convinced her that she should do it. She started getting very, very excited because she realized that if she actually noticed experimentally that parity was violated this was a huge result.

She and her husband had actually left China 20 years before in 1936, and for the last 20 years they had not gone back to their homeland. They had planned a return trip on a cruise liner and they were ready to go on their trip back to China for the first time in 20 years when Wu realized that she really needed to stay home and do her experiment. She let her husband go back to China without her, she stayed at home, did an experiment watching the decay of certain atomic nuclei and she was able to go back over the Christmas holiday to Columbia and report to Lee and Yang that she had shown that parity is violated by the weak interactions.

This is something that was certainly worth a Nobel Prize and sadly Lee and Yang won the Nobel Prize, but Wu did not. It was absolutely a change in how we think about how the fundamental laws of physics work. Parity, taking the mirror image of something is something that seems like it should be a symmetry, but it is not. Certain clever people said well maybe parity is not a symmetry and maybe charged conjugation, changing particles for antiparticles is also not a symmetry, but maybe if you combine them you get a symmetry, maybe if you change particles for antiparticles and then take the mirror image, CP would be a good symmetry of the world.

Well you can kind of guess how this is going to go. The answer is no. In 1964, Jim Cronin and Val Fitch did an experiment where they showed that CP was also violated. They also used the kaon system looking at different decays of neutral kaons. CP is also not a symmetry of nature so you might guess well what about CPT. What if I take an experiment change all the particles for antiparticles, take the mirror image and then run it backwards in time, maybe that's a symmetry of nature.

Depending on your level of cynicism you might say no I think that none of these are symmetries of nature. I don't even think that the CPT combination is going to be a good symmetry. It turns out that there is actually a theorem that you can prove that under certain very general assumptions the combination CPT must be a symmetry of nature. These are not very strong assumptions, basically locality, the things that happen don't instantly affect things that happen there, and invariance under moving at different speeds and moving in different directions, the certain very basic features of physics that we think are true and if those features are true then it must be the case that CPT is a good symmetry of nature.

Theorems are slippery beasts. It is always easy to convince yourself that their assumptions on which they are based are absolutely true and then to find out later that they are not, but still as far as we can tell experimentally CPT, that combination, seems to be a perfectly good symmetry of nature. What that means is if CP is violated, if you don't get the same answer when you change particles for antiparticles and take the mirror image, but CPT is not violated, than T must be violated. That makes sense. It's kind of like you're multiplying the violations by each other to cancel out. CP is violated, T is also violated, that's why CPT can be conserved and this is why nobody expected that CPLEAR would not find T violation. We already knew that CP was violated. We have a theorem that says that CPT must be preserved therefore it's an implication that time reversal invariance should not be respected by the weak interactions. That was the prediction of our theory; that's exactly what they showed at the experiment.

The moral of this story is that even though the weak interactions do give you a difference between going forward in time and going backward in time, it is a matter of definition whether or not you want to say what we call in words time reversal is T or whether what we call in words time reversal is the whole combination, CPT. If I define what I mean by time reversal to be first change particles with antiparticles then take the mirror image, then run the movie backward in time, that is a symmetry of nature. In fact, this is a theorem. If the laws of physics are reversible, if Laplace was right, then the information we have about the world now predicts what will happen in the future and retrodicts what happens in the past, there is always some kind of symmetry relating time reversal. There is always some kind of symmetry that says if I do enough to the system I can run the movie backwards and it will give me the same result as running the movie forwards.

For our purposes in these lectures, the important lesson is that even though there is some technical sense in which the weak interactions violate time reversal invariance that does not explain the arrow of time. Our macroscopic arrow of time, why we can remember yesterday but not tomorrow, why causes precede effects, why we grow older as time goes on, those are all due to irreversibility, those are not due to reversible processes like the weak interactions. If your friends ask you yes it is true that the weak interactions of particle physics do tell the difference between moving forward and backward, but no they don't, explain the arrow of time.

Time in Quantum Mechanics
Lecture 8

Quantum mechanics is a model of physics that was developed in the first half of the 20th century and came to replace the classical Newtonian view. It describes the world on both very small and very large scales, although on those very large scales, Newtonian mechanics also works. When we look at things that are very small, however, the predictions of quantum mechanics deviate from the expectations of Newtonian mechanics. Quantum mechanics is the correct theory of reality as far as everything we know about the universe is concerned, so we need to understand what the implications of this theory are for our understanding of the arrow of time.

Folktales about Quantum Mechanics
- Quantum mechanics is more complicated that just a theory of quanta—discrete packets. Quantum mechanics says that some things come in discrete packets, such as individual photons in an electromagnetic wave, but other things do not, such as time.

- In quantum mechanics, it's also true that we can predict only the probability of outcomes; we cannot predict with 100% certainty that any particular outcome will ever attain.

Quantum v. Classical Mechanics
- The laws of physics in classical mechanics are deterministic. They tell us what will happen and what did happen, and information is conserved along the way. Quantum mechanics is a much richer theory.

- The way quantum mechanics differs from classical mechanics is in what can be observed. In classical mechanics, we don't worry much about what we can observe. The state of the system is defined by its position and velocity.

- In quantum mechanics, we don't define a system by giving positions and velocities. In fact, there are no such things as positions and velocities. Instead, the world is described by something called a wave function.
    - This is basically a set of pieces of information that answers the following question: If we were to observe a certain thing, what is the probability that we would get a certain answer?

    - For example, I know that I have an orange, but I don't know exactly what state the orange is in. The orange has a wave function. If I look for where the orange is, the wave function can tell me the probability that I see it in one place or that I see it in some other place.

    - It is not just that quantum mechanics doesn't let us know where the orange is; quantum mechanics says there is no such thing as where the orange is. It's not that our measurements aren't good enough; it's that quantum mechanics doesn't let us go from the state of the universe to a prediction with a perfect match. There are always probabilities that get in the way.

- In quantum mechanics, we say that objects exist in superpositions of classical properties. In other words, there really is no position of the orange. It's not in one position or the other; it's in both. The wave function tells us how much of the superposition is here and how much is there and, therefore, the probability we have of seeing it in one position versus the other.

- The wave function can be a positive or a negative number. This means that we can contribute to the wave function in different ways, and they can interfere with each other. A positive contribution can cancel a negative contribution.

- There is no connection between this way of thinking and classical mechanics. There are no negative numbers for the probability of seeing something in classical mechanics. There are no negative numbers for probability in quantum mechanics either, but the "machine" we use to make the probability—the wave function—can be negative.

The Remarkable Consequences of Quantum Mechanics

- When we think about reality in this way—that it's made of wave functions, not positions and velocities—remarkable consequences follow.

- Some of those consequences are as simple as the fact that certain things are quantized. For example, in an atom, the energy levels that an electron can be in come only in discrete, quantized pieces. You cannot get any possible energy for the electron around an atom; you can get only certain possibilities.

- The major difference between quantum mechanics and classical mechanics is in the relationship between what exists and what you can see.
    o In classical mechanics, you can see everything there is. If you look hard enough, you can measure the position and velocity of every particle. Quantum mechanics embodies the wave function, but you don't see the wave function.

    o For example, the orange could have one position or another, but it doesn't have either one until I look at it. The orange has a wave function that describes the probability that I will see it.

    o When I look at the orange, that is when reality collapses. Before I looked at it, it was really in both positions—that's what the wave function is—but when I look at it, I will only ever see the orange in one position.

    o This is a bizarre feature of quantum mechanics: that reality seems to change when you look at it. But it's not true that our interactions with the world bring reality into existence. The laws of physics are still being obeyed, but the laws themselves are disturbing to us. They say that the rules by which physical objects evolve are different when we are looking at them and when we are not looking at them.

The Measurement Problem

- The rules of quantum mechanics tell us that things evolve in one way when we are not looking at them and another way when we look at them. This prompts us to ask: What counts as an observation? When does it become the case that reality collapses into one configuration?

- One of the puzzles here is the fact that wave function collapse seems to be irreversible, which means that information seems to be lost. If I look at the orange and I see it in one position, I don't know where it came from; I don't know what the wave function of the orange was before I looked.
  - There is a huge number of possible wave functions, all of which have some probability that I will see the orange in one position, and once I see it, that's where it is. I cannot reconstruct what the wave function was before I did my measurement.

  - That sounds like true irreversibility, an arrow of time, and this is why when we say that the fundamental laws of physics are reversible, we have to add a footnote saying, "except for quantum mechanics."

- In the 1960s, three physicists, Yakir Aharonov, Peter Bergmann, and Joel Lebowitz, argued that the reason we think there is time-asymmetry in wave function collapse is that we're asking a time-asymmetric question.
  - If we asked a symmetric question, we would get a symmetric answer, even in quantum mechanics. In other words, quantum mechanics doesn't impose time-asymmetry on the world; the world imposes time-asymmetry on quantum mechanics.

  - The lesson for us is that the arrow of time is not explained by the time-asymmetry of quantum mechanics. The time-asymmetry of quantum mechanics is explained by the arrow of time.

Interpretations of Quantum Mechanics

- The fact that information doesn't seem to be conserved in quantum mechanics has driven a philosophical field of interpretation.

- The leading interpretation, known as the Copenhagen interpretation, was developed by the founders of quantum mechanics in the 1920s and 1930s. This interpretation simply says that there are two different ways for a quantum mechanical system to evolve: what happens when you're not looking at it and what happens when you look at it.
    - According to the Copenhagen interpretation, the collapse of the wave function is real and it is irreversible. The collapse occurs when a macroscopic classical system, such as a person or a cat, interacts with a small-scale quantum mechanical system.

    - For a number of physicists, the Copenhagen interpretation is basically a stopgap measure; it's good enough as long as we don't do sufficiently careful experiments to distinguish between what happens on small scales and what happens on large scales.

- The leading alternative to the Copenhagen interpretation is the many-worlds interpretation, according to which there is no separate way of evolving when you are looking at a system versus when you are not looking at a system. The world evolves in only one way, and that is in accordance with the rules of quantum mechanics.
    - The many-worlds interpretation takes seriously the fact that the observer is a quantum mechanical system. Not only does an orange have multiple positions it could be in, but the observer has different possibilities, and they are also described by a quantum mechanical superposition.

    - When I look at the orange, my wave function interacts with the wave function of the orange, and I evolve into a superposition myself. There is a version of me that sees the orange in one place and another version of me that sees the orange in another place, and these two versions exist in separate worlds.

- o Every time two quantum mechanical systems interact with each other, the wave function splits rather than collapsing. It splits into alternative versions of reality, all of which are equally real.

- According to the many-worlds interpretation, there is one wave function for the whole universe, and it evolves according to one equation. It is true that that wave function describes many universes, but it is still the same amount of information contained in the same wave function.

- The evolution of the wave function is completely deterministic. In fact, Erwin Schrödinger wrote the equation that tells us how the wave function evolves. The reason information seems to be lost when wave functions collapse is that we don't know where we are in the wave function.

- The application of this story to reality seems outlandish at first, but the more we look at it, the more logical it becomes and the better it fits with other things we think are true about the universe.

Suggested Reading

Carroll, *From Eternity to Here*, chapter 11.

Greene, *The Fabric of the Cosmos*.

Questions to Consider

1. What is a quantum state? What are the different ways it can evolve?

2. Why does the process of observing a quantum system seem to pick out a direction of time? How does this relate to the second law?

Time in Quantum Mechanics
Lecture 8—Transcript

We have alluded several times now to quantum mechanics, the theory of the world, the model of physics, that was developed in the first half of the 20th century and applies to the really tiny world of subatomic particles. Quantum mechanics is really a way of doing physics, a way that we placed the Newtonian classical world and even though I have said it and other people have said it, it's not really completely accurate to say that quantum mechanics describes the subatomic world. The truth is quantum mechanics describes the whole world, the small and the very, very large. The good news is that in the very, very large scales Newtonian mechanics also works.

When you go through your life, when you go from here to the moon in a rocket ship or if you just get in your car and go shopping, you don't need to know quantum mechanics, but that doesn't mean that quantum mechanics isn't applying. It just means that you're in a realm where the predictions of quantum mechanics are the same as the predictions of the Newtonian mechanics so Newtonian mechanics gets you through the day. Quantum mechanics is not just a theory of the very small, it's only that when we look at things that are very small the predictions of quantum mechanics deviate from the expectations we would have from Newtonian mechanics and it is quantum mechanics that gets it right. Quantum mechanics is the correct theory of reality as far as everything we know about the universe is concerned, so we need to understand what the implications of that theory are for our understandings of the arrow of time.

The first step is to figure out that there're certain folk beliefs about quantum mechanics that are not correct. Certain things that you will hear if you linger on the wrong street corners about what quantum mechanics says that we want to remove from your mind. The first one is the idea that quantum mechanics is a theory of quanta, that is a quantum is just a discrete packet of something, quanta are just more than one quantum, and you might get the impression from the name that quantum mechanics says that everything comes in discrete packets. That is simply false.

Quantum mechanics is a little bit more complicated than that. It says that some things come in discrete packets. For example, if you have an electromagnetic field, a wave in an electromagnetic field, then you look at it, if you look at it carefully enough you see individual particles called photons. There are other things in quantum mechanics that are not discrete at all. Time, as far as we can tell, is not at all discrete in quantum mechanics so the name quantum mechanics is a little bit misleading. Do not let it force you into thinking that it implies things really come in discrete jumps.

The other potentially confusing thing about quantum mechanics is the role of uncertainty. There is something called the Heisenberg Uncertainty Principle, which is absolutely central to understanding the implications of quantum mechanics. Quantum mechanics makes the following very careful statement that if you want to predict the future outcome of an experiment based on the current state of the world, you can only predict the probability of getting a certain outcome. You cannot predict with 100 percent certainty that any particular outcome will ever attain. It's not that you are not good enough to do it, quantum mechanics says that it cannot be done. This leads to a sort of hand wavy uncertainty kind of thing where you say well quantum mechanics doesn't let us say what will happen.

Quantum mechanics says what will happen extremely precisely. It is just that the language it speaks is in probabilities, but those probabilities are not up for grabs. They are predicted to extremely high accuracy by quantum mechanics. It turns out that our understanding of particle physics based on quantum mechanics is the most precise theory ever invented. Quantum mechanics predicts probabilities. That doesn't mean that everything in quantum mechanics is simply uncertain.

To understand how quantum mechanics works, it is useful to contrast it with classical physics. When we talk about Newtonian mechanics, or classical mechanics, these mean the same thing. It is the vision of the world that was given to us by Isaac Newton and it's a vision of the world that was expanded on by Newton's followers in subsequent years, but it was still always within the framework of classical mechanics. When Maxwell figured out how to marry electricity and magnetism together his theory of electromagnetism was still a classical theory. When Einstein built the Theory of Relativity, the special theory and the general theory, those were still classical theories.

Quantum mechanics is a different way of doing physics that does not fit with how we think of electromagnetism according to Maxwell or relativity according to Einstein. You have to start from scratch and you can get things right, but you have to work in a different language. The language of Newton was what we talked about in the context of Laplace. You give me the state of the world right now, by which I mean the position and velocity of every particle of the world, and I can use the laws of physics to predict what will happen in the future and to figure out what happened in the past, to retrodict the past.

The laws of physics in Newtonian mechanics are deterministic. They tell me what will happen and what did happen and information is conserved along the way. Now even in Newtonian mechanics we might have uncertainties in our predictions because we have uncertainties in our measurements. If you do not know exactly the positions and the velocities of everything in the world, then you cannot make precise predictions, but there was always the possibility that if your measurements were better, your predictions would be better, and your measurements could in principle be infinitely good and then your predictions would be infinitely good.

Quantum mechanics is a different theory and a much richer one. Quantum mechanics, the way that it really differs from classical mechanics is in the role of what you can observe. In classical mechanics what you can observe wasn't even really something we worried about very much. The state of the system was defined by its position and its velocity where the position and velocity of all of its parts, and you could measure the position and the velocity of all of its parts. There would be some uncertainty if your measurement apparatus was not very good, but you could improve it just by making more and more precise measurements. The difference between quantum mechanics and classical mechanics is that just because something exists doesn't mean you can observe it.

In quantum mechanics, you don't define a system by giving positions and velocities. There are no such things as positions and velocities. In quantum mechanics, what the world is described by is something called a wave function. It is also called a quantum state or a quantum amplitude. It's basically a set of pieces of information that answers the question, "If I were to observe a certain thing what is the probability I would get a certain answer?" For example, I know that I have an orange, but I don't know exactly what state the orange is in. The orange has a wave function. The wave function will tell me if I look for where the orange is, the wave function says what is the probability that I see it in one place or that I see it in some other place.

This is not a matter of our ignorance. It is not just that quantum mechanics doesn't let us know where the orange is, the point is that quantum mechanics says there is no such thing as where the orange is. It's not because we are not good enough, it's because quantum mechanics doesn't let us go from the state of the universe to a prediction with a perfect match. There are always probabilities that get in the way.

The way that we talk about this is to say that objects exist in super positions of classical properties. In other words, there really is no position of the orange. There is a possibility that we would find the orange here and a possibility that we would find the orange there, but it's not in either one position or the other. It is in both. It is in a super position of both possibilities. The wave function tells me how much of the super position is here, how much is there, and therefore how much probability I have of seeing it in one position versus the other.

The other great thing about the wave function is that it can be a negative number, just like a wave on the ocean means sometimes the water is above its typical height, sometimes it's below it. A wave function can be a positive number or a negative number. That means that you can contribute to the wave function in different ways and they can interfere with each other. A positive contribution can cancel a negative contribution. This leads to really interesting phenomena in particle physics where, for example, you can add a new particle to your theory that gives some other particle a new way to interact and the net effect is that the probability of interaction goes down because that new addition was a negative number in the wave function. We get the probability from the wave function by squaring it. You square -1, you get +1, so the probabilities are always positive, but the wave function can be positive or negative.

There is no connection between this way of thinking in classical mechanics. There are no negative numbers for the probability of seeing something in classical mechanics. There are no negative numbers for the probability in quantum mechanics either, but the machine you use to make the probability, the wave function, can be negative. You have to square it to get what the probability is.

When you put all of this together, when you think about reality in this way, that it's made of wave functions, it is not made of positions and velocities, remarkable consequences follow. Some of those consequences are as simple as certain things are quantized. For example, in an atom the energy levels that an electron can be in come only in discrete quantized pieces. You cannot get any possible energy for the electron around an atom, you only get certain possibilities. That has crucial implications for how we live our lives.

For example, if you close your eyes it gets dark. This is not a surprise to you. You say well there is no light coming in, but if you look at it carefully there is a tremendous amount of light even in your closed eye because you are a warm body, you are giving off infrared radiation. The amount of infrared radiation inside your eye is much greater than the amount of light you get from the external world, but because it's lower energy per photon because the infrared light has longer wavelengths and less energy per particle. Even though there's a lot of infrared photons in your eye they're not enough to make you see anything and that's because of quantum mechanics. To get the chemistry in your eye to be excited and to say I'm seeing something requires individual photons of high energy. You can see visible light, you can't see infrared light. It wouldn't matter to you whether your eyes are open or not, you would be blinded all the time. You can thank quantum mechanics for the fact that we can see the world and apprehend it and everything goes dark when you close your eyes.

Now as far as we know, this quantization does not extend to space and time itself. There might be theories in the future that do quantize space and time. One of the problems with quantum mechanics is that we have not yet been able to reconcile it with relativity, not special relativity, that was Einstein's first theory, but his next theory general relativity, which we will talk about in great detail, his theory of gravity and the warping and changing of space and time. General relativity doesn't play nice with quantum mechanics and the search for a theory of quantum gravity is one of the major goals of modern physics. If we get there it may be the case that time and space become quantized. But, it may also be the case that they do not. We just have to say that we don't know the answer to this. The name quantum mechanics is just not a very good name.

The major difference between quantum mechanics and classical mechanics is in the difference, the relationship between what exists and what you can see. In classical mechanics, you can see everything that there is. If you look hard enough you can measure the position and velocity of every particle. In quantum mechanics, what there is is the wave function. But, what you see when you look is something different, you don't see the wave function. For example, that orange that we had that could've had one position or another; it doesn't have either one until I look at it. The orange has a wave function that describes the probability I will get a certain answer when I look at it. Before I look at the orange, it is truly in a superposition; the wave function describes it being in more than one place. It is when I look at it, when I make that observation, that the reality of the orange collapses to one possibility. You see the orange in one position and not in a superposition. What you see is different than what there is.

This is a bizarre and disturbing feature of quantum mechanics that reality seems to change when you look at it. This is something that has led to all sorts of bad popular science saying that our interactions with the world bring reality into existence. It's not like that. It is still the laws of physics are being obeyed. The problem is the laws of physics are really disturbing to us. They don't quite make sense to us. We're saying that the rules by which physical objects evolve are different when we are looking at them and when we are not looking at them. That is not an exaggeration; that is how quantum mechanics is taught in every undergraduate quantum mechanics course.

Many of us, including myself, are unsatisfied with this way of looking at things. We think that this is not how reality works, this is just a reflection that we do not yet understand it. This is called the measurement problem in quantum mechanics. The rules of quantum mechanics as they are written down say that there is one way that things evolve when we are not looking at them, another way that they evolve when we look at them. But then you ask yourself what counts as an observation, when am I observing it. What if it's a robot that is observing it, what if it is my cat that is observing it, what if it is just a rock that falls on it? When does it become the case that reality collapses into one configuration? We call this collapse literally the collapse of the wave function so the question is when do wave functions collapse, what counts as a measurement.

The short answer is we don't know, we don't have a reliable answer to the measurement problem, we don't know what the correct way of thinking about it is. One of the puzzles that makes us wonder is the fact that wave function collapse seems to be irreversible. If it is true that the orange is spread out in its wave function, it could be here, it could be there, it could be there, but when I look at it I see it in one location. Then if I look away and look back again it is still in that location so it was spread out before I looked, after I looked it is localized into one position. What this means is that information seems to be lost. If I look at the orange I see it in one position, I know it is there now, I know if I look again it will still be there, but I don't know where it came from. I don't know what the wave function of the orange was before I looked. There is a huge number of possible wave functions, of possible quantum states, all of which have some amplitude, some probability, that I will see the orange there and once I see it that's where it is. I cannot reconstruct what the wave function was before I did my measurement.

That sounds like true irreversibility. That's sounds like an arrow of time and this is why when we talk about the fact that the fundamental laws of physics are reversible, we often have to add a little footnote saying except for quantum mechanics that's not so clear. We can ask ourselves the question is this once again quantum mechanics' fault or is this our fault for not understanding quantum mechanics as well as we should.

Way back in the 1960s, this question was investigated very carefully by three physicists: Yakir Aharonov, Peter Bergmann, and Joel Lebowitz. And the question they asked is, is the time-asymmetry of wave function collapse really the fault of quantum mechanics or is it our fault for setting up the question in a certain way? Of course, I'm stating this in a leading fashion, what they argued was that it is our fault, it is not the fault of quantum mechanics. They said that the reason why we think there is time-asymmetry in wave function collapse is because we're asking a time-asymmetric question. If we asked a symmetric question we would get a symmetric answer even in quantum mechanics.

What Aharonov, Bergmann, and Lebowitz, ABL for short, what they said was think about how we usually work quantum mechanics. We prepare a system that has some wave function that the probability for being seen in different positions and then we observe it and we record the answer to our observation. In other words, there is an initial condition followed by an observation. What if we try to do that symmetrically, so first setting it up and then observing is one way of doing it. We could also set it up observe and then set it up again. This sounds weird. What we're saying is we have a state for the system to be in, we observe it, and then after the observation we insist that it be in some particular state so we prepare it, we observe it, and then what we call post-selection, we only include those times that the final state of the system is something that we chose ahead of time just like the initial state.

This is not something that we do ordinarily because we don't have access to the future. We have access to the past so we set things up in the past and then we make an observation. What ABL found is that if we insist on asking questions that are symmetric with respect to the past and the future then the formula for predicting the probability of quantum mechanical experiments is itself symmetric. In other words, quantum mechanics treats conditions set up before the experiment and conditions set up after the experiment in precisely the same way. There's no intrinsic time direction in quantum mechanics, it's just that we ask questions that presume a direction of time from the start.

You might be a little bit leery about this, how do you, after all, set up a condition after you've done the experiment? Well the answer in quantum mechanics is you just measure the object twice, you measure it in what you call the measurement, then you measure it later on and if the state at the end is not the state you want you throw away that experimental result. You set up the particle, for example, with some position or some wave function, you evolve it a little bit, you observe it, you let it evolve more, and then you insist that it be in some position at the end of the day. You can do that by measuring its position, by doing this experiment many, many times and when it was at the end of the day in the position you wanted it you keep that experimental result. If it is anywhere else you throw it away, you only keep the experimental results that agree with your final boundary condition.

These experiments have been done. People have checked the formula derived by ABL and they found that it agrees with experiment. The conclusion is that quantum mechanics does not impose time-asymmetry on the world. The world imposes time-asymmetry on quantum mechanics. The reason why wave function collapse seems to be time-asymmetric to us is because we have asked a time-asymmetric question. We set up initial conditions and then make an observation rather than making an observation and then imposing final conditions.

Why do we do that? Why is it that we always set up the system and then observe it rather than the other way around. The answer comes not from quantum mechanics, but from our knowledge of the world. Remember we remember what the world was like in the past. We do not remember the future so if we set something up and then let time pass and then do the experiment, when we do the experiment we can still remember how we set up the apparatus. We cannot do that for the future because we do not have a memory of the future. Why don't we have a memory of the future, ultimately because the entropy was lower in the past. This is an assertion that I have made before, we have not justified it yet, we'll have a whole lecture in the future where you will learn why the reason you remember the past and not the future is because entropy used to be lower in the past. The future is harder to predict than the past was because in the past the entropy was smaller.

The lesson of this is that the arrow of time is not explained by the time-asymmetry of quantum mechanics. The time-asymmetry of quantum mechanics is explained by the arrow of time. The reason why wave function collapses irreversible is the same reason why the melting of an ice cube is irreversible. It is because we know more about the past than we know about the future. This doesn't mean that there's nothing to worry about. We still need to ask about what about conservation of information. It's not like the ice cube where information didn't seem to be conserved because we can't keep track of it right. We cannot see the position momentum of every water molecule in the glass of water, but in the quantum mechanical system it seems like we can. We have a single particle or a single orange we know everything there is to know about it, why can't we just keep track of all the information in the quantum system as time goes on?

This is a very good question and again this is a source of controversy among people who work in quantum mechanics as professional physicists. The naïve answer is that in quantum mechanics information is not conserved. If I look at an orange and I see it in a particular place I don't know what the wave function was before. It appears to be gone. This whole controversy is what drives the philosophical field of interpreting quantum mechanics so people argue over which is the right way to think about quantum mechanics, what is the right interpretation to impose upon this weird way of thinking about the world.

The leading interpretation, in the sense that most people who do physics accept that it is probably true, is called the Copenhagen interpretation. It is what was developed by Niels Bohr and Werner Heisenberg and other of the founders of quantum mechanics back in the 1920s and '30s. The Copenhagen interpretation says something very simple. It says that there are simply two different ways for quantum mechanical systems to evolve. There is what happens when you're not looking at it and what happens when you look at it. The collapse of the wave function according to the Copenhagen interpretation is real, it is true, and it is irreversible. Collapse happens when a macroscopic classical system like a person or a cat or a robot interacts with a small-scale quantum mechanical system.

The problem of course is this interpretation makes no sense whatsoever so I said that most working physicists believe in it, but I don't think that they should believe in it. What you will find is that in the subset of physicists who think hard about quantum mechanics nobody believes in the Copenhagen interpretation. It is basically a stopgap measure. It's sort of a kludgy way of thinking about the world that is good enough as long as we don't do sufficiently careful experiments to distinguish between what happens on small scales and what happens on large scales.

Remember I started the lecture by saying that quantum mechanics is not just a theory of the very small, it is a theory of the whole world including the very large. Quantum mechanics should apply to me and you as well as it applies to individual electrons. If that is true then the Copenhagen interpretation cannot be right. The Copenhagen interpretation relies on the fact that you and I are not quantum mechanical, but we are classical things, but no one really believes that the real world is made of small particles that obey quantum mechanics and then when you take all those small particles and put them together to make a human being they suddenly stop obeying quantum mechanics.

The Copenhagen Interpretation doesn't have answers to questions like why is the behavior of the world different when we look at system versus when we don't. What really counts as looking, does a virus interacting count as looking. What is the dividing line between the classical world and the quantum world? There is another interpretation of quantum mechanics; there are many actually, but the leading alternative to the Copenhagen interpretation is called the many worlds interpretation.

The many-worlds interpretation says that there is no separate way of evolving when you are looking at a system versus when you are not looking at as system. There is only one way the world evolves and that is in accord with the rules of quantum mechanics. The differences is that the many worlds interpretation takes seriously the fact that you, the observer, are a quantum mechanical system, not just the thing you're looking at, but the observer is just as quantum mechanical as everything else, so not only does an orange have a position it could be in or multiple different positions, not only does an electron have the possibility of being found in different places in the universe, but you the observer have different possibilities and they are described by a quantum mechanical super position just as much as the state of the electron is.

What does that mean? What that means is that when I as a classical observer interact with the quantum mechanical particle, the right way to treat it is not to treat me classically, but to think about my wave function. My wave function interacts with the wave function of, let's say, the orange. The simple minded way that we talked about it before was I look at the orange and it could've been in two positions, but I will only ever see it in a single position. What happens according to the many worlds interpretation is that I look at the orange and I evolve into a super position myself. There is a version of me that sees the orange in one place; there's another version of me that sees the orange in another place, and these two versions live in separate worlds. They live in separate universes so this is a theory of multiple universes. Every time that two quantum mechanical systems interact with each other the wave function doesn't collapse, it splits. It splits into alternative versions of reality, all of which are equally real.

Now this story sounds a little outlandish. People don't like the many-worlds interpretation both for good reasons and bad reasons. One of the bad reasons is that they say that is an awful lot of universes out there. Every time we observe a quantum mechanical system a whole new universe comes into existence. The reason why that is a bad way of thinking about it is that the world is not weighted down by the number of universes that it has. The many-worlds interpretation says that there is one wave function for the whole universe and it's simply evolves according to one equation, one wave function, one equation, that is the simplest possible theory we can write down.

It is true that that wave function describes many, many universes, but it is still the same amount of information contained in the same wave function. That's one of the benefits of the many-worlds interpretation. Another benefit is that the evolution of the wave function is completely deterministic. There's an equation, Schrödinger's equation written down by Erwin Schrödinger another one of the giants of the early days of quantum mechanics that tells you how the wave function evolves. In the many worlds interpretation the evolution of the universe is just as deterministic and reversible as it was for Laplace and Newton.

If you tell me the wave function of the universe now and you tell me Schrödinger's equation, I can evolve that wave function forward in time, I can evolve it backward in time and no information is lost. You might ask why it seems that information is lost when wave functions collapse and that is because I don't know where I am in the wave function. There are different copies of me in the many different universes that are produced; that is why information seems to be lost. It's not that entropy is increasing, it's not that the fundamental laws of physics are truly irreversible, it's just that my ignorance increases because I don't know where I am.

This is a complicated question whether or not this fairy story can actually apply to reality. It seems outlandish on the face of it, but the more you look at it the more logical it becomes and the better it fits in with other things we think are true about the universe. For our present purposes, the important thing is that understanding quantum mechanics reveals that there are true mysteries to which we don't know the answer. Richard Feynman once famously said, "It's not that hard to understand general relativity, but I can safely say that nobody understands quantum mechanics." These issues of interpretation are very real, but for our immediate purposes all we need to know is quantum mechanics does not explain the arrow of time in our universe. It does not explain why entropy increases toward the future, it is easy to explain why entropy increases. The real mystery is why entropy was lower in the past. Quantum mechanics doesn't help with that. To understand that we're going to have to think a little bit more carefully.

Entropy and Counting
Lecture 9

Thus far in the course, we have talked about time and measuring time, and we've said that the real mystery of time is the arrow of time, the fact that the past is different from the future. We have also claimed that the reason the past is different from the future is that entropy increases. In this lecture, we will dive into what entropy really is—give the rigorous modern definition of it—and explain what we do and do not understand about how entropy works.

Boltzmann's Introduction of Atoms

- Our modern understanding of entropy comes from the Austrian physicist Ludwig Boltzmann in the 1870s. What Boltzmann added to the previous work of Carnot and Clausius was the idea that matter is made of atoms. This is a tremendously powerful idea. It means that ice, water, and water vapor are not different things but different arrangements of the same fundamental set of things.

- Of course, even in a very small collection of macroscopic material, there is a large number of atoms. The number that scientists use to indicate a macroscopic amount is Avogadro's number, which is the number of carbon atoms in 12 grams of carbon ($\sim 6 \times 10^{23}$).

- Avogadro's number is an indication that we can throw away information. That is the key to going from the world of atoms to the macroscopic world. It's called "coarse graining."
  - This is basically the idea that we can't possibly keep track of what every atom is doing in any tiny macroscopic object, so we need to keep track of less-than-perfect information.

  - Laplace taught us that if we wanted to know exactly what was going to happen, we would have to know the position and velocity of every atom, but that is completely impractical in the real world. Instead, we average; we take typical features of the whole collection of atoms in some macroscopic object.

o These typical features include the density of the material (the number of atoms in any one volume), the pressure, the temperature (the average kinetic energy), and heat (the total thermal energy). Boltzmann realized that entropy, like these other thermodynamic properties, could be thought of as a property of the arrangement of atoms.

Boltzmann's Formula for Entropy

- Consider a box of gas with a partition in the middle of it that has a tiny hole. Most of the time, the atoms in the gas bounce around inside the box on either the right side or the left, but occasionally, an atom will pass from one side to the other through the hole.

- If we start with all the atoms on the left and we wait long enough, the box of gas will equilibrate, just as heat equilibrates when we bring a hot object close to a cold object. Intuitively, it's clear to us that entropy increases during this process.
    o Now imagine that we have 2000 particles of gas, all on the left side of the box. Every atom has a 1% chance per second of going through the hole; it doesn't matter whether the atom is on the left or the right. What happens in this circumstance?

    o Obviously, if all the atoms start on the left, they will gradually diffuse to the right until they even out. After only about 200 seconds, we can't tell the difference between the left and right sides of the box. Clearly, entropy has increased, and clearly, this is an irreversible process.

- Boltzmann tried to describe this process quantitatively. He tried to figure out how many combinations there would be of atoms in one configuration versus another and show that our intuitive notion of high entropy corresponds to many different arrangements of a particular configuration.

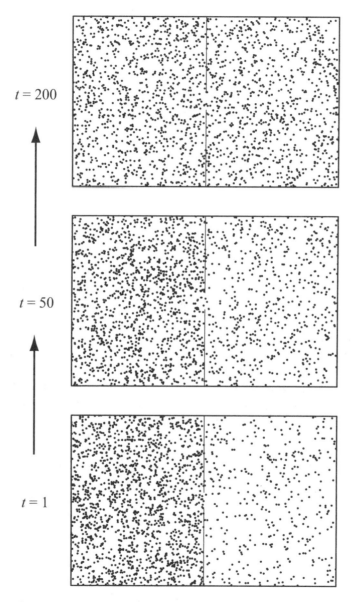

Entropy increases over time as the distribution of particles equilibrates.

- o If we try to count the number of arrangements of atoms, we see that it quickly becomes astronomical. The maximum number of arrangements with 1000 atoms on the right and 1000 on the left is greater than 2×10^{600}.

- o The number of arrangements is small when there is an imbalance between left and right and large when there is a balance. Our intuitive notion is that entropy is higher when things are balanced and lower when it is unbalanced.

- Boltzmann developed the idea that entropy is really just counting the number of arrangements. In other words, entropy tells us the number of microscopic arrangements of a set of atoms that look indistinguishable to our macroscopic perception.

- Boltzmann's formula is that entropy is proportional to the logarithm of the number of indistinguishable states. This formula is $S = k \log W$, where S is entropy; k is Boltzmann's constant, a proportionality constant; and W is the number of ways we can rearrange things.

Understanding Logarithms

- A logarithm is basically the power to which you would raise the number 10 to get the number you want. The logarithm of 10 = 1. Why? Because in order to get 10, you raise 10 to the power 1, $10^1 = 10$. The logarithm of 100 is 2 because $10^2 = 100$. The logarithm of 1000 is 3 because $10^3 = 1000$ and so forth. The logarithm of 1 is 0 because $10^0 = 1$.

- Taking the logarithm of a power of 10 is easy, but we can also define the logarithm for every number in between. Basically, the logarithm is a function. It's small when the number you're taking the logarithm of is small, and it's negative when the number is less than 1.

- Notice that the logarithm grows slowly. The logarithm goes from 0 to 1 to 2 as the number we are taking the logarithm of goes from 1 to 10 to 100. The logarithm tells us the number of ways we can rearrange something, but it tells us in a more compact and manageable form.

- Recall that in our box of gas, the lowest-entropy configuration had 2000 particles on the left and 0 on the right. The number of combinations that looked like that was 1 and the logarithm of 1 = 0, so the entropy of that configuration is 0. It's the most orderly configuration we can have.

- With 1999 atoms on the left of our box and 1 atom on the right, the number of combinations was 2000, and the logarithm of 2000 is 3.3. With 1998 atoms on the left and 2 on the right, the number of combinations was almost 2 million, and the logarithm of that is 6.3. The highest number of combinations was 2×10^{600}, and the logarithm of that is just 600.3.

- Remember that before Boltzmann, Clausius had given us a notion of entropy that was very specific about exchanging heat. Boltzmann wanted to give a much more general definition, but he still wanted his general definition to match with Clausius's specific definition in the same arrangement.
    - One of the properties of entropy as it was already understood is that if we have two systems, each with a certain entropy, and we put them together to make one larger system, the total entropy is the sum of the individual entropies.

    - But Boltzmann wanted to define entropy in terms of the number of arrangements, and when we connect two systems, we open up many more ways to arrange the individual particles. The number of arrangements is not the sum of the individual arrangements but the product.

    - We take the number of arrangements we had in one system and multiply it by the number of arrangements in the other system. If we try to define entropy as the total number of ways we could arrange the atoms, that would give us an incorrect formula.

- The magic of the logarithm is that it fixes this problem. It has this wonderful property that the logarithm of a product is the sum of the logarithms. The logarithm of 10 × 10 is the logarithm of 10 + the logarithm of 10, which is 1 + 1, which is 2.

- Boltzmann realized that putting the entropy proportional to the logarithm of the number of arrangements preserves the property that the entropy of two systems added together is the entropy of each system added to the entropy of the other system.

Coarse Graining

- The microstate of a system is the exact Laplacian state. It is what we need to predict the future and to retrodict the past. If we know the microstate, then the evolution of the system would be perfectly reversible.

- But if we look at a glass of water with an ice cube in it, we don't see the exact microstate of every atom inside the glass of water. We know that it is a glass of water; it has a certain temperature; and maybe it has an ice cube. We call that information the macrostate.

- In the process of coarse graining, we define ahead of time what it is that we can macroscopically observe, and two arrangements of atoms that look the same according to our macroscopic observations are put into the same macrostate. We take all the possible states we could have and chunk them up into microstates that look the same macroscopically. The set of all those microstates is the macrostate.

- What Boltzmann's entropy tells us is that the entropy of an individual configuration of particles is the logarithm of the total number of other configurations in the same macrostate. We coarse grain by making some observations; we calculate how many particles and how many arrangements look the same, take the logarithm, and that's the entropy.

Summing Up Boltzmann

- Once we have Boltzmann's program in place, we have an understanding of how entropy works at a quantitative level. We take all the ways that a state can be—all the microstates of some system—and we divide up that space of states into macrostates. We calculate in each macrostate how many microstates there are, we take a logarithm of that number, and that is the entropy. Given that understanding, the fact that entropy increases makes sense.

- With Boltzmann's definition of entropy, high entropy means there are more ways to arrange things in a certain way and low entropy means there are fewer ways. Given a system that has low entropy, if we just let it evolve, it will naturally move to a higher-entropy configuration simply because there are more ways to be high entropy than to be low entropy.

Suggested Reading

Albert, *Time and Chance*.

Carroll, *From Eternity to Here*, chapter 8.

Questions to Consider

1. Imagine you have two liquids that tend to unmix themselves when you try to mix them together. What can you conclude about the behavior of the molecules from which they are made?

2. Can you think of some different ways we could coarse-grain all the states of gas molecules in a box? Why do some coarse-grainings seem more appropriate than others?

Entropy and Counting
Lecture 9—Transcript

We are now one-third of the way through the course. It is a good time to sort of take stock, figure out what we have learned so far and where we are going. The truth is that so far what we have really been doing is laying the groundwork. We talked about time, what time is. Is it just as real in the past and the future. We have talked about measuring time with clocks and calendars and we said that the real mystery of time is the arrow of time, the fact that the past is different from the future. We have also claimed that the reason why the past is different from the future is because of entropy, because of the Second Law of Thermodynamics which says that toward the future entropy increases, toward the past it was lower. We have investigated a couple of possible explanations for why there is an arrow of time and said that they are not the right explanation. It is not because time reversal is violated by the weak interactions of particle physics and it's not because of the collapse of the wave function being irreversible either. It has to come down to entropy.

In this lecture, we really start answering the question, not just setting it up. We are going to dive into what entropy really is, give the rigorous modern definition of it and explain what we do understand and what we do not understand about how entropy works. Our modern understanding of entropy comes from the 1870s, that is not very modern, but it's still the right answer as far as we currently know. It comes from the Austrian physicist Ludwig Boltzmann.

The idea of entropy predated Boltzmann. It came from Carnot and Clausius, the second law and the word and the notion of entropy, what Boltzmann added to the discussion was the idea that matter is made of atoms. Atoms have a really important role in how we think about stuff in the universe. You have all seen the periodic table, if only because it was hanging in the room when you took chemistry back in high school, the list of all the different chemical elements. The idea of atoms goes all the way back to ancient Greece. People like Democritus said the world is made of discrete objects called atoms, so when chemists in the 19th century realized that elements could be divided into their smallest possible pieces, they naturally gave the word atoms to those smallest possible units of an element and that is what you see in that periodic table; the elements of chemistry, hydrogen, zirconium, iron, and so forth.

These days we would say that even the atoms can be divided into smaller particles, the protons, neutrons, and electrons. The protons and neutrons are made of quarks and gluons, but none of that is absolutely necessary for Boltzmann and his understanding of entropy. All we need is that matter consists of discrete little pieces. This is a tremendously powerful idea. It means that when you see ice, when you see water, when you see water vapor, these are not different things, these are different arrangements of the same fundamental set of things. Since there's only really electrons, protons, and neutrons, everything you see around you, every form of matter is just a different arrangement of those substances.

When we talk about the different properties of different kinds of matter, we have to explain that in terms of the different arrangements of the underlying pieces, different arrangements of the atoms and the particles. This was the task that physicists found themselves faced with in the 1800s when the chemists began to convince the physicists that there were such a thing as atoms. The chemists and the physicists teamed up to invent what was called kinetic theory trying to understand thermodynamics, the science of heat and fluids and motion in terms of atomic theory, in terms of the idea that all of these different substances were collections of atoms.

What that means is that what you might think of as an absolute law could just be a statistical law. It could just be that probably the atoms will do a certain thing. The point is that even in that very small collection of macroscopic material, there is a large number of atoms. The number that scientists use to indicate a macroscopic amount is Avogadro's number. It's the number of carbon atoms in 12 grams of carbon. Twelve grams is not very much, it is a good sized pinch of sand, but in that grain of sand there is something like 6×10^{23} atoms or molecules. That's a huge number, 6 followed by 23 zeros. It's a giant number, far bigger than ones that we use in our everyday lives.

Avogadro's number is the indication that we can throw away information. That is the key to going from the world of atoms to the macroscopic world.

It's called coarse graining. It is basically the idea that because in any tiny macroscopic objects, not to mention any big macroscopic object, we can't possibly keep track of what every atom is doing. Therefore we need to keep track of less than perfect information. We don't write down the position and the velocity of every atom. Laplace taught us that if we wanted to know exactly what was going to happen, that is what we would have to do, but it is completely impractical in the real world. Instead we average, we take typical features of the whole collection of atoms in some macroscopic object.

We talk about the density of the material, which is the number of atoms in any one little volume. We talk about the pressure of a fluid or a gas; that is just the force. When the atoms hit some boundary those atoms push on the boundary and that's the pressure that you feel at the boundary. We talk about the temperature of a material; that is the average kinetic energy, the energy of motion, because all the atoms are jiggling back and forth. High temperature means the atoms are jiggling fast, low temperature means they are jiggling slowly.

Then we talk about the heat; that is just the total thermal energy. You take the temperature and you include all of it over the whole volume; that is how you get the heat. What you notice about all of these ideas is that they're all properties of the atoms. Density, temperature, heat, these are not separate substances. There is no heat fluid. Back in the early days of the 19th century they thought there was a heat fluid; they called it caloric, but atomic theory says that they were not correct, but all of these different things are properties of the atoms and so it is obvious to ask, is entropy also a property of the atoms.

That's what Boltzmann put together. He realized that entropy, just like the other thermodynamic quantities could be thought of as a property of the underlying atomic description. If you like think of grains of sand, just imagine we blow up our atoms until they are the size of grains of sand and imagine that some of those grains are blue and some of those grains of sand are red. If they are in separate piles or if they are in one box, but all the blue grains are on one side and all the red grains are on the other side, that is a low entropy configuration. There is the blue sand and the red sand. If you mix it together it is clear that the entropy goes up. It's clear that's an irreversible process. You mix all the sand together and what happens is that instead of blue and red grains from a distance everything would just look purple. If you are far enough away your eye does not perceive the individual grains of sand. When they are all mixed together it would look like one mass of purple sand.

We know that if you look very, very closely the individual grains still maintain their identities. Every grain is either blue or red. The grains have not become purple, but because we have mixed them together the substance now looks purple. There's a property of the sand that now is all mixed together is high entropy, but when we increase the entropy by mixing it we did not change the substance, we merely rearranged. That is underlying Boltzmann's idea that entropy can be thought of as a property of the arrangement of the atoms, not as a substance all by itself.

To see this at work in a real situation, consider the example we will be using over and over again, which is a box of gas, but a box of gas with a partition. You have a box of gas divided in two, there is a dividing thing in the middle of it, but there is a tiny hole in the partition so most of the time the atoms in the gas just bounce around inside whichever side they are on, the left side or the right side. Rarely, but occasionally an atom will go right through the hole and pass from one side to the other. The nice thing about this example is it is perfectly obvious how it will behave. You don't need science, you don't need equations, but think very hard and figure out what the box of gas will do.

Imagine that you have this partitioned box of gas and you start with all of the gas on one side so there is a little hole in the middle, particles from the one side, let's say the left-hand side, will gradually leak over to the right-hand side. If you start with all the atoms on the left, no atoms on the right, and you wait long enough the box of gas will equilibrate just like the heat equilibrates when you bring a hot object close to a cold object. This empty box next to a full box equilibrates until there is half the gas on the left-hand side, half the gas on the right-hand side. It is intuitively clear to us that entropy increases during this process.

Why is that true? Because we imagine that that is the direction in which the gas will naturally evolve. The gas will even itself out until it's 50/50, left and right, and once it's there it will not naturally go back. You can imagine it going back, but it is not going to naturally happen. That's an irreversible process. Entropy has gone up.

We can consider this a little bit more quantitatively I told you we are going to be taking things seriously in this lecture. We are going to work our way up to Boltzmann's equation defining the entropy. Imagine you have this box of gas and to make things definite, imagine there are 2000 particles of gas. This is obviously more than a handful, but it is much, much less than Avogadro's number. This is much, much less than a truly macroscopic amount, but what we will see is this is more than good enough to see the effects of entropy and the averaging process of having so many molecules. You start with 2000 atoms or molecules and you put them all on the left-hand side of the box, and let's just say to make things definite, that every atom has a 1 percent chance per second of going through the hole, of flipping sides from left to right. It doesn't matter whether the atom is on the left or on the right. If it is on the left it has a 1 percent chance of going to the right, if it is on the right it has a 1 percent chance of going to the left.

What happens in this circumstance? Obviously if you start with all the atoms on the left they will gradually diffuse to the right until they even out. If you're the kind of person who likes to write little computer programs, you can easily program this yourself, draw the picture and watch what happens. After only about 200 seconds, after a few minutes of these atoms going back and forth you can't tell the difference between the left-hand side of the box and the right-hand side of the box. It looks like there is the same amount of atoms on either side. Clearly the entropy has gone up, clearly this is an irreversible process so what we would like to do is to describe that process quantitatively.

We do not want to just draw pictures of it and look at it happening. We want to write down equations that we can solve and predict what will happen in any circumstance. That is exactly what Boltzmann did. He tried to figure out how many combinations there would be of atoms in one configuration versus another and showed that our intuitive notion of high entropy corresponds to many different arrangements of a particular configuration. For example, in our box of gas with two sides and 2000 atoms, let's say we only keep track of whether an atom is on the left-hand side or the right-hand side. We are not keeping track of the specific position and momentum of every atom, it's just like the real world. We can more or less see how full the left-hand side is, more or less see how full the right-hand side is, but we do not see every atom.

Our task is to count how many arrangements look one way versus another and then apply to that counting some formula that lets us calculate the entropy. Let's actually do that. How many ways are there to have all the atoms on the left and none of the atoms on the right? The answer is very simple. There is one way. You know that all the atoms are on the left, none of the atoms are on the right. That's all you need to tell me. There is no other way to do that. It becomes interesting when we have 1999 atoms on the left and 1 atom on the right. How many ways are there to do that? Well basically you start with the 2000 atoms on the left and you take 1 of them, you move it to the right. How many ways are there to take 1 atom? There are 2000 different ways because there are 2000 different atoms you could've chosen. Therefore, there are 2000 ways to get 1 atom on the right, 1999 on the left. You see that there already many more ways to get 1 on the right and 1999 on the left than there were to have zero on the right and 2000 on the left. That is the trend we're going to see. As the numbers become closer on the left than right, the ways that you can do it, the number of combinations that look like that increase very rapidly.

Let's ask what happens if you have 2 atoms on the right and 1998 atoms on the left. You might think well there were 2000 ways to choose 1 atom and there's 1999 ways to choose another atom so the number of arrangements is 2000 × 1999. Now that is cheating a little bit because you have 2 atoms and it didn't matter which one you chose first and which one you chose second. The right number of arrangements is (2000 × 1999) / 2 because it doesn't matter whether you took one atom first or the other atom first. The answer is almost 2 million different arrangements. It's 1,999,000 different arrangements, much larger than the 2000 arrangements you had with 1 on the right and 1999 on the left.

Don't worry, I'm not going to do this for three or four or five atoms on the right, the point is that there is a formula, there is a mathematical procedure for counting the number of arrangements. You are not surprised to learn that the maximum number of arrangements has 1000 atoms on the right, 1000 atoms on the left. You might be surprised to learn that the number of such arrangements is greater than 2×10^{600}; that is 2 followed by 600 zeros, an enormous number. Just for calibration purposes, the total number of particles in the visible universe is only 10^{88}, so 10^{600} is a number that we would never come into contact with in our everyday lives, but that is the number of ways you can have 1000 atoms in one box and 1000 atoms in the other.

Then as you continue to move atoms from the left to the right, the number of arrangements goes down again. Once you have no atoms on the left and 2000 on the right, that's again a unique configuration. There's only one way to do that. The number of arrangements is small when there is an imbalance between left and right. The number of arrangements is large when there is a balance between left and right. Our intuitive notion is that the entropy is higher when things are balanced, lower when it is unbalanced.

That gave Boltzmann the idea that what the entropy is really doing is just counting the number of arrangements. That is to say the entropy tells us the number of microscopic arrangements of a set of atoms that look indistinguishable to our macroscopic perception. We do not see every individual atom, all we see is the total number of atoms on the left and the total number of atoms on the right. Boltzmann says that if you count the number of arrangements that tells you the entropy. He doesn't say that the number of arrangements is the total entropy—that is too good to be true. Boltzmann's formula is that the entropy is proportional to the logarithm of the number of indistinguishable states. That is to say Boltzmann gives a formula for the entropy in terms of the number of ways we have to rearrange the system which don't change its macroscopic properties. This formula is S, the entropy gets the letter S because E is already taken up by energy. The entropy is $S = k \log W$.

This is such an important formula it is engraved on Boltzmann's tombstone, which you can visit in Vienna. I always tell my physics students that Boltzmann has achieved every physicist's dream. He has an equation on his tombstone and they should be thinking in their physics career, what is the equation that will someday go on my tombstone? Boltzmann's equation S

= $k \log W$ is not that hard to understand. S is the entropy, k is Boltzmann's constant. It's just a constant of nature. It's sort of a proportionality constant out there that translates from the number of states to the entropy of the system. W is Boltzmann's notation for the number of ways we can rearrange things and so W in our box of gas was 1 or 1999 or 2×10^{600}, etcetera, and log means the logarithm. It's a way of going from the actual quantity to a more manageable way of dealing with that quantity.

We have to understand what a logarithm is if we are really going to wrap our heads around Boltzmann's formula. Boltzmann is telling us that the entropy is proportional to the logarithm of the number of rearrangements. The logarithm is basically the power to which you would raise the number 10 to get the number you want, so that probably is not immediately clear if you are not already familiar with logarithms. The point is that the logarithm of 10 is equal to 1. Why? Because in order to get 10 you raise 10 to the power 1, 10^1 equals 10. The logarithm of 10 is therefore 1. The logarithm of 100 is 2 because 10^2 is 100. The logarithm of 1000 is 3 because 10^3 is 1000 and so forth. The logarithm of 1 is zero because 10^0 is 1.

It is easy if you're taking the logarithm of a power of 10 like 1000 or 10,000 or whatever. The logarithm of 1 million is 6, no problem at all. But, you can also define the logarithm for every number in between. You can take the logarithm of 2, of 5, of Pi, whatever you want. Basically the logarithm is a function. It is small when the number you're taking the logarithm of is small. In fact, it's negative when the number is less than 1. The logarithm of 1/10 is -1 because 1/10 is 10^{-1}. The logarithm is small when x is small, it is large when x is large, but you'll notice that the logarithm grows more slowly. We multiply 10 by 10 by 10 by 10, we only add to the logarithm; we go from zero to 1 to 2 as the number we are taking the logarithm of goes from 1 to 10 to 100. The log tells us the number of ways we can rearrange it, but it tells us in a more compact and manageable form.

Boltzmann's equation says if you want to know the entropy of something count the total number of combinations that look the same, take the logarithm and then multiply by this constant of nature, which I am going to call Boltzmann's constant. He didn't call it Boltzmann's constant, but in his honor that is what we call it now. So, go back to our box of gas, remember the lowest entropy configuration has 2000 particles on the left, zero on the right. The number of combinations that look like that is 1, the logarithm of 1 is zero so literally the entropy of that configuration is zero. It's the most orderly configuration you can have. Its entropy is the lowest number you can ever get, zero.

When there was 1 atom on the right and 1999 on the left, there were 2000 combinations like that. The logarithm of 2000 is 3.3. When there was 1998 on the left and 2 on the right, you had almost 2 million combinations and the logarithm of that number is 6.3. Once you get all the way up to 1000 on the left, 1000 on the right, absolutely equal, highest entropy configuration, you had 2×10^{600} combinations. The logarithm of that is just 600, 600.3 to be precise. So, you see how much easier it is to write down the logarithm than to write down the actual number of combinations. This is a huge number of combinations, when you have 1000 on the left and 1000 on the right, 2×10^{600}, once you take the logarithm of that it is just 600.3. That is much easier to deal with.

To be honest, the reason why Boltzmann says that the entropy is the logarithm of W of the number of microscopic arrangements, rather than W itself is not just because the logarithm is smaller and therefore easier to deal with. Remember Boltzmann came along when entropy had already been defined. Clausius had given us the notion of entropy. He had a definition of it that was very specific about exchanging heat. Boltzmann wanted a much more general definition, but he still wanted his general definition to match with Clausius's specific definition in the same arrangement. One of the properties of entropy as it was already understood is that if you take two boxes of gas or two objects, one box has a certain entropy and the other box has another entropy, and you put them together to make one larger system, the total entropy is the sum of the entropy in the first thing plus the entropy of the second thing. That was something that was already known about entropy. The entropy of the total is the sum of the individual entropies.

But, Boltzmann wanted to define the entropy in terms of the number of arrangements and when you take one box and connect it to another box you open up many more ways to arrange the individual particles. Now you can move particles between the two boxes so the number of arrangements is not just the sum of the total number you had before, it's the product. You take the number of arrangements you had in one box and you multiply it by the number of arrangements you have in the other box. If you tried to define the entropy as the total number of ways you could arrange the atoms, that would give you a wrong formula for the entropy of adding two things together. It would multiply them rather than add them.

The magic of the logarithm is it fixes this problem. The logarithm has this wonderful property that the logarithm of a product is the sum of the logarithms. Consider the logarithm of 100. The logarithm of 100 is 2, 10^2 is 100. We can also write it as the logarithm of 10×10 because 10×10 is 100. The log of a product is the sum of the logarithms. That means that logarithm of 10×10 is the logarithm of 10 plus the logarithm of 10, which is $1 + 1$, which equal 2. It is the right formula. What Boltzmann realized was that putting the entropy proportional to the logarithm of the number of arrangements preserves the nice property that the entropy of two systems added together is the entropy of each system added to the entropy of the other system.

There are a lot of details here that we're not going into. For example, I said that the logarithm is the number to which you raise 10 to to get the number you want. You can choose any number to do the raising. You can take the logarithm base 2, that's the number you raise 2 to to get the number you want. You can take the logarithm base 12, whatever you want. In science we usually use either base 2 or base E, this weird irrational number, Euler's constant that is very convenient for calculus and physics, but for this course we are just going to take the logarithm base 10 because we all know that 10 to the power of 3 is 1000 and so forth.

Now a crucial role in all of these considerations is that we are taking microscopic configurations of atoms and treating them the same. We are using the phrase over and over again, a certain arrangements of atoms look the same macroscopically. We say that means they are in the same macro state so what we are doing is we are distinguishing between the microstate of a system, which is the exact Laplacian state. The thing you would need if you had all of the positions and momenta to predict the future and to retrodict the past, that is the microstate. If you knew the microstate then the evolution of the system would be perfectly reversible, you would be able to use the laws of physics if you were Laplace's demon to go forward and backward in time.

But we don't know the microstate. You look at a glass of water with an ice cube in it, you do not see the exact microstate of every atom inside the glass of water. You know that it is a glass of water, it has a certain temperature, maybe there's an ice cube in there. We call that information the macrostate. We define ahead of time what it is we can macroscopically observe and two arrangements of atoms that look the same according to our macroscopic observations we put into the same macrostate. This process is coarse-graining. It says we take all the possible states you could have and we chunk them up, we chunk them up into microstates that look the same macroscopically. We call the set of all those microstates the macrostate.

What Boltzmann's entropy tells us is that the entropy of an individual configuration of particles is the logarithm of the total number of other configurations in the same macrostate. We coarse-grain by some observations, we calculate how many particles, how many arrangements look the same, take the logarithm, that's the entropy. Now, there is obviously a step here that requires some thinking. How do you do the coarse-graining? How do you decide what I can macroscopically see and not? What if there is some alien who comes along who has much better vision than I do and is able to see the positions and the velocities of every molecule in the glass of water with the alien coarse-grain differently. It seems a little arbitrary.

Imagine that you are playing pool one day, playing billiards, and you've scattered the balls across the table and a friend of yours walks by and says wow that's an amazing configuration of billiard balls. You look and you say I don't see anything special about it. It's just randomly scattered across the table. Your friend says yes but the particular place where the 8-ball is and the 3-ball is how unlikely it was that it would end up in precisely that configuration. You would not be impressed by your friend's purported amazement. You know that they would not have predicted ahead of time that that particular arrangement of billiard balls was special.

Every individual arrangement of billiard balls on the pool table is individually unlikely, but we treat different configurations the same if they look more or less the same to us. We would say yes that was a good break and all the balls are scattered across the table or that was pretty lame and they're all clustered around one pocket or something like that. We decide ahead of time what we can measure macroscopically. We use that definition to do the coarse-graining.

For example, when it comes to atoms, it is not simply an accident that when we look at a box of gas or a glass of water we can see things and measure things like the temperature and the density and so forth. This is not arbitrary. You could imagine picking other features of that collection of molecules, but they would be harder to measure. The way we coarse-grain the real world depends on what is actually easy for us to measure. Even though aliens might be better at measuring some things than we are, they would not measure different things than us. That is because it's the laws of physics that basically give us what is easy to measure. If you build a machine it would be very, very hard to build the machine that would actually be able to look at the position and velocity of every molecule of water in a glass of water. It is not difficult to build a machine that measures the density or the temperature and so forth. The way that we coarse-grain the real world is very natural given how the laws of physics work and how we actually measure things in reality.

Once this is all in place, once you take Boltzmann's program and take it seriously we have an understanding of how entropy works at a quantitative level. We take all of the ways that a state can be, all the microstates of some system, a box of gas or a glass of water, all of the different ways that the individual molecules can be arranged. We divide up that space of states, that set of all possible configurations, into macrostates. We say these look alike to me, these look alike to me answer so forth. We calculate in each macrostate how many microstates are there, we take a logarithm of that number and that is the entropy. Given that understanding, it makes sense to us now why entropy increases.

Boltzmann's definition of entropy is that high entropy means there are more ways to arrange things that way, low entropy means there are fewer ways. Therefore, given a system that has a low entropy, if you just let it evolve, let it change its state over time, without tricking it, without conspiring or setting it up in some subtle way, it will naturally move to a higher entropy configuration simply because there are more ways to be high entropy than to be low entropy. In fact, there are many, many, many more ways to be high entropy than to be low entropy, so Boltzmann explains for us why entropy increases toward the future. There are simply more ways to be that way. He also makes the difficult very explicit, why was the entropy lower in the past? There are fewer ways to be low entropy. So, this puzzle why was the entropy lower in the past becomes not just puzzling but a real conundrum. We are going to have to work hard to offer an explanation for this phenomenon.

Playing with Entropy
Lecture 10

In the last lecture, we learned what entropy really is: The world around us is made of atoms that have many different ways of arranging themselves, but when we look at things in the universe, we do not see the individual atoms. There are many different ways that atoms look the same to us. Boltzmann taught us that entropy is just a logarithm of the number of ways to arrange the atoms so that they look macroscopically indistinguishable. In this lecture, we will compare Boltzmann's ideas about entropy with those of Clausius and Carnot; we'll then refine our definition and look at some of its philosophical implications.

Boltzmann v. Clausius

- Boltzmann's version of the second law of thermodynamics is that entropy increases because there are more ways to be high entropy than to be low entropy. We need to make sure that this idea of entropy is compatible with the formulations of Carnot and Clausius.

- According to Clausius, if we put together two boxes of gas with different temperatures, the temperatures will even out. With Boltzmann, we saw that if we opened a small hole between one box that was full of atoms and one box that was empty, the density of atoms would generally tend to even out. We want to check that that is also true for temperature, and we can do so with a fairly easy thought experiment.
    o Temperature tells us how fast individual atoms are moving. With a high-temperature box and a low-temperature box, we can think of fast- and slow-moving atoms as different-colored grains of sand.

    o In the boxes of gas with different temperatures, there are more arrangements with an equal number of hot atoms in the left and right boxes and an equal number of cold atoms in the left and right boxes; therefore, the highest-entropy state is one in which the temperature is the same.

- Boltzmann's reasoning gets us back to Clausius's version of the second law of thermodynamics: Temperatures tend to come to equilibrium. Boltzmann has unified the entropy of heat with the entropy of mixing. His one formula lets us understand both circumstances.

Boltzmann v. Carnot

- Carnot's interest in building the perfect steam engine led him to invent a cycle—a way of running a steam engine—that was maximally efficient. He realized that this maximally efficient steam engine requires a reversible process but that most real-world steam engines were irreversible.

- We would like to use Boltzmann's language of entropy to distinguish between efficient and inefficient steam engines. Another way of phrasing the question is: Can we get useful work out of a certain arrangement of atoms? A box of gas has energy it. Can we use that energy for some purpose?
    o For this experiment, we start with a thin piston dividing a large cylinder, and we reconstruct our two boxes of gas. If we put all the gas on one side of the piston and the other side remains empty (a vacuum), we know that the piston will push away from the gas. It will move in the direction of the vacuum and expand the volume of the region that has the gas in it.

    o In other words, the atoms moving around do useful work; they push the piston in one direction. We are extracting energy from the gas because as we expand its volume, we are cooling it down. We have lowered the temperature of the atoms.

    o This is a low-entropy configuration, and we can do useful work with it. It's not that energy is created or destroyed; we are just taking energy out of the atoms and putting it into the piston.

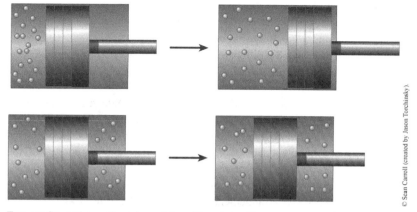

Two configurations of gas in a piston: They have the same energy, but one is low entropy and the other is high entropy.

- Now imagine the same amount of energy but in a different configuration. Consider the same gas atoms with the same temperature and the piston in the middle of the cylinder, but put half the atoms on one side of the piston and half on the other side. The total energy is the same as we had before, but the distribution of atoms is different.
  - With the same amount of gas on either side of the piston, there is no net force, and the piston doesn't move. Even though the same amount of energy is in the piston now as before, we cannot extract that energy.

  - The reason for this, of course, is that the energy is in a high-entropy form. High entropy is an even distribution of gas atoms on both sides of the piston, and even though the energy is present, we cannot take it out.

- Boltzmann's way of looking at entropy, which says that entropy will be highest when there is a large number of arrangements, recovers Carnot's insight that low-entropy energy can do useful work and high-entropy energy cannot.

Refining Entropy

- We have said that entropy measures the disorderliness of a certain configuration of stuff. However, the true statement of entropy is that it is the logarithm of the number of configurations that look macroscopically the same.

- To get this notion exactly straight, consider again gas in a large box, a room-sized box. Imagine that the whole atmosphere in the room is squeezed into a tiny cube in the middle, just 1 mm on a side. This is not literally a cube but a randomly chosen configuration of all the gas atoms in the room.
  - Such a cube is a very low-entropy configuration. There are very few arrangements of the atmosphere in a room that, just by chance, have all the atoms squeezed into a 1-mm cube.

  - We could also consider a random arrangement of the atmosphere in the room squeezed into a 10-cm-tall version of the Statue of Liberty. This is also a very low-entropy configuration, but it is a much higher-entropy configuration than the 1-mm cube, simply because it is larger. There are more configurations that look like a 10-cm-tall Statue of Liberty than look like a 1-mm cube.

- It's important that we don't confuse complexity or simplicity with low entropy and high entropy. We also shouldn't get the idea that high entropy means more things are mixed together. What we really care about is the number of configurations.

Philosophical Implications of Boltzmann's Entropy

- The most important implication of Boltzmann's way of thinking about entropy is that the second law of thermodynamics is not really a law; it's a statistical statement about probabilities.
  - The reason entropy increases is that there are more ways to be high entropy than to be low entropy. Thus, it's probable that a random configuration that isn't already at maximum entropy will tend toward maximum entropy.

- Although it's improbable, entropy can decrease. With a box of gas in equilibrium—its highest-entropy state—there will occasionally be fluctuations that decrease entropy. However, the period of time it would take for a noticeable decrease in entropy to occur is probably much longer than the age of the universe.

- Once we accept the idea that entropy increasing is probable but not necessary, we can do experiments that intentionally decrease entropy.
  - Imagine we have a box of gas that starts in a low-entropy state, such as all the gas in one corner, and we let it evolve so that all the gas spreads out—entropy increases. Laplace's demon, which can see the position and momentum of every atom, can simply reverse the velocity of every atom from the end configuration to the initial configuration.

  - Because the laws of physics are reversible, the box of gas will experience an exactly backward evolution, from high entropy to low entropy. This is a gross violation of the second law of thermodynamics, but of course, it wouldn't happen spontaneously; we have arranged it to happen as a thought experiment.

  - However, any tiny deviation will spoil the experiment. We might know the position and velocity of all the atoms of gas in the box, but any disturbance of even one atom will change its precise velocity and position and spoil the experiment. Entropy will not decrease because we would be unable to accurately arrange all the velocities and positions of atoms in the box.

  - Even if we can imagine an isolated box of gas with decreasing entropy, once that box starts interacting with the outside world, the velocities and positions of the atoms in the box will be disturbed and entropy will no longer decrease.

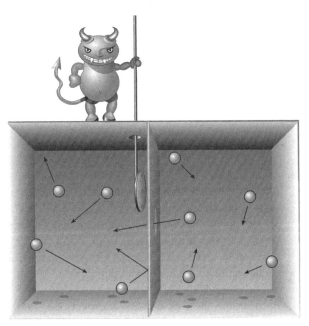

Maxwell's demon decreases entropy by regulating the flow of high- and low-entropy particles.

Maxwell's Demon

- James Clerk Maxwell, known for unifying electricity and magnetism, proposed his own thought experiment to investigate the statistical nature of the second law in relation to Boltzmann's ideas about entropy.

- Consider again our box of gas with a partition in the middle at thermal equilibrium. Maxwell's demon stands on top of this box with a switch he can use to allow some atoms to go from right to left and others to go from left to right in the box. In particular, he allows the high-velocity atoms to go from left to right and the low-velocity atoms to go from right to left, but not vice versa.

- The demon does not disturb the velocity of any of the atoms, but because of his work, we end up with a box that has a high temperature on the right and a low temperature on the left. Again, we have violated Clausius's formulation of the second law. Heat seems to have flowed from thermal equilibrium out of thermal equilibrium.

- Obviously, the demon himself somehow creates entropy, but it is far from obvious how he does so. Over the years, people showed that the demon's door could operate without increasing entropy and that he could observe the atoms without increasing entropy.

- In 1960, Rolf Landauer, a physicist at IBM, made a model in which he showed that the demon could observe the atoms without increasing entropy by simply recording their measurements to make the experiment a reversible process. He showed that irreversible computations are the ones that generate entropy.

- It wasn't until 1982 that Charles Bennett, a computer scientist at IBM, finally solved the problem in a satisfactory way. What Bennett realized, using Landauer's ideas, is that Maxwell's demon can keep track of only a finite amount of information.
  - To convince ourselves that the atoms are going only one way and not the other, the demon must constantly update his information about where the atoms are. He needs to keep track of a potentially infinite amount of information.
  - That's possible as long as he can erase the previous entries in his notebook, but Bennett saw that erasing information is irreversible. Once something has been erased, we don't know where the information went.

Suggested Reading

Carroll, *From Eternity to Here*, chapter 8.

Von Baeyer, *Warmth Disperses and Time Passes*.

Questions to Consider

1. Imagine you are a screenwriter, and your science fiction movie script includes a character who truly lives backward in time. Can you invent a plausible dialogue that character would have with someone experiencing the ordinary arrow of time?

2. Can you think of a Maxwell's demon–like experiment that lets you decrease entropy in a box without increasing it anywhere else? (Answer: No, but it's fun to try.)

Playing with Entropy
Lecture 10—Transcript

Now we know what entropy really is. The world around us is made of atoms. Those atoms have many different ways of arranging themselves, but when we look at the things in the universe we do not see the individual atoms. There are many different ways that look the same to us and Boltzmann taught us that the entropy is just a logarithm of the number of ways to arrange the atoms so that they look macroscopically indistinguishable. It is the number of indistinguishable microstates within a macrostate.

Basically in the last lecture we ate our vegetables. We learned about logarithms. We got the definition right. We considered an example and calculated some numbers. For the rest of the course, it is all going to be about the main course and the dessert. No more logarithms and very few numbers. All you need to remember is that a high entropy state is when there are many arrangements that look that way. A low entropy state is one where there are few arrangements that look that way.

If you are shuffling a deck of cards there are many ways for the cards to be randomly shuffled. There is only one way for them to be perfectly ordinarily arranged. If you are playing pool, there are few ways for all the balls to be racked together. There are many ways for them to be scattered across the table. If you have a glass of water there are few ways to have ice and warm water. There are more ways to have all of the water be in the same state. Entropy tends to go up, according to Boltzmann, simply because there are more ways to be high entropy than to be low entropy. That is his version of the Second Law of Thermodynamics.

We also, of course, have the other formulations of the Second Law of Thermodynamics. Our first task is to make sure that Boltzmann's definition of entropy helps us to recover the previous ideas that we had. We had Carnot's version of a maximally efficient steam engine, the one that was absolutely reversible and we also have Clausius's notion that if you take two objects and put them together at different temperatures the temperatures will even out. We should check that Boltzmann's way of thinking about entropy is compatible with these preexisting ways.

Let's think about Clausius. He says you have one box of gas at one temperature and you have another box of gas at another temperature. If you put them together they will equilibrate, they will come to the same temperature. Now Boltzmann when we were investigating his idea we used one box of gas and another box of gas, but instead of different temperatures we had one box that was full of atoms and the other one that was empty with a small hole in between, we found that the atoms would generally tend to even out. We would like to do the same thing for the heat that is in the boxes that Clausius was considering.

Note that Clausius's definition wouldn't tell us anything about the example we looked at in the last lecture. Clausius talked about the transfer of heat, but when we have a box at constant temperature, just with a higher density on one side than the other, then Clausius is silent. He is not saying anything. The temperature is the same on both sides.

Boltzmann tells us what we know to be true, which is that if you have one box with a lot of atoms they will generally leak to the other box until the density is more or less the same. We want to check that that is also true for the temperature and it is a pretty easy thought experiment to do. Remember temperature is just telling us how fast the individual atoms are moving. In each box if you have a high temperature box and a low temperature box, think of the fast moving atoms as red grains of sand, the slow moving atoms as blue grains of sand. There are more arrangements where the sand grains are mixed than ones where they are kept segregated in the different boxes. In other words, in the boxes of gas with different temperatures there are more arrangements where there are an equal number of hot atoms in the left box and the right box and an equal number of cold atoms in the left box and the right box, and therefore the highest entropy state is one in which the temperature is the same.

Boltzmann's reasoning gets us back to Clausius's version of the Second Law of Thermodynamics, temperatures tend to come to equilibrium. What Boltzmann has done is to unite, to unify the entropy of heat with the entropy of mixing. We have one formula that is on Boltzmann's tombstone that lets us understand both different circumstances. That is a classic example of progress in science when you can understand two different phenomena using one single unified picture.

What about Carnot's picture. It sounds a little different. Remember Sadi Carnot was interested in building the perfect steam engine so he invented a cycle, a way of running a steam engine that was maximally efficient. He realized that to get this maximally efficient steam engine you would have to have a process that was reversible and that most real world steam engines were irreversible we would now say that the entropy is increasing. What we would like to do is to think of entropy in Boltzmann's language and use it to distinguish between efficient and inefficient steam engines, or another way of saying it is can we get useful work out of a certain arrangement of atoms. Useful work is stuff that we can do something with, we can push a vehicle or we can lift something from the ground to a higher altitude. That is what we would like to be able to do. A box of gas has energy in it. The question is can we use that energy for some purpose. Can we make it do useful work?

Consider a piston, not a big, thick piston, but a rather thin piston that is in a big cylinder. Basically we start with our piston in the middle of the cylinder and we are reconstructing our two boxes of gas. We have one side of the cylinder and the other side of the cylinder divided by the piston so what we're going to do is we're going to put gas into that cylinder, into that chamber, and we are going to ask can it do useful work. First, imagine that we put all of the gas on one side of the piston so we have one side of the piston that is full of gas, the other one is completely empty. Well what is going to happen? And again, we don't even need to use science or equations, we can just think about it intuitively.

If one side of the piston is empty, if it is a vacuum, and the other side of the piston is a gas, it has atoms moving around and pushing on the piston exerting pressure then we know what's going to happen. That piston is going to push away from the gas. It is going to move into the direction of the vacuum and expand the volume of the region that has the gas in it. In other words, those atoms moving around do useful work. They push the piston in one uniform direction. We are extracting energy from the gas because as we expand its volume, just as in a refrigerator, expanding the volume of the gas cools it down. We have lowered the temperature of the atoms. In that configuration that's a low entropy configuration and we can do useful work with it. It's not that energy is created or destroyed, we are just taking energy out of the atoms and putting it into the piston using it to move our car along the highway.

Now imagine that the same amount of energy, but in a different configuration. Take the same gas molecules, the same number of molecules with the same temperature, but have the piston in the middle of the cylinder and put half of the molecules on one side of the piston, half of the molecules on the other side so the total energy is the same as we had before. It is just that the distribution of atoms is different. But, notice what happens. Now that the same amount of gas is on one side of the piston as on the other side, there is no net force. The piston doesn't move. We can push or pull the piston, but that requires work. That would take work away from us. Even though the same amount of energy is in the piston now as was before we cannot extract that energy out. The reason, of course, is because the energy is in a high entropy form. High entropy is an even distribution of gas molecules on both sides of the piston and even though the energy is there, we cannot take it out.

In other words, Boltzmann's way of looking at entropy, which says that the entropy will be highest when there is a large number of arrangements, when the gas molecules are evenly spread one side and the other, recovers Sadi Carnot's insight that low entropy energy can do useful work, high entropy energy cannot. That reinforces the idea that we had that entropy is one way of measuring the uselessness of energy. The total amount of energy is conserved, but it can go from a useful form, a low entropy form, to a useless form, high entropy. That is what happens when you burn fuel. When you burn gasoline in your car, you're not creating energy, you're releasing it and you are converting it into a higher entropy form. That is why you can burn fuel, but you cannot un-burn it.

If you look at the other side of lectures I have done for the Teaching Company about dark matter and dark energy, you will learn about dark energy, which is most of the energy in our universe today. It is absolutely uniformly spread through space. Something like 73 percent of the total energy of the universe is in this form of mysterious cosmological stuff called dark energy. We know it's there. It is making the universe accelerate. Our astronomical observations say that it is there, but the sad story is that, because it is absolutely perfectly uniform throughout space, it is in its highest entropy state. Dark energy is maximally useless. You would like to say well there is all this energy in the universe, can we somehow use it to do something and the answer is no. Dark energy is like fuel that has already been burned. It is energy in its most useless configuration.

Back to our everyday world, now that we understand the definition of entropy, Boltzmann gave it to us, and we understand that we recovered the Second Law of Thermodynamics, we can clean up some of the sloppy ways that we have been talking about entropy. For example, you will often hear and I will often say it because it's almost completely true that entropy measures the disorderliness of a certain configuration of stuff, the randomness or the haphazardness. The reason why that's almost true is because entropy measures disorder when all other things are created equal, but sometimes not all other things are created equal.

The true statement of entropy is not that it is the disorderliness, the true statement of entropy is it is the logarithm of the number of configurations that look macroscopically the same. Whenever you're confused about entropy, go back to the equation that's on Boltzmann's tombstone; that is the right definition. To get this notion exactly straight, consider again gas in a box, let's say a big size box, a box the size of the room that you are in right now. Imagine that the whole atmosphere in the room you're in were squeezed into a very tiny cube right in the middle of the room, just 1 mm on a side. Now, I am not imagining that there is literally a box that's holding it together, just imagine that happens to be a randomly chosen configuration of all the gas molecules in your room. It's a loud configuration, but it is a very, very low entropy configuration. There are very few arrangements of the atmosphere in your room that just by chance have all the molecules squeezed into a 1 mm across cube. That's very, very low entropy.

We could also consider a random arrangement of the molecules of atoms of air in your room, which are squeezed into a miniature version of the Statue of Liberty 10 cm high. Just by chance, all the air molecules in your room miniaturize a little form of the Statue of Liberty including the crown and the torch and all that about 10 cm tall. That is also a very low entropy configuration. However, when you plug in the numbers having all the air molecules in the room form a little Statue of Liberty 10 cm tall is much higher entropy than having them all in a little cube or even a little sphere just 1 mm across. Even though the shape of the Statue of Liberty is more specific, even though it is more complex and has more details, because that Statue of Liberty that we are hypothesizing is simply bigger there are just more ways to arrange the atoms to look like that than to squeeze them all into 1 mm across. There are more configurations that look like a 10 cm Statue of Liberty than look like a 1 mm cube. So, the entropy is higher, that Statue of Liberty is more likely even though it is more complex. Don't mix up complexity or simplicity with low entropy and high entropy.

Don't even mix up the idea that high entropy means that things are more mixed together. We all know that oil and water don't mix. If you take milk and stir it into your coffee they naturally mix together. If you take oil and water and stir them together they might mix for a little while, but their natural evolution is to separate, so you want to ask yourself why do oil and water behave in the opposite way to milk and coffee. It seems to violate the Second Law of Thermodynamics. What you have to do is actually think about the individual properties of oil molecules and water molecules. It turns out that oil molecules are hydrophilic. They like the water molecules and they like to stick them into a very particular configuration so when one oil molecule is next to water it freezes into a particular configuration. What that means is that if all of the oil is mixed with all of the water the total number of arrangements is actually smaller than when the two are segregated. When the oil and water are segregated from each other, the oil is in whatever configuration it wants, the water is in whatever configuration it wants, when they're mixed together they stick and there are fewer such configurations.

I am driving home this example because over and over again what you really should care about is, are there more configurations or less configurations? Again, entropy is not about complexity or simplicity. If you want to know what is higher entropy you plug in the number of configurations that look the same. Now often you won't know. You will have something like a cloud of gas that wants to make a galaxy and you don't know how to count how many configurations look that way, but you can also work the definition backwards. If you know that a certain procedure, a certain process in nature happens in one way, but not the other, for example, a giant gas cloud or dust cloud forms a galaxy, but galaxies do not dissolve into giant clouds of gas and dust, you know that the way that the system wants to go is in the direction of higher entropy. So, a galaxy is higher entropy than an undifferentiated smooth cloud of gas and dust. That is a simple way of realizing whether something is higher or lower in entropy.

Now, we can get to some of the deep philosophical implications of Boltzmann's way of thinking about entropy. The biggest implication is that the Second Law of Thermodynamics is not really a law. It's a really good idea. It is a statistical statement about probabilities. Boltzmann says that the reason why entropy goes up is because there are more ways to be high entropy than to be low entropy and what that means is that it is probable that a random configuration that isn't already at maximum entropy will tend towards maximum entropy, will tend towards increasing the logarithm of the number of states.

What that means is that occasionally, improbably, but if you wait long enough it will happen, entropy will decrease. The Second Law is not an absolute law. In fact, if you take a box of gas that is in equilibrium, a box of gas that is in its highest entropy state, let's say in our example where there are 1000 atoms on one side and 1000 atoms on the other side, certainly if you wait there will occasionally be fluctuations where there are 1002 atoms on one side and 998 on the other and vice versa. What that means is if you keep track of exactly the number corresponding to the entropy of a box of gas, if it's sitting at its maximum, if it's in equilibrium, there'll be fluctuations, just as the random motions of the molecules the entropy will go down a little bit and then it will relax back to equilibrium.

It turns out that this is actually relevant to the functioning of life. In certain macroscopic biological molecules, you do not just see what they want you to do in the most probable way, you look at the fluctuations and that is incredibly important for how cells actually function. This is cutting edge biology trying to match the statistical understanding of fluctuations to how molecules work in our bodies. Now in the real world all around us, we are not in equilibrium, we are not a box of gas with highest possible entropy. We are many, many, many molecules, it's not just 2000, it's a lot more molecules in the real world than Avogadro's number 6×10^{23} so it is true that the entropy of something in the real world could go down, but the probability of it happening is incredibly tiny.

If you have let's say 100 particles in a box moving back and forth and you want to know how long would it take for a sizable, noticeable decrease in entropy, it's much longer than the age of the universe is the most probable answer. Because the number of atoms, the number of molecules, the number of moving parts in the universe is so large, even though Boltzmann says that it's only probable that entropy will increase. It is so overwhelmingly probable that it is practically a law. Nevertheless it was hard for people to accept this. In the 1870s, all through the 1890s, when Boltzmann and other people like Maxwell and Gibbs and Thompson were pushing this idea that atoms helped us understand entropy they met with enormous resistance. After all this was the Second Law of Thermodynamics we are talking about. The First Law says that energy is conserved. No one was going to suggest that that was only approximate. That is absolute. Energy just is conserved and now you are saying that entropy increasing is not absolutely conserved, that met a lot of resistance.

Furthermore, once you enter that idea into your head that the entropy increasing is just probable, not necessary, you realize that you could do experiments which intentionally decreased the entropy. Imagine you have a box of gas which starts in a low entropy state, like all the gas is in one little corner, and you let it evolve so that all the gas spreads out, the entropy goes up. Imagine that you are Laplace's demon so you can see every atom, you can see its position and its momentum. Take the configuration you are in at the end of that experiment when you have a high entropy box of gas and just reverse the velocity of every single atom. Send every atom and molecule backwards along the path it came. What will happen? Because the laws of physics are reversible the box of gas will evolve exactly backwards the evolution that it did going from low entropy to high entropy. In other words, if you arrange it that way, which is certainly a thought experiment kind of thing, you cannot possibly do this in the real world, but if you could you would arrange a box of gas that decreased in entropy even though it was a closed system. In gross violation of what we think of as the Second Law of Thermodynamics.

Now, it is important to emphasize this doesn't happen by itself spontaneously. You could arrange it to happen and even arranging it to happen is again only a thought experiment. The point is that any tiny deviation would spoil it. Imagine you had 10^{23} molecules of gas in your box and you keep track of the momentum, the velocity, and the position of every single molecule and you delicately arrange it so that the entropy will go down. Just before you start your experiment the cat that is in your laboratory goes and touches one of those molecules and disturbs it from the precise velocity that you gave it, just that one disturbance of one molecule is enough to completely spoil the experiment. The entropy would not go down because you have not accurately arranged all of the velocities and positions of the atoms in your box. It is like shooting an arrow at a target that is millions of miles away, any tiny deviation from the perfect accuracy means you'll miss the target completely.

Even if you could imagine doing it, what this means is there's no danger that in the real world you would ever meet a person who lives backward in time. There is no danger that Merlin from the once and future king or Benjamin Button from the movie or F. Scott Fitzgerald's story could actually possibly live in the real world. Even if you can imagine a box of gas isolated whose entropy goes down, once that box of gas started interacting with the outside world our interactions would disturb the velocities and the positions of the atoms in that box and the entropy would stop going down. Therefore there is no danger that in the real world we'll meet aliens or people close by or artificial intelligences that remember the future and don't remember the past. The idea of consciousness is something that is absolutely tied up with the increase of entropy in the arrow of time and because it is so delicately arranged it is absolutely universal. Everything we will ever meet in the real world experiences the arrow of time in the same direction.

This possibility of a thought experiment, the possibility of arranging something carefully to decrease in entropy was very intriguing to the physicists of the 19th century. Remember they thought that entropy increase was absolute. Kinetic theory, atomic theory suggested that it was not absolute and therefore they worried that they were missing something. They worried that maybe they should be fixing there atomic theory so that the increase of entropy was not just probabilistic, that it really became absolute.

Along these direction there was a thought experiment proposed by James Clark Maxwell, the guy who was most famous for unifying electricity and magnetism, but also did a lot of work on kinetic theory and thermodynamics. Like Boltzmann, Maxwell believed in atoms. He believed that Boltzmann's ideas about entropy were on the right track so he was not a competitor. He wanted to believe in the answer, he was sympathetic, but he was puzzled about the statistical nature of the Second Law and so he proposed this thought experiment that people should think about to get their ideas exactly right.

Maxwell's thought experiment was the famous idea of Maxwell's demon. I believe that Maxwell actually did call it a demon, unlike Laplace. The idea in the 19th century was that demons were these little things that were intelligent and a little bit mischievous so that they got in our way. Maxwell's demon sits on top of our box of gas with a partition in the middle so there's a box of gas with a bunch of molecules on the left, a bunch of molecules on the right, a partition in the middle with a little hole in it so atoms can pass back and forth. The difference is that Maxwell's demon stands on top of the box and he looks at the molecules and he has a little switch that he can turn on and off. What Maxwell's demon does is to let some molecules go from right to left, other molecules go from left to right. In particular, he is choosy, he lets the high velocity atoms go from left to right and the low velocity atoms go from right to left, but not vice versa.

If that could happen what you would get after the demon has worked for quite awhile, he has not disturbed the velocity of any of the atoms, he just lets them through or doesn't let them through. But, because of the work of the demon who started with equal temperatures on both sides and at the end you have a high temperature on the right and a low temperature on the left. You have violated Clausius's formulation of the Second Law. Heat seems to have flowed from thermal equilibrium out of thermal equilibrium. This was a challenge to kinetic theory and atomic theory and its understanding of how entropy worked. This comes from 1871, so this is right at the beginning of when people were thinking about these ideas and they were very worried about it.

It's kind of obvious what the solution to Maxwell's demon conundrum has to be, but it is very far from obvious how it actually works. The obvious thought is that somehow the demon itself creates entropy. Remember the problem is that in the box of gas the demon stands on it can decrease the entropy of the gas inside the box by letting molecules go one way but not the other. To do that the demon has to do something, it has to look at the molecules, it has to observe them, it has to open and close the door so you worry that maybe you are just not being careful enough, you are not keeping track of the activities of the demon. If you could you would realize that the demon is actually generating heat or entropy or something like that.

Literally for decades people thought hard about the Maxwell's demon thought experiment, trying to see how the entropy really went up, really obeyed the Second Law of Thermodynamics. The first idea was that well when the demon opens and closes the door the door slams shut and it makes noise or something like that thereby increasing the entropy of the world. People quickly showed that that was not the right answer. You can design a little door that the demon controls that moves almost imperceptibly up and down so you don't increase any entropy.

It was then suggested that well maybe when the demon looks at the molecules, that process of observation increases the entropy of the universe and a famous physicist named Leo Szilard, who actually played an important role in the atomic bomb project, in 1929 thought about the Maxwell's demon problem. He showed that the demon could observe the molecules again without increasing any entropy. The effect of the demon looking at the molecules and keeping track could be made as pointless, as affectless as you want. It does not need to have an important influence on what the molecules do. That's 1929, almost 60 years after the original experiment and we were still wondering about it.

The next idea came in 1960 when Rolf Landauer, a physicist at IBM made an explicit model where he showed that you can measure the velocity of the atoms in the box reversibly. In other words, what we want to do is not increase entropy, so you want to let the demon look at the molecules without increasing their entropy, to absolutely prove it you just do it in such a way that you can undo it, make it a reversible process. What Landauer was able to show that Szilard was not was that the way to make the measurements reversible was to simply write them down in a notebook. You need if you're the demon to have a way of keeping track of the information.

This actually played a crucial role in the development of what we now call information theory. Landauer was able to show that irreversible computations are the ones that generate entropy. He made a connection between calculation and computation and the increase of entropy in the universe, what we now call Landauer's principle. It says that irreversible computations increase entropy, reversible computations do not. Taking 2 + 2 and adding them together to get 4 increases the entropy of the universe.

But Landauer was not able, of course, to convince us that Maxwell's demon really did increase the entropy of the universe. In fact, he worried us. He seemed to be telling us that you could be Maxwell's demon without increasing the entropy. It wasn't until 1982, over a century after Maxwell proposed the idea, that Charles Bennett who, who is a computer scientist also working at IBM, finally solved the problem in a way that many of us think is satisfactory. What Bennett realized, again using ideas from information theory, is that since Landauer showed Maxwell's demon needs to have a notebook eventually that notebook is going to run out of pages. Now this sounds a little silly, but the point is that a finite size Maxwell's demon can only keep track of a finite amount of information.

To really convince yourself that the molecules will only going one way and not the other the demon needs to constantly be updating his information about where the molecules are. He needs to keep a potentially infinite amount of information. That's okay as long as he can erase the previous entries in his notebook. What Bennett realized is that it's erasing information that is the irreversible stuff. Maxwell's demon can look at the molecules, get the information that's reversible, the demon can open the door and he can let heat flow in one direction and not the other; that's also perfectly fine, but if he wants to keep doing this for a long period he needs to erase the information in his memory and it's erasing information that increases the entropy of the universe. That makes sense to us. Increasing entropy means doing something irreversible, writing something down is reversible, but erasing something is not. Once you've erased it you don't know where it went.

I want to mention one final thing about Boltzmann. He met with resistance from many physicists about his ideas that entropy was statistical not absolute. He also struggled with depression all of his life. He committed suicide ultimately in 1906, which was ironic because only in 1905 Einstein and other people put together perhaps the strongest evidence yet that atoms were real and they really existed. A lot of people speculate that it was the resistance to his ideas that led him to suicide, but that is not true. He simply was depressive. He struggled with that all of his life, but the important thing to remember is that Boltzmann is as influential as any of the physicists that we know, as Einstein, as Newton, as Galileo. His role in putting together a statistical view of the world is absolutely central to how we think about the universe today.

The Past Hypothesis
Lecture 11

You should be convinced by now that Boltzmann's understanding of entropy is the correct one and that it can be reconciled with our previous understandings of the second law of thermodynamics from Carnot and Clausius. In this lecture, however, we'll see that there are some loose ends. It's true that Boltzmann's explanation gives us the right answer under certain assumptions, but these assumptions are not unproblematic. As we'll see, Boltzmann helps us understand why entropy will be higher in the future but not why it was lower in the past.

A Space of States

- To begin, let's imagine that we have a set of medium-entropy microstates in one macrostate and a larger set of high-entropy microstates in a larger macrostate. Keep in mind here that a state evolves through time, which means that every state lies on a unique trajectory. Given the laws of physics and given the state we are in, the complete definition of a state is that it is all the information we need to predict the future and reconstruct the past.

- Every state is on a trajectory, and trajectories never end. That is a consequence of the reversibility of the laws of physics. Trajectories do not stop evolving, nor do two trajectories come together or one split apart. Every trajectory is unique and will continue to evolve uniquely forever.

- This space of states that we are thinking about is full of trajectories passing through it, with every state on a unique trajectory. We know that the entropy now in the universe is not as high as it could possibly be or as low as it could possibly be, so we are, in some sense, in a medium-entropy state right now.

- Boltzmann points out that if we follow all the medium-entropy states, some of them will evolve into other medium-entropy states, some of them will go to higher entropy, and some of them will go to lower entropy. Remember, it's not an absolute law that medium entropy must go to higher entropy; it is possible to go to low entropy. But if we count, we will find that many more medium-entropy states will evolve to high entropy than to low entropy

A Typical Microstate?

- This picture raises two puzzles, the first of which relates to the statement that if we pick a typical trajectory, the entropy will increase.

- We have many trajectories in the medium-entropy macrostate, this is basically our current world—all the different arrangements of the particles that make us up. Boltzmann says that to predict what happens next, we randomly choose a typical microstate within this macrostate and ask what typically happens. If many more of the microstates increase in entropy, it's probable that entropy will increase.

- But who's to say what is typical? How do we know that we are on one of the trajectories that goes up rather than going down? The answer is that we don't. We have made an assumption that our trajectory will take us to higher entropy.

- The idea behind this kind of assumption goes back to Laplace and his principle of indifference: Every allowed possibility should be treated as equally likely.
    - Laplace's principle tells us that if we want to predict the behavior of the universe, it is acceptable to consider what happens to a typical microstate; we simply count the microstates that do one thing versus another thing and turn that counting into a probability. If we don't know what's going to happen, we assume that every alternative is equally likely.

    - It's true that if you knew the microstate of the universe, you could flip a coin and know whether it would come up heads or tails. But we don't know the microstate of the universe;

therefore, we say that it's 50/50. We assign equal weight to undistinguishable alternatives.

- o Of course, there is only one universe, and it does have a microstate, so the justification for this assumption is a little bit shaky. Nevertheless, most people are happy to say that our microstate is probably typical within the macrostate we're in.

Lohschmidt's Reversibility Objection

- A larger problem with our picture of microstates goes back to the origins of Boltzmann's ideas. This problem is known as Lohschmidt's reversibility objection, named for Boltzmann's mentor, the Austrian physicist Josef Lohschmidt.

- Lohschmidt's reversibility objection basically comes down to a simple fact about the evolution of microstates within a space of all possible states, namely, that the trajectories don't begin or end. As a simple consequence of that fact, there are just as many trajectories along which entropy is decreasing as there are trajectories along which entropy is increasing.

- Boltzmann's statement is also true: If we start with medium entropy and let it evolve, entropy will usually increase. But how can we reconcile this with the statement that there are just as many trajectories in which entropy is decreasing as increasing? The answer comes down to a matter of which question we are asking.

- Boltzmann phrased his question in a particular way: Start with a medium-entropy macrostate and ask what happens to it. But that question is already sneaking in some directionality of time. Lohschmidt says that we could just as easily ask: Where did the medium-entropy state come from? What is a typical past of a state that is now in medium entropy? This is exactly the same kind of calculation, but with the arrow of time reversed.

- Because the laws of physics are reversible, the features of the past are the same as the features of the future. Therefore, for all the states that have medium entropy, many more of them came from high-entropy states in the past than came from low-entropy states. If we are going to take seriously Boltzmann's claim that most medium-entropy states increase in entropy as time goes on, then we must take seriously Lohschmidt's point that most of these medium-entropy states came from a higher-entropy past.

- Boltzmann gave us a convincing argument that starting from our medium-entropy configuration, it is natural for entropy to increase. There are more ways to be high entropy than to be low entropy. The same argument says that it is natural to come from a higher-entropy past because there are more ways to be high entropy than to be low entropy in the past, as well.

- Ultimately, Lohschmidt's reversibility objection is valid. If all we have to work with are underlying laws of physics that are symmetric with respect to past and future and Boltzmann's statistical ideas that we chunk up the space of states into macrostates that look the same to our macroscopic observation, we do not derive a different behavior for the future than we do for the past. We need to add something to Boltzmann's machinery, an extra assumption that is explicitly asymmetric with respect to past and future.

The Past Hypothesis
- This extra piece of machinery, the past hypothesis, is the assumption that the entropy of our observable universe was, not long ago, much lower than it is today. Further, the universe was the kind of low-entropy configuration that, under natural thermodynamic evolution, would get us to where we are today. There are many ways to be low entropy, but there is a specific low-entropy state that naturally grows into the universe we see.

- The past hypothesis for the universe talks about what we would call conditions near the Big Bang. Obviously, Boltzmann and Lohschmidt didn't know about the Big Bang, but they knew that the correct way to justify the second law of thermodynamics within

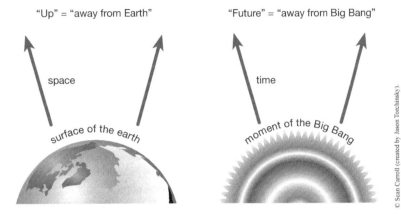

Just as space gets a direction because we live in the vicinity of an influential object, the Earth, time gets a direction because we live in the vicinity of an influential event, the Big Bang.

Boltzmann's framework was to add this assumption that explicitly broke the symmetry of past and future.

- The past hypothesis teaches us that the arrow time—the real difference between the past and the future—is not a matter of atomic theory, statistical mechanics, or anything like that. It's not deeply engrained in the nature of time; it is a feature of our environment, a feature of the universe in which we find ourselves.

- Given this assumption, Laplace's principle of indifference, and the definition of entropy as the logarithm of the number of microstates in a macrostate, Boltzmann gives us a convincing argument for why the entropy of the universe tomorrow should be higher than it was today. He doesn't give us an argument for why it would have been lower yesterday.

- The past hypothesis tells us that the reason the entropy of the universe was lower yesterday than it is today is that it was even lower the day before yesterday. The reason it was even lower the day before yesterday is that it was even lower the day before that, and so on, back 13.7 billion years to the Big Bang.

- We can't derive the history of the universe as we know it on the basis of purely dynamical grounds. We have to put in an assumption about the boundary condition in the past. There is no matching assumption about the boundary condition of the future. We get the answers right by assuming that there is nothing special about the future; it will be a typical state. This is where the imbalance between past and future comes from. We make a past hypothesis, but there is no future hypothesis.

A Future Hypothesis?

- Craig Callender, a philosopher of physics, has asked us to imagine what a future hypothesis would be like. His seemingly bizarre thought experiment involving Fabergé eggs collecting in your dresser drawer is actually no more bizarre than to imagine that there is a past condition in which the universe was in some very specific low-entropy macrostate among all the possible configurations it could have been in.

- The past hypothesis seems to be necessary and true if we want to understand the past, but it doesn't come from any underlying theory. It must come from cosmology. In other words, it's the conditions near the Big Bang—13.7 billion years ago—that are what we refer to when we talk about the past hypothesis.

- With that in mind, we understand the statement that the arrow of time is like the arrow of space. The arrow of space here on Earth is not something intrinsic to the laws of physics; it is a feature of our local environment, distorted by the gravitational field of Earth. The past hypothesis tells us that the arrow of time is the same. It doesn't come from the laws of physics but from our local environment— the observable universe—influenced by the Big Bang.

Suggested Reading

Albert, *Time and Chance*.

Carroll, *From Eternity to Here*, chapter 8.

Price, *Time's Arrow and Archimedes' Point*.

Questions to Consider

1. Can you imagine inventing new laws of physics—not like those in our world—that naturally have entropy increasing in one direction of time and decreasing in the other?

2. What are some other examples of the principle of indifference that we use to calculate probabilities? Do you think this kind of principle is warranted?

3. Following Callender's thought experiment, can you think of some other interesting future boundary conditions and how they would manifest themselves without violating the microscopic laws of physics?

The Past Hypothesis
Lecture 11—Transcript

Hopefully by now, you are convinced that Boltzmann's understanding of entropy is the right one. We are able to think about the world in terms of atoms, arrangements of atoms that look macroscopically the same to us and explain why entropy increases. Entropy because there are more ways to be high entropy than to be low entropy. We are able to recover our previous understandings of the Second Law from Carnot and Clausius in terms of efficiency of engines and heat equilibrating naturally over time.

In this lecture, we are going to admit that there are problems. We're going to reveal the dirty laundry. It is true that Boltzmann's explanation gives us the right answer under certain assumptions, but these assumptions are not unproblematic. We really need to interrogate why the world works exactly in this way; what are the loose ends that need to be cleared up. There is no real reason to be coy about it, there is one loose end that is the biggest one. Boltzmann helps us understand why entropy will be bigger in the future. He fails at helping us understand why entropy was lower in the past. This is one of those puzzles that wasn't a puzzle before the correct ideas came along.

Before Boltzmann to the extent that there was something called the Second Law, it was a law. The law says that entropy increases over time and therefore a consequence of that law is that entropy was lower in the past. What Boltzmann says is that the Second Law is not autonomous. It is not separate from the rest of the behavior or matter. The Second Law is something that we can derive if we know what atoms do. When you look into the derivation that Boltzmann gives us you find that it is easy to explain why entropy goes up toward the future, but it is very hard to explain why entropy goes down to the past.

The question in front of us is do we just need to work harder, we have to come up with some clever explanation as to why the understanding of the world in terms of atoms would make the entropy of the universe smaller in the past or is our theory incomplete. Do we need to add something to it? And we'll find that in fact yes we do need to add something to Boltzmann understanding to truly understand why the Second Law is true and has always been true in the history of the observable universe.

To get there, think about a set of states. When we think about entropy ala Boltzmann we think about some system, whether it is a glass of water or the air in a room or the whole universe, and we think about the abstract space of every single configuration that system could be in, the set of all possible states. We could be thinking about it classically ala Isaac Newton in terms of momentum and positions, that is often the language that we will use. We could be thinking about it in terms of quantum mechanics, in terms of wave functions. We don't use that language as often, but it works equally well. It's the same kind of story. We have the set of all possible states and then we divide it up. We divide the microstates up into sets of macrostates and a macrostate is a collection of many microstates all of which look the same to us.

The entropy is basically the volume, it is really the logarithm of the volume of that macrostate, how many microstates are in it. For some large set of states, we have a set of low entropy microstates in one macrostate. Let's imagine we have a set of medium entropy microstates in another macrostate, and we have a larger set of high entropy microstates in a bigger macrostate. The important thing to keep in mind is that a state evolves through time and that means that every state lies on the unique trajectory. Given the laws of physics and given what state we are in the whole definition of a state is that it is all the information we need to predict the future and reconstruct the past.

Every state is on a trajectory. It evolves toward the future, it came from somewhere in the past, and trajectories never end. That is a consequence of the reversibility of the laws of physics. Trajectories do not stop evolving, nor do two trajectories come together or one trajectory split apart. Every trajectory is unique and it goes on forever as long as time lasts. Maybe there is some ambiguity at the Big Bang depending on what the universe did then, but at least since the Big Bang the trajectory of the universe has simply been evolving and will continue to be evolving uniquely.

This space of states that we are thinking about is full of trajectories passing through it and every state lies on a unique trajectory. If you like, in every point in this space there is an arrow going into it saying which state you came from to get there and there is an arrow going out of it saying where you will go in the future. Every single point has an arrow coming in and an arrow going out, whether it's high entropy, low entropy, or whatever. What Boltzmann asks us to do is to think about what we call the medium entropy macrostate, that is to say there is some lower entropy set of states, there are some higher entropy set of states, and there are some medium states that are just right.

We know that our entropy now in the universe is not as high as it could possibly be or as low as it could possibly be, so we are in some sense medium entropy right now. Boltzmann points out that if you follow all of these state, all of these medium entropy states, some of them will just evolve into other medium entropy states, some of them will go to higher entropy, and some of them will go to lower entropy. Boltzmann knows that's true. It is not an absolute law anymore that medium entropy must go to higher entropy. It is possible to go to low entropy.

But, Boltzmann says we should simply count. We should compare the number of medium entropy states that go to high entropy versus those that go to low entropy. The counting is very easy. In fact, it's very difficult to represent it accurately in a picture because the total number of states is much huger than we can draw, but the upshot is the same. Many more medium entropy states evolve to high entropy than to low entropy. If you pick, you have randomly chosen a medium entropy state it is more likely its entropy will increase than will decrease. We can simply count the number of trajectories where entropy goes up versus going down. There is more where the entropy goes up than when the entropy goes down starting from a medium entropy state.

Now there are two big puzzles that this picture raises and one puzzle is kind of minor and annoying and most people ignore it. The other puzzle is big and profound and we're going to be worrying about for the entire rest of the course. Let's first get to the minor and annoying puzzle. That is the part of the statement where I said in this medium entropy macrostate most trajectories increase in entropy. In other words, if we pick a typical trajectory the entropy will go up. The question is who says we should pick, randomly, a trajectory? In other words, we have many, many trajectories in the medium entropy macrostate that in other words this is basically our current world; all the different arrangements in our world of the atoms and particles that make us up, we don't know the specific positions and momentum of all of them, we know the macroscopic features.

Boltzmann says that to predict what happens next you choose randomly a typical macrostate, a typical microstate within this macrostate and you ask what typically happens. If many, many more of the microstates increase in entropy, you say it is probable that entropy will go up. But, who is to say that we are typical? Who is to say that we are on one of the trajectories that actually does go up rather than going down? Even though we don't know the microstate of the universe, we think that there is something called the microstate of the universe. It could be true that we just get unlucky, that we are on a trajectory that will take us to lower entropy. How do we know that's not true? The answer is we don't know, in some abstract sense, you can't prove that it's not true. It is an assumption. The assumption actually predates Boltzmann.

The idea behind this kind of assumption goes all the way back to Laplace himself. We have talked about Laplace's contributions in terms of determinism and predicting the future and Laplace's demon, but he is also one of the pioneers of probability theory. Laplace thought about what it means to be typical, what it means to be randomly chosen. He invented what we now call the principle of indifference. Basically the principle of indifference says that when you have some information, but not all of the information, let's say you know the macrostate of the universe, but you don't know the microstate of the universe, the principle of indifference says you should treat every allowed possibility as equally likely. Every alternative is created equal.

In other words, Laplace's principle of indifference tells us that if we want to predict the behavior of the universe it is okay to consider what happens to a typical microstate, just simply count the microstates that do one thing versus another thing, and turn that counting into a probability. Everybody makes this assumption all the time. If you don't know what's going to happen you assume every alternative is equally likely. If you flip a fair coin, it is true that if you knew the microstate of the universe, you would predict that coin is going to come up either heads or tails then you would know which one. You don't know the microstate of the universe, therefore you say it's 50/50. You're assigning equal weight to undistinguishable alternatives.

Of course, there is only one universe. The universe does have a microstate so the justification for this assumption is a little bit shaky. Nevertheless, everybody makes it and even though it is a problem we would like to be able say which microstate we are on and to use that to somehow justify using Laplace's principle of indifference, this is not something that a lot of people worry about. Most of us are happy simply to say our microstate is probably typical within the macrostate we're in and to reason on that grounds. That is what we are going to be doing for the rest of this course.

That is not the only problem. That is not, in fact, the important problem that we will be talking about. The important problem we will be talking about goes right back to the very first days of Boltzmann's ideas.

Boltzmann had a mentor back in Austria named Josef Lohschmidt. He was a slightly older Austrian physicist and he was a friend of Boltzmann. He was not in any sense a competitor. He was not trying to tear him down, but just like Maxwell with his demon. Lohschmidt worried a lot about the foundations of this statistical point of view on entropy and statistical mechanics. Lohschmidt invented an objection to Boltzmann's version of the Second Law, which we now call Lohschmidt's reversibility objection.

Lohschmidt's reversibility objection basically comes down to a very simple fact about the evolution of microstates within this space of all possible states, namely that the trajectories don't begin or end. Every trajectory you are on, came from somewhere and goes somewhere. As a simple consequence of that fact, there are just as many trajectories along which entropy is decreasing as there are entropy increasing trajectories. Boltzmann's statement is true if you start with medium entropy and let it go you will usually increase in entropy, but there are just as many trajectories in which the entropy is decreasing as increasing. You might ask, how can we reconcile these two true statements?

It is a matter of which question we are asking. Boltzmann phrased his question in a very particular way. Start with a medium entropy macrostate, ask what happens to it. But, you see that's already sneaking in some directionality to time; you start with something and then ask what happens. Lohschmidt says I could just as easily ask the question, where did the medium entropy states come from? In other words, instead of saying, Where do the medium entropy states go to in the future? I could say, What is a typical past of a state that is now medium entropy? Of course, it is exactly the same kind of calculation, but with the arrow of time reversed.

Crucially because the laws of physics are reversible, the features of the past are the same as the features of the future. Therefore, the consequence of this is that for all the states that have medium entropy many more of them came from high entropy states in the past than came from low entropy states in the past. If you are going to take seriously Boltzmann's claim that most medium entropy states increase in entropy as time goes on, then you should take seriously Lohschmidt's point that most of these medium entropy states came from a higher entropy past.

The problem is nobody believes that's true. We do not think that is how the universe actually works. Boltzmann, if we take him seriously, managed to give us a very convincing argument that, starting from the universe today, starting from our medium entropy configuration that we find ourselves in, it is natural for entropy to increase. There are more ways to be high entropy than to be low entropy. It is not an absolute law, it is just a probability. The same argument says that it is natural to come from a higher entropy past because there are more ways to be high entropy than to be low entropy. He can explain the future, he cannot explain the past.

Now Boltzmann didn't believe that the past was higher entropy, he believed, just like everybody else, that the Second Law has been true throughout the history of the observable universe. But, the manifest implication of his ideas was different, it implies that the entropy was higher in the past. Now this is something that is worth sort of emphasizing and dwelling on because this is the first implication of Boltzmann's ideas that is not at all intuitive. It makes sense to us that if you put gas in a box and most of it is on one side and not the other side it tends to even out. Put a hot thing next to a cold thing, their temperature tends to equilibrate. All of that makes sense to us.

But because we live in a world with such a pronounced arrow of time, we tend to think about the future and the past differently. We tend to think that our current situation has some sort of tendency to have been going from lower entropy to higher entropy. When Boltzmann says, "I can predict on the basis of statistics that the future will be higher entropy," we nod our heads and go, ah, yes, that seems to make very much sense. When Lohschmidt says that exactly the same logic predicts that our current macrostate came from a higher entropy past, we shake our heads and go no that cannot possibly be right. But, the logic is exactly the same so we want to know, how can we escape this conclusion? What we want to do is to accept Boltzmann's logic that says entropy will increase toward the future, but reject the fact that it implies it was higher in the past.

It is interesting to note that historically even Boltzmann himself fell prey to the difficulty we have in treating the arrow of time fairly. Remember way back in the beginning lectures we talked about the view from no end from the idea of standing outside the universe, not assuming there is an arrow of time and just contemplating all the different ways the time could evolve. Now Boltzmann himself when he was dealing with the Second Law, he proved a theorem called the H-theorem, which amounted, essentially, to proving the statement that the entropy must always increase in time. To get the technical connection right, H was minus the entropy and he proved that it should always go down. I don't know why that was a clever thing to do.

Essentially Boltzmann's H-theorem is a statement that entropy tends to increase and he didn't think that he put any assumption in there about time-asymmetry. He thought that he was just making reasonable assumptions about statistics, but later people who dug into the dirty details of the axioms Boltzmann used to prove his H-theorem realized there was a subtle assumption that did sneak time-asymmetry into it. Boltzmann assumed that when particles were bouncing into each other that the velocities and direction in which the particles were moving were uncorrelated with each other. If you have some box of gas, particles bouncing around, where one particle is moving doesn't know what other particles are doing. That sounds completely innocuous. This is the assumption of molecular chaos, he called it.

The problem with this assumption is that, once you start the evolution, even if particles start uncorrelated once they bump into each other, they are suddenly correlated. The two particles that are moving apart are moving apart from the same place because they just bumped into each other. If you start your box of gas with an assumption that the motion is uncorrelated it instantly breaks the next moment. By putting that assumption in at the beginning, you are sneaking in the arrow of time from the start.

This was the challenge that people faced. They wanted to derive the arrow of time, not just assume it from the start. What Boltzmann had bequeathed was a set of machinery that didn't have an arrow of time built in. It could explain entropy going up toward the future, but it also explains entropy going up toward the past, which nobody thought was true. The challenge was could you use these time-symmetric underlying laws of physics to derive a time-asymmetric conclusion. The answer is no.

Lohschmidt was right. It was not that he was making some mistake or that Boltzmann wasn't careful enough. Lohschmidt's reversibility objection is absolutely valid. If all you have to work with are underlying laws of physics that are symmetric with respect to past and future, and Boltzmann's statistical ideas that you chunk up the space of states into macrostates that look the same to our macroscopic observation, you do not derive a different behavior for the future than you do for the past. You need to add something to that machinery, you need to add an extra assumption, and you need to add an extra assumption that is explicitly asymmetric with respect to past and future.

That extra assumption is what we call the past hypothesis. The past hypothesis is the assumption that the entropy of our observable universe, not too long ago, was much lower than it is today. In fact, a little bit more carefully the past hypothesis assumes that the early universe was in a configuration which was, number one, low entropy, and number two, was the kind of low entropy configuration which, under natural thermodynamic evolution, would get us to where we are today. The point is that there are many ways to be low entropy, but there's a specific low entropy kind of thing that naturally grows into the universe we see.

For example, if you imagine a box of gas and you took a snapshot of it and you noticed that all of the gas was sort of concentrated in the vicinity of one corner of the box of gas. Now according to statistics that could just be an unlikely fluctuation, but you might say well that would make sense to me if all of the gas in the box started in a very, very concentrated form right in the corner and I'm just observing it as this expanding and filling the box. That is a sensible explanation for what you have observed. That is a version in the box of gas of what we called the past hypothesis for the universe.

The past hypothesis for the universe talks about what we would call conditions near the Big Bang. Now obviously Boltzmann and his friends, Lohschmidt, et cetera, they didn't know about the Big Bang, but they knew that the correct way, or at least they eventually realized, the correct way of justifying the Second Law of Thermodynamics within Boltzmann's framework, was to add this assumption that explicitly broke the symmetry of past and future. The point is that the past hypothesis teaches us that the arrow time, the real difference between the past and the future is not a matter of atomic theory, statistical mechanics, or anything like that. It's not deeply engrained in the nature of time, it is a feature of our environment, a feature of the universe in which we find ourselves. It is a statement about the stuff in the universe and what it was doing in the past.

One way of thinking about it is Boltzmann, given his assumption, given Laplace's principle of indifference, and given the definition of entropy is the number of microstates or the log of the number of microstates in a macrostate, he gives us a convincing argument for why the entropy of the universe tomorrow should be higher than it was today. He doesn't give us an argument for why it would have been lower yesterday. The past hypothesis says the reason why the entropy of the universe was lower yesterday than it is today was because it was even lower the day before yesterday. The reason why it was even lower the day before yesterday is that it was even lower the day before that and so on 13.7 billion years all the way back to the Big Bang.

This is a remarkable fact. We can't derive the history of the universe as we know and love it, on the basis of purely dynamical grounds. We have to put in an assumption about the boundary condition in the past. There is no matching assumption about the boundary condition of the future. We get the answers right by assuming that there is nothing special about the future. The future will be a typical state. We are evolving toward higher and higher entropy that will keep going. You don't need to say anything about the future. That's where the imbalance comes from. That's where the difference between past and future comes from. We make a past hypothesis; there is no future hypothesis.

Now just to give you an idea what a future hypothesis would be like, remember there is no future hypothesis that we actually think is part of physics, we only think that there is a past hypothesis, but philosopher of physics Craig Callender has asked us to imagine what it would be like. He says imagine you go to a fortune teller, one who is always right, and the fortune teller gives you bad news. She says you are going to die within a couple of weeks, but before you die all of the world's Imperial Faberge eggs are going to collect in the dresser drawer in your bedroom.

Now you don't like the fact that she says you are going to die, but you're not worried too much because you say well I will just prevent the Faberge eggs from collecting in my dresser drawer. These are these beautiful Russian works of art. There are about 42 of them scattered throughout the world's museums. It shouldn't be too hard to not collect them in my dresser drawer. What happens is, despite your best efforts, they keep accumulating there, these Faberge eggs. There is a delivery that was supposed to go to a museum, but the address got written a little bit sloppily and instead it gets delivered to your house and there's an egg. You get scared so you throw the egg out the window, but it happens to hit a lamppost and careens off back into your bedroom and lands in your dresser drawer. Things like this happen over and over again and gradually one by one your dresser drawer fills up with Imperial Faberge eggs.

The important thing about this thought experiment is that no laws of physics are broken along the way. Just like it does not break the laws of physics to take a scrambled egg and turn it back into a regular egg. It is unlikely, but it could happen. Likewise, this future boundary condition that says all of the eggs land in your dresser drawer can be achieved without breaking any laws of physics, just breaking the laws of probability, unlikely things keep happening.

Now you say to yourself, that is kind of a silly thought experiment. It's bizarre to imagine that there's some condition in the past that is so specific, there might be many ways to get there, but it says something specific about what the world will be like two weeks from now. The reason why it is, nevertheless, worth thinking about is because it is actually no more bizarre than to imagine that there is a past condition that says the world is in some very specific macrostate, some low entropy macrostate, some macrostate that is very, very small numbers among all the possible configurations we could have been in.

The past hypothesis seems to be necessary and true if we want to understand the past. It doesn't come from the underlying theory. It's not atomic theory, Newtonian mechanics, classical mechanics, quantum mechanics, particle physics, none of those things justify or explain the past hypothesis. It has to come from cosmology. In other words, it's conditions near the Big Bang, 13.7 billion years ago, that are what we refer to when we talk about the past hypothesis. The universe started with low entropy.

It is an interesting historical question, why didn't Boltzmann invent the Big Bang. Just to get the history right, Boltzmann died in 1906, the controversies over entropy were mostly 1870s, 1880s, 1890s, the Big Bang didn't come along until after Einstein invented general relativity in 1916. The idea of the Big Bang wasn't really developed in details until the 1920s, but you can ask yourself given that Boltzmann and his friends should have known that the success of their program for understanding thermodynamics relied on an assumption that the universe started with a low entropy, should they have invented something like the Big Bang? Should they have been able to say the universe started in a condition like that and has been evolving ever since? They did not say that, so it is an interesting historical thought experiment, but maybe that's what they should have thought of.

With that in mind, we understand the statement that we made long ago that the arrow of time is like the arrow of space. The arrow of space here on Earth is not something intrinsic to the laws of physics, it's just because we are standing on the Earth. The gravitational field of the Earth distorts our local environment; it creates a direction, namely down, that we all agree on what that direction is. It's not universal. We don't agree on in between us here and someone on a continent on the other side of the world. It is a feature of our local environment. The past hypothesis tells us that the arrow of time is the same way. It doesn't come from the deep down laws of physics. It comes from our local environment. Now, by local we mean the whole observable universe, but just like the Earth is an influential object that creates an arrow of space, the Big Bang is an influential event that creates an arrow of time.

The next challenge we have is to be cosmologists to explain why the early universe had such a low entropy. We need to do that, not just to understand cosmology itself, but ultimately to understand why we can go from eggs to scrambled eggs, but not from scrambled eggs to eggs.

Memory, Causality, and Action
Lecture 12

With Boltzmann's idea of entropy and the past hypothesis, we now understand why there is an arrow of time. But remember that there is more than one version of the arrow of time. In addition to melting ice cubes in a glass of water, the fact that we can remember yesterday but not tomorrow is an example of the arrow of time. We've claimed that these kinds of examples also reduce to the fact that entropy is increasing. In this lecture, we will try to justify that grandiose claim, to connect the human aspects of the arrow of time to the physical aspects of increasing entropy.

The Past Hypothesis in Everyday Life
- The past hypothesis creates an imbalance between the past and the future; it gives us extra information about the past.
  - The past hypothesis modifies Laplace's principle of indifference in that it doesn't allow us to assume that we're equally likely to be in any of the microstates contained within some macrostate.

  - We happen to be in one of the very specific microstates in which, if we evolve them into the past, entropy would decrease.

- This information is deeply ingrained in how we think. If you're walking down the street and you see a broken egg on the sidewalk, you can retrodict the past of the

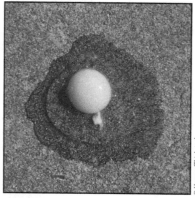

We don't look at the macrostate of the broken egg and imagine that it is from some random microstate that we can't identify; we imagine that the microstate of the egg was one that came from a lower-entropy past.

egg much more readily than you can predict its future because of the past hypothesis.

- ○ You don't look at the macrostate of the egg and imagine that it has some random microstate that you don't know. You imagine that the microstate of the egg was one that came from a lower-entropy past.

- ○ You know that the way the real world creates broken eggs is first by making unbroken eggs. That is the likely evolution if you want to connect the past hypothesis—the low-entropy past of the Big Bang—to the current situation, where there is a broken egg lying on the sidewalk.

- This is where our asymmetry of epistemic access comes from. We are able to say more about the likely past given our current information than about the likely future because we know more about the past. We have a boundary condition in the past at the Big Bang of very low entropy. There is no boundary condition of the future.

- We think that the macroscopic information we have about the egg is all we need to talk about the past of the egg and the future of the egg, but if that were true, the set of things we could say about the egg's future would be the same as the set of things we could say about the egg's past.

- It is not just the knowledge of the state of the egg that lets us say it was unbroken. We pick out the possible trajectory of the history of the universe along which the egg was unbroken when we say that that's what we construct the past to be. That selection is based on the idea that there is a past hypothesis that fixes the universe to start in a state of very low entropy.

- This is why there is a directionality to our knowledge of the universe. If we didn't know about the past hypothesis, we wouldn't be able to remember the past any more than we think we can remember the future.

- We tend to think that the past is real and fixed, but there is a tension with that way of thinking and our belief that the underlying laws of physics are reversible.
    - At the deep level of our best understanding of the laws of nature, there is no more reality or fixedness to the past than there is to the future. The real difference between the past and the future is not how real or settled it is but how much we know about it.

    - We can infer much more accurately about the past because we know not only the current macrostate but also the past hypothesis. When it comes to predicting the future, all we have is the current macrostate. There is no future hypothesis.

Memory
- Let's think about an abstract notion of memory, the idea that we have something in the present world that tells us something about the past world. This is some artifact or record—a photograph, for example—that testifies to something that was really true in the past.

- Imagine that you have a photograph of yourself taken with John Lennon back in the 1970s. Why is it that this photograph is evidence that you actually met John Lennon? What reason do we have to say that just because you have a photograph, that event probably happened in the past?
    - Let's talk about the photograph in the same language that we talked about the egg on the sidewalk. What will happen to the photograph in the future? We cannot predict the exact future state of the photograph, but we know that it will eventually decay.

    - What about the past of the photograph? The reason there is a photograph now is that a real event took place and someone took a picture of it. But why do we treat the past of the photograph differently from the future of the photograph?

- o The answer is the past hypothesis. Just as the most likely future of the photograph is to decay and have its molecules scattered throughout the universe, if it weren't for the past hypothesis, the most likely way for that photograph to come into existence would be by randomly becoming assembled out of molecules throughout the universe.

- o Note that the way we justify using the photograph as evidence of the event is to say that if the event hadn't happened, the photograph wouldn't be likely. But we are not asking how likely is the photograph given the event or not given the event; instead, we're asking, given the photograph, how likely is the event?

- o Given the photograph today—given some information about the macrostate of the current world—what can you conclude about the past? The answer is that without the past hypothesis, you could certainly not conclude that the event actually happened.

- o To go from the photograph today to the actual event would generally lead you to a sort of random collection of things happening unless you could narrow down the trajectory of the universe by also including a low-entropy boundary condition in the past. It is the past hypothesis that lets us believe our memories, that lets us take for granted that what we think is documentation of past events is actually reliable.

Cause and Effect

- Another aspect of the arrow of time is that causes precede effects. But if we believe that the laws of physics are reversible, we can explain any moment in the history of the universe in terms of any other moment.

- Why is it that, in our phenomenological real-life world, we think that causes happen first and events happen second?
    - o Given pieces of broken glass on the floor, we don't think about every possible way for the molecules of glass to arrange themselves in the form of a broken window.

- We imagine there was a past low-entropy boundary condition and given the past hypothesis, then the most likely way to get pieces of broken glass from the window on the floor is to have an unbroken window that something broke.

- What we call the cause of the broken window on the floor is the event that connects the unbroken window to the broken window. It is the past event that is lower entropy that needs to be connected to the future event that tells us that the cause must have happened before the effect.

- It is ultimately because we know more about the past (because of the past hypothesis) that what we call causes always precede what we call effects.

Free Will

- The same logic applies to the human ability to make choices, what we call free will. We think we have the ability to decide what to have for dinner tomorrow night. We don't think we have the ability to decide what to have had for dinner last night.

- But again, the laws of physics treat the two events the same. There is the future, and there is the past, and they are connected by the laws of physics to the present. If we knew the microstates, there would be no difference between predicting the future and retrodicting the past. But we don't know them. We have less information.

- Of course, the reason we think we have free will about the future and not about the past is the past hypothesis. The past hypothesis fixes enough about what happens in the past that we don't think we can affect it because there is no future hypothesis. The set of things that are open to us in the future is very large. We conceptualize that large set of possibilities in terms of choices, in terms of free will.

- Determinism—the idea of starting with the current state of the universe and predicting what will happen in the future—has a bad reputation because it is confused with fatalism. But the determinism of the laws of physics isn't the same as fatalism.

- o Physics tells us that if we knew the microstate, we could say what our fate might be in the future, but we will never know the microstate—we can't.

- o Physics can be determined yet not fatalistic exactly because there is no simple future boundary condition.

The Effects of the Past Hypothesis

- The arrow of time is not just a physics problem. Everything about how we live our lives relies on the arrow of time in an absolutely intimate way. The arrow of time, in turn, relies on the combination of the past hypothesis and our incomplete knowledge of the current state of the universe.

- This is not the usual way we think, but it is the right way of matching our best description of reality with the everyday reality in which we live. It is a reconciliation of the fundamental laws of physics with what we do in everyday life.

Suggested Reading

Albert, *Time and Chance*.

Carroll, *From Eternity to Here*, chapter 9.

Price, *Time's Arrow and Archimedes' Point*.

Questions to Consider

1. If entropy were higher in both the past and the future, would it be possible to have a reliable record of events in the past?

2. Why do we generally say that causes precede effects?

3. Do you think conservation of information robs us of free will? What would be the consequences of that view for how we live our lives?

Memory, Causality, and Action
Lecture 12—Transcript

By now we understand why there is an arrow of time. Given Boltzmann's understanding of entropy, there are more ways for entropy to be increased toward the future and given the past hypothesis there is a reason to believe that entropy was smaller in the past. Now admittedly, the past hypothesis is just an extra hypothesis tacked on that says the entropy was small, and what we would call the Big Bang, 13.7 billion years ago, but it is just one extra added ingredient to the theory. Once we assume that 13.7 billion years ago the entropy was very low. The entropy has been increasing ever since. It is nowhere near as high as it can get. It would take a lot longer than 13.7 billion years for the universe to go all the way to thermal equilibrium, so we live in the aftermath of that low entropy beginning. That is why entropy has been increasing ever since the universe started.

However, you remember that there is more than one version of the arrow of time. Sometimes when we're talking about the arrow of time, we use examples like melting ice cubes in a glass of water or scent of a perfume spreading through the room or mixing cream into coffee, but other times when we talk about the arrow of time we use examples like I can remember yesterday, I cannot remember tomorrow. I can make a choice about what happens tomorrow, but I cannot make a choice about what I had done yesterday.

What about these other examples of the arrow of time? The claim we made earlier on was that they all reduce to what we call the thermodynamic arrow of time, the fact that entropy is increasing. This is the lecture in which we try to justify that incredibly grandiose claim. We try to connect the human aspects of the arrow of time to the physical aspects of entropy increasing and that we should admit from the start it can be a little bit disconcerting. It's a little bit spooky and alien to us to take features of human life that we think of as absolutely intrinsic and necessary and try to explain them in terms of laws of physics that are basically based on statistics.

We will say a lot of things that sound outlandish, but that is what happens when you try to reconcile the way we go through our everyday lives with what physicists have learned about how nature works at a deep level. Remember that Newtonian physics, as interpreted by Laplace, and as continues all the way through quantum mechanics and more modern versions of physics, supports the idea of eternalism or at least the idea that there is nothing special about the present moment. We have the moment we call now, but if you believe that there are laws of physics they connect the moment we call now to the future and the past. At least if you believe either classical mechanics, which you shouldn't, or the many worlds interpretation of quantum mechanics, which at least has a chance of being right, if you knew everything about the present state of the universe, you could predict the past and future with perfect accuracy.

In any version of quantum mechanics, if you knew the present state of the universe exactly, you could at least predict the future in explicit probabilities. You would really know a lot about what could possibly happen. However, we don't know everything. We don't have perfect information about the present microstate of the universe. That is what we imagine Laplace's demon would have, but we certainly don't think that we have it. It's not just sort of a technical problem. We will never have explicit knowledge of the complete microstate of the universe.

In the real world, when we talk about the future, when we talk about trying to predict what will happen in the future, we need to work in a world of incomplete information. Now that we have understood Boltzmann's ideas about entropy and macrostates constructed from microstates that look macroscopically the same, we can be very, very explicit about what it is we do know and what it is we don't know. Well we know, well what we talk as if we know, is the macrostate of the current universe, that is to say we have some incomplete information about the present state. If we model the universe as a box of gas or a glass of water, we know things like the temperature and the density, we know some course-grained features of reality at the present time. We also know the past hypothesis, we know, we will talk about this later, but the only sensible way to proceed in life is to go under the assumption that the past hypothesis is true.

We assume that both, we have current information about the macroscopic configuration of stuff in the universe, and we also know that sometime in the past, 13.7 billion years ago, the entropy was much, much lower. It's the imbalance that we have extra information about the past versus the future that, of course, gives us the arrow of time. In other words, the past hypothesis modifies Laplace's principle of indifference. Remember the principle of indifference said that if you're in some macrostate you assume that we're equally likely to be in any of the microstates contained within that macrostate. Given the macroscopic information, we could be in any of the microstates compatible with that macroscopic information.

The past hypothesis says that is not quite right. Within our macrostate, within all the possible microstates that look like the universe currently looks, we are not equally likely to be in any of the microstates, we happen to be in one of the very specific microstates which, if you evolve it into the past, the entropy would decrease. That is a very tiny fraction of all of the macrostates, all of the microstates within our macrostate, most of them have the property that entropy would increase if you followed them to the past. We don't know which microstate that is, but it is one that had a much lower entropy in the past. That is the modification of Laplace's principle of indifference that is given to us by the past hypothesis. It picks out a tiny fraction of the microstates among all those within our macrostates.

This information is incredibly useful to us as we go through our lives even if it's so ingrained in how we think that we don't actually access it consciously. Let me give you a very, very specific example of how we use the knowledge of the past hypothesis every day, even though we don't put it in those terms. Imagine that you are walking down the street and you see on the sidewalk a broken egg. There are a lot of eggs in these lectures because they're great illustrations of how entropy works, so you're lucky enough that when you're walking down the sidewalk there's a broken egg lying on the sidewalk. You see it splattered in some particular configuration.

What you are observing is the macrostate. You're not observing every single molecule in the egg, its position and its momentum, you have some coarse-grained features. There's a bit of shell. There's a bit of egg yolk, etcetera. That is current information; that is information about the universe right now; that is what you have access to, not complete information, some macroscopic coarse-grained information about the current universe. What you want to know is what will happen in the future and what did happen in the past. You want to make predictions and retrodictions based on the information you have.

What can you say about the likely future of this broken egg lying on the sidewalk? The answer is you can't say the one true thing that will happen, you don't have enough information. Even if you have other macroscopic information about the world around you, the weather and the number of cars on the streets and so forth, it is still not enough information to say with 100 percent certainty what will happen to the egg. It is possible that it will just sit there for the next several days, maybe it will get really hot outside and the egg will fry right there on the sidewalk. Maybe somebody will come up and clean it up. Maybe the rain will come and wash the egg away. Maybe a dog or a cat will walk by and eat the egg. There are many different possibilities. All of them are compatible with the laws of physics and with our knowledge of the macrostate.

It's even possible that the particular microstate of the egg will cause it to leap up and un-break. It will Humpty Dumpty itself back into the form of an unbroken egg. That is not likely. It is an extremely small number of microstates that would behave that way, but there is some probability for that to happen even if it's really, really small. What we see is the future of the egg is pretty open. There are a lot of things that are consistent with what we know.

Now let's ask, what was the past of the egg? What was that egg doing 24 hours ago? Right now, we see that it is broken on the sidewalk. It looks to us what we would describe as fairly fresh, so what we say is we don't know where the egg was 24 hours ago, but we're essentially certain that it was an unbroken egg. There is a huge imbalance between how we talk about the future of the egg and how we talk about the past of the egg. In the future, many things are possible, many futures are open to the egg. In the past, we think that there was an unbroken egg. The egg did not begin life messily scattered around the ground. It did not assemble itself randomly from molecules and atoms moving to the universe just to form a broken egg. There is an unbroken egg that broke.

Why is it that we can be so much more specific about the past of the egg than the future of the egg? The answer is the past hypothesis. We don't look at the egg, look at its macrostate and imagine that it is just some random microstate that we don't know which one. We imagine that the microstate of the egg was one that came from a lower entropy past. We know that the way that the real world creates broken eggs is first by making unbroken eggs. That is the likely evolution. If you want to connect the past hypothesis, the low entropy past of the Big Bang, to the current situation where there is a broken egg lying on the sidewalk.

That is where our asymmetry of epistemic access comes from. We are able to say more about the likely pasts given our current information than about the likely future because we know more about the past. We have a boundary condition in the past at the Big Bang of very low entropy. There is no boundary condition of the future. This way of thinking is alien to us. We think that the egg itself, the macroscopic information that we have looking at it, is all we need to talk about the past of the egg and the future of the egg. That's not true. We're cheating when we think that. If that were true the set of things we could say about the egg's future would be the same as the set of things we could say about the egg's past. It is not just the knowledge of the state of the egg that lets us say it was unbroken. We pick out the possible trajectory of the history of the universe along which the egg was unbroken when we say that that's what we construct the past to be. That is based on the idea that there is a past hypothesis that fixes the universe to start in a state of very low entropy.

This is why there is a directionality to our knowledge of the universe. If we didn't know about that we wouldn't be able to remember the past anymore than we think we can remember the future. We tend to think that the past is fixed. The past is real, the past happened. It's not something that we can alter by our current actions, but there is a tension with that way of thinking and our belief that the underlying laws of physics are reversible. If you believe that the underlying laws of physics or Newtonian mechanics or quantum mechanics, they treat the past and future the same. At the deep level, at the deep level of our best understanding of the laws of nature, there is no more reality or fixedness to the past than there is to the future.

The real difference between the past and the future is not how real it is, not how settled it is, but how much we know about it. The difference between the past and future is that we can infer much more accurately about the past because we not only know the current macrostate, we also know the past hypothesis. When it comes to predicting the future, all we have is the current macrostate. There is no future hypothesis. The point is that the past hypothesis illuminates the actual past. There is one particular past. There is one particular future, it's a matter of how well we know it. Because we have the past hypothesis, we know the past relatively well. Because there is no future hypothesis, the future seems up for grabs.

Let's see a little bit more specifically how this works in the case of memory. Now when I talk about memory you tend to think of your actual memory in your actual brain. We'll talk about the brain in a later lecture. Right now let's think about the abstract notion of memory, the idea that we have something in the present world that tells us something about the past world. We have some artifact, some record, some historical document, a photograph for example, that tells us, that testifies to something that was really true in the past. We think we can have photographs or records of the past, we don't think we can have photographs or records of the future. Why is that?

Let's look at a particular example. Let's imagine that you have a prized possession, a photograph that was taken back in the 1970s when you ran into John Lennon. Your friend was there with a camera; they took a picture of you shaking hands with John Lennon back in the 1970s, you've held on to this photograph as evidence of this moment that was a great moment in your life. You got to meet one of your favorite artists and musicians. Our question is, why is it true that that photograph is evidence that you actually did meet and shake hands with John Lennon? We are not questioning that it is evidence. I agree that that photograph is good evidence. This is before Photoshop or anything like that. We are assuming you didn't doctor the photograph. But, what right do we have to say that just because you have a photograph that thing actually probably happened in the past?

The image that is represented in the picture is something that is a true feature of the past. We wouldn't say that about the future. Let's talk about the photograph in the same language that we talked about the egg lying on the ground. What will happen to the photograph in the future. It's a document, it's a physical collection of atoms. The entropy of that document is going to gradually increase as time goes on. It's not a closed system perfectly, but it interacts gently with the environment, the entropy of the world is going up the photograph will fade with time. It will get yellow, it will crumble, it might by accident get damaged. Someone could spill coffee on it. It could get ripped. As the future goes forward, we cannot predict the exact future state of the photograph, but we can see that it will gray and decay and eventually crumble away. Ten thousand years from now this photograph will not be around, even in a couple of years it might very well not be around if it's subject to a lot of disturbances.

What about the past of the photograph? Of course, we assume that the reason why there is a past image, why there is a photograph now is because there was a real event and we took a picture of it. But, what right do we have to treat the past of the photograph differently from the future of the photograph? The answer is the past hypothesis. Just like the most likely future of the photograph is to decay and have its molecules scattered throughout the universe. If it weren't for the past hypothesis the most likely way for that photograph to come into existence would be by randomly becoming assembled out of molecules throughout the universe.

Now, nobody thinks that this is what happened. I am not trying to suggest that there's any reason to believe that your photograph is not good evidence of the actual event. I am sure that it is. But we have to understand how the logic works. The usual way we would put the logic is not right. You would say two things. You would say if I hadn't shaken hands with John Lennon then it's very unlikely that I would have the photo. If I didn't actually shake hands then of all the ways the molecules in the universe could have arranged themselves it is very, very unlikely they would've arranged themselves into a picture like this. Furthermore, the second thing you would say is if I did shake hands it makes perfect sense that I have the photo. It is easy to take a picture and that picture survives until today.

The way that you would justify using your photograph as evidence of the event is to say that if the event hadn't happened the photograph wouldn't be likely and if the event hadn't happened the photograph would be likely. All of these are true; none of them are relevant. That is not the question we are asking. The question we are asking is not given the event or not given the event how likely is the photograph. We are saying given the photograph how likely is the event?

Given the photograph today, given some information about the macrostate of the current world, what can you conclude about the past? The answer is that without the past hypothesis you could certainly not conclude that the event actually happened. To go from the photograph today to the actual event would generally lead you to sort of random collection of things happening unless you could narrow down the trajectory of the universe by also including a low entropy boundary condition in the past. It is the past hypothesis that lets us believe our memories, that lets us take for granted that what we think is documentation of past events is actually reliable.

This way of thinking has deep repercussions. You can take that little part of the lecture and keep playing it over and thinking about it. Hopefully you'll come to understand that this is all a consequence of thinking about the microstates, macrostates, and what is probable and improbable in the whole set of things the universe can do. It goes beyond simple memory and reconstruction, it also works toward the future. Toward the past the helpfulness of the past hypothesis is it gives us a memory, but toward the future we want to ask about cause and effect.

Remember we talked way back when about the fact that one aspect of the arrow of time is that causes precede effects. You say the window broke because the kids outside hit a baseball that went through the window. It would be bizarre and insane sounding to say the kids hit a baseball because the window was going to break and they had to break it somehow. That is not how cause and effect works. Causes happen first, effects happen next. Why is this true? Why do causes come first and then effects after? After all, if you believe the laws of physics are reversible, I can explain any moment in the history of the universe in terms of any other moment. I can explain the kids playing baseball in terms of the broken window. I can say the state of the universe right now that has the window broken there, the kids scattering around, and everything that I know about the universe if I knew the microstate, I could go backwards and say ah yes there are the kids out there.

Why is it that, in our phenomenological everyday real life world, we think that causes happen first and the events happened second? The reason is because given the broken window lying on the floor we don't just think about every possible way for the molecules of glass to arrange themselves in the form of a broken window on the floor. We also imagine there was a past low entropy boundary condition and given that past hypothesis then the most likely way to get a broken window on the floor is to have an unbroken window in the window that something broke. What we call the cause of the broken window on the floor is the event that connects the unbroken window to the broken window.

It is the past event that is lower entropy that needs to be connected to the future event that tells us that the cause must have happened before the effect. It is ultimately because we know more about the past because of the past hypothesis that what we call causes always precede what we call effects. This way of thinking goes even deeper than that. Causes and effects, windows breaking, knocking over glasses of wine, and so forth, that's fine, but the same logic applies to your human ability to make choices, what we call your free will. We think that you have the ability to decide where to go to have dinner tonight or tomorrow night. We don't think that you have the ability to decide what to have had for dinner last night. Again, the laws of physics treat the two events the same. There is the future, there is the past, they are connected by the laws of physics to the present.

If you knew the microstates there would be no difference between predicting the future and retrodicting the past. But, we don't know that. We have less information. You will not be at this point surprised to learn that the reason we think we have free will about the future and not about the past is because of the past hypothesis. The past hypothesis fixes enough about what happens in the past that we don't think we can affect it because there is no future hypothesis. The set of things that are open to us in the future is very, very large. We conceptualize that large set of possibilities in terms of choices, in terms of free will.

Philosophers argue, up to the present day and I'm sure into the future, about whether or not free will is real. That is not the argument that I'm trying to adjudicate right now. There are people called compatabilists who think that the laws of physics do govern everything and nevertheless we should talk in the language of free will because we don't know what will happen next even though the laws of physics tell us what will happen next if we know the microstate. We don't know the microstate and therefore the way to speak about the world is as if we were making choices. There are also libertarians in the philosophical sense, not the political sense, a philosophical libertarian is one who believes that our human choices overcome the laws of physics, that you are not determined by the laws of physics.

I am not one of those because I am a physicist, but I do believe that free will is the right way to talk. There is a third group who says that free will is wrong. Free will does not exist because the laws of physics tell us what is going to happen in the future. Personally, even though I believe in the laws of physics, even though I believe that the future is determined at least probabilistically if we know the microstate of the universe, I don't mind talking about free will. I believe that it's okay to say we have free will because we don't know what the microstate is. We have the macrostate of the universe.

When we use the language of free will we are not describing the individual microstates of the universe. We are not talking about atoms and molecules. Just like when we talk about the melting of ice in a glass of water, we can talk about that in the language of the molecules of water or we can talk about it the thermodynamic language of fluids and temperatures and densities. We don't talk about it in both languages at once, you don't talk about the temperature of the water and the position and velocity of every molecule. You use one vocabulary or the other.

Likewise I would say that you can use the vocabulary of atoms and microstates to describe what you as a human being will do and in that language there is no free will. There is only Schrödinger's equation and the laws of physics. But, when we don't talk that language, which makes sense because we don't have all that information. We talk about human beings in a different vocabulary and that vocabulary can include the possibility of making choices only, of course, making choices that are compatible with the laws of physics and our current macrostate, but there are many such choices.

This is a very rich area of not only physics and philosophy, but also neuroscience, the study of when we actually do make choices is one that is a hot topic in people who are looking at the brain. Consciousness is not a clean thing. It is kind of a mess that turns out you can make choices before you know that you are making choices. Nevertheless for our purposes what matters is knowing the macrostate leaves choices open to us, just knowing what you can possibly know about the universe right now does not fix the future.

Thinking this way reveals that determinism, the idea that starting with the current state of the universe and predicting what will happen in the future, has a bad reputation. People confuse determinism with what we might call fatalism. Fatalism is the idea that, I am doomed to have something happen to me in the future. I am destined to struggle with poverty, or I am destined to be bored with my job or I am destined to kiss a frog and turn him into a handsome prince, something like that, some gross future boundary condition. That is what fatalism is. The determinism that lurks in the laws of physics is not like that because the determinism of the laws of physics says, yes, if you knew the microstate then you could say in the future you will be kissing frogs, working in a dead end job, or winning the lottery, whatever it is that will happen, but you don't. You never will. You don't know the microstate, you can't. The determinism of the deep down laws of physics cannot possibly affect how you, personally, think about the future.

It can be determined yet not fatalistic exactly because there is no simple future boundary condition. That is not, as far as we know, how the laws of physics work. So, the right way to think about determinism is compared, contrasted with the usual way people think about predicting the future. When we think about predicting the future, we think about something like what happened in *Macbeth* where Macbeth talked to the three crones stirring a cauldron that was bubbling and the crones made a prediction he was going to die when the forest marked on him et cetera, etcetera. Making a very specific prediction, whether it is Faberge eggs or kissing frogs or losing your kingdom, that's what we think of when we think of determinism.

The real world is more like a little annoying kid who says I know exactly what is going to happen to you and you say well, tell me what is going to happen to me. The little annoying kid says I can't tell you and then you say alright I am going to ignore you, I am just going to go on and do something, and then you do something and the kid says I knew that was going to happen. This is annoying, but it doesn't change how you live your life. The universe knows what is going to happen to you next, but you don't know, nobody else knows, there is no realistic sense which the determinism of the laws of physics affects how we should think about our future choices.

What it all means is that the arrow of time is not just a physics problem. It is not just a story about mixing different kinds of sand or heating up boxes of gas or anything like that. Everything about how we live our lives relies on the arrow of time in an absolutely intimate way. The way that we conceptualize the past versus the future, what we remember, what we predict, what we plan for, how we do things, make choices, affect the future, and live our lives toward a future oriented way, all of this relies on the arrow of time, which in turn relies on the combination of the past hypothesis and our incomplete knowledge of the current state of the universe. This is not the usual way we think about it. This is, however, the right way of matching our best description of reality with the everyday reality in which we live. You might think it is a little bit creepy or off-putting, it is not the usual way we think, but I think it is actually quite marvelous and wonderful that we can reconcile the fundamental laws of physics with what we do in our everyday life.

Boltzmann Brains
Lecture 13

Boltzmann's story of what entropy is and why it tends to increase is compelling: Entropy goes up because there are more ways to be high entropy than to be low entropy. But to make the past work, we have to add this extra ugly ingredient, the past hypothesis. People have argued about this ingredient ever since Boltzmann came up with the idea. What does the past hypothesis really mean, and where does it come from? Our larger goal is to try to find a theory that explains the past hypothesis, and we start in this lecture by looking at a particular explanation that didn't work out.

Poincaré's Recurrence Theorem and Zermelo's Objection

- The 19th-century mathematician and physicist Henri Poincaré gave us the recurrence theorem: If there are particles in some system that can take on only a finite number of states or can move only within a bounded region, then whatever state we find that system in at one point in time will repeat itself an infinite number of times toward the future.

- This doesn't sound particularly surprising. If there are a finite number of things you can do and you have an infinite amount of time to do them, then you will keep doing the same thing over and over again. But when you think about the implications of this idea for the real world, it becomes disconcerting.
    - If we put an ice cube in a glass of warm water, the ice will melt; entropy will increase. But if we leave the glass of water on the table for an arbitrarily long period of time, Poincaré's recurrence theorem says that the ice cube will re-form.

    - In other words, Poincaré says that if we wait long enough, entropy will spontaneously decrease. In fact, his theorem says that the lower-entropy configuration will certainly happen. Fundamentally, this is simply a consequence of reversibility.

- The recurrence theorem of Poincaré was turned Boltzmann by a German mathematician named Ernst Zermelo. According to Zermelo, if motion is periodic, anything that goes up will eventually go back down and then go up again.
    o We can't prove that entropy will increase if we can prove that entropy will eventually decrease and then increase again, going on forever.

    o Zermelo's recurrence objection is stronger than Loschmidt's reversibility objection: If entropy is increasing now, the recurrence theorem says that at some time in the future, it will certainly decrease. Thus, it can't be shown that there is a tendency for entropy to go one way or the other. We haven't really established a true arrow of time.

    o The simple answer to Zermelo's objection is that in the visible world—the observable part of reality—entropy was, in fact, very low in the observable past. There's a period of time in the universe that we have access to—about 13.7 billion years—but that period of time is much shorter than the timescale over which, according to the recurrence theorem, anything interesting would recur.

Boltzmann and the Anthropic Principle

- Because Boltzmann worked in a world that was governed by Newtonian physics, he thought of space and time as absolute and eternal. He saw the universe as an infinitely big space scattered throughout with infinitely many particles that were in thermal equilibrium most of the time.
    o In Newtonian physics, time doesn't begin or end; in an infinitely long time, the particles would come to thermal equilibrium. But Boltzmann also knew that thermal equilibrium doesn't last forever; it fluctuates.

    o If we wait long enough, arbitrarily large fluctuations to lower-entropy configurations will eventually happen. It may take a

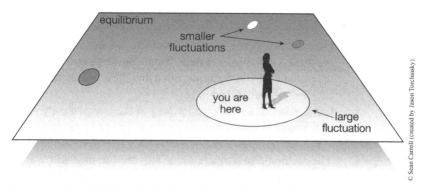

In a thermal equilibrium Newtonian universe, there are times and places where a large fluctuation happens, large enough to form a galaxy, we can live in these multiverses.

very long time, but in this Newtonian eternal universe, we have forever to wait.

- Boltzmann then made an argument based on the anthropic principle, the idea that we human beings don't get to see the typical parts of a big universe—a universe where there are many different conditions in many different places.
  - If typical parts of the universe are inhospitable to the existence of life, then we are going to find ourselves only in those regions that are hospitable.

  - The anthropic principle says that our job is not to construct a theory of cosmology in which the entire universe is hospitable to us. Our job is to construct a theory in which those regions that are hospitable look like the universe in which we actually live.

  - Boltzmann's strategy was to take this eternal Newtonian universe with fluctuating entropy and apply to it the anthropic principle.

- Even today, the anthropic principle has a somewhat shady reputation. It can seem empty. It's obviously true that we live where we can live, but that's irrelevant unless we live in a universe where we can live in some places and not others.

- o For example, there is a lot more volume in the space in between the planets in our solar system than on the surface of any one of the planets. Should we be surprised that life arose on the surface of the Earth rather than in between the planets? Of course not; life would just not survive for long in the interplanetary space of the solar system.

- o The anthropic principle may or may not be the right way to think about the universe, but it is certainly an important part of cosmology. If we live in a universe that is big enough to have very different regions, then the anthropic principle tells us why we live here rather than somewhere else.

- Boltzmann said that most of the places in this large universe—most of which is in thermal equilibrium—cannot support life. But there will be fluctuations—regions of space in which the molecules randomly find themselves in a low-entropy configuration. If we wait long enough, there will be large fluctuations, and if we wait for a very long time, there will be a fluctuation large enough to form an entire galaxy.

- Without knowing about the Big Bang or the existence of other galaxies, Boltzmann suggested that there could be random, smoothly distributed gas and dust in the universe, and if we wait for eternity, it will simply happen that the random motions of that stuff will collect into something that looks like our galaxy. Outside that region, it will not look like our galaxy, but here, it will.

- In Boltzmann's scenario, entropy does not increase forever. It can decrease locally because of some fluctuation, and then it will increase again. The overall evolution of entropy is actually symmetric; it goes up and down equally often. Is Boltzmann going to predict that entropy increases just as much as it decreases? That's not the right way to think about it.

- As we argued, the direction of time in which entropy was lower is the direction we remember (the past). The region of time in which entropy is higher is the direction where we can make choices (the future). It doesn't matter what coordinates you use; what matters is that we always define the past to be the direction in which entropy was lower.

- If entropy fluctuates, there are two possible places we can live: while it's going down or while it's going up. People living in either place would define the past as the minimum point of entropy. This scenario is potentially compatible with the data.

- Boltzmann's idea that the random evolution of chaos ultimately creates the universe is compelling, but in the end, his hypothesis fails. This was pointed out in the 1930s by Sir Arthur Eddington, an astrophysicist in England, who noted that small fluctuations in entropy are much, much more likely than large fluctuations.
    - Boltzmann's scenario makes a prediction, namely, we make whatever we want in this Newtonian universe via the tiniest fluctuation possible. If we want to make a star with some planets, we make the star and the planets; we would not make billions of other stars or an entire galaxy.

    - If we want to make a person, we wouldn't need to make the Earth. We could just fluctuate random molecules into the shape of a person. Of course, that's very unlikely, but it's much more likely than fluctuating the entire Earth with many, many people on it.

    - If we are in an eternal Newtonian universe with random fluctuations, the easiest way to make an apple pie—contrary to what Carl Sagan said—is just to fluctuate the apple pie.

Boltzmann Brains

- What does Boltzmann's scenario say about you and the past hypothesis? In this eternal universe where things fluctuate into existence, it's immensely more probable that you fluctuated into existence with your memories randomly out of the surrounding chaos than that your memories actually happened.

- In fact, you don't even need you. A thinking person is defined by his or her brain and central nervous system. If that's the case, then we don't need to fluctuate a whole body into existence; we can just fluctuate a brain with whatever memories or thoughts we want it to have. This horrible implication of Boltzmann's scenario is called Boltzmann brains.

- If we live in chaos—if the universe simply fluctuates—almost all observers in that universe would be disembodied brains. The fact that we are not disembodied brains means that we don't live in that universe. But how do we know that we're not Boltzmann brains?
    o The simple and short argument against that statement is this: If you believe that you're in some city and outside of your current location, the city exists, Boltzmann's scenario says that the impression of that city is much easier to put into your brain by itself than actually creating the city. Thus, once you leave where you are, there shouldn't be any city there.

    o The real reason that Boltzmann wasn't right is something called cognitive instability. If you fluctuate into existence randomly, all of your impressions about physics, math, logic, and philosophy are most likely to also have fluctuated randomly into your brain.

    o If this scenario is true, you can never have any rational justification for believing that the scenario is true. You can never have any reason to believe the thoughts you think you have if you live in this universe. That's why it's called cognitive instability. Once you believe it, you realize you have no rational reason for believing it.

Suggested Reading

Carroll, *From Eternity to Here*, chapter 10.

Questions to Consider

1. What kind of collection of matter do you think it takes to make a functioning "brain"? Can something that fluctuates randomly into existence be just as alive and conscious as something that evolves and ages normally?

2. What is the best argument that we are not Boltzmann brains? What's the best reason to think that the universe as a whole is not a random fluctuation from a higher-entropy state?

Boltzmann Brains
Lecture 13—Transcript

To remind you of our goal here, just remember that we are trying to understand the mysteries of time. In particular, the mystery of the arrow of time, why the past is different from the future. We have put together a story that is hopefully pretty convincing. There is good news and bad news in this story. The good news is we understand why entropy increases toward the future. Ludwig Boltzmann explained that entropy is really a way of counting how many different ways you can rearrange the microscopic constituents of a system, which keep it macroscopically looking the same. As you go toward the future, entropy goes up just because there are more ways to be high entropy than to be low entropy. The bad news is we don't understand, just on the basis of Boltzmann's counting, why the entropy was lower in the past. To make that come true, we need to introduce a new ingredient, the past hypothesis.

We call it a hypothesis, but it is actually just a true fact. It is something that must be correct about our universe. It certainly fits all of the data. It is the idea that at some time in the observable past, we think now 13.7 billion years ago, the universe started in a very low entropy state. That is the past hypothesis. Once you do that you explain both why entropy was lower in the past and why entropy will be higher in the future.

There is another bit of good news, which we tried to argue for in the last lecture, which is that if you believe the story about entropy increasing, the Second Law of Thermodynamics, you can argue that that one fact underlies all of the differences between the past and the future. The fact that you remember the past and not the future, the fact that you can make choices and affect the future, but not the past. The whole story we like to think fits together, but there is an obvious thing that we would like to do. There is that past hypothesis that we just put in there as an extra ingredient, it's a little bit ugly compared to the rest of the story.

Boltzmann's story of what entropy is and why entropy tends to increase is very, very compelling. Once you believe in atoms, once you understand what Boltzmann is trying to do, it just all makes sense. Of course, entropy goes up. There're more ways to be high entropy than to be low entropy. But then to make the past work, you had to add this extra ugly ingredient, the past hypothesis. This has been something that people have been arguing about from the moment that Boltzmann came up with the idea to the present day. What does the past hypothesis really mean, where does it come from?

There're different attitudes you can take toward that question. One attitude absolutely is maybe it is just like that, maybe there is no bigger explanation for why the early universe had a low entropy, maybe it is a brute fact. That is not something we are going to disprove. It is absolutely on the table, but it's clearly unsatisfying. There is something about the low entropy past of the universe that doesn't seem natural to us. It doesn't seem that, if you were just to start a universe, it would be so special in that way. Remember if low entropy means there are very few arrangements of a system that looked that way, what we are saying is that the very earliest moments of the universe could have been many, many configurations and were a very, very specific kind. That's what the Big Bang was like. That seems to be a feature that cries out for an explanation.

Our bigger goal is going to be to try to come up with a theory that explains the past hypothesis that says yes, that was a natural outcome. It was not just arbitrary. It could not have been any other way. We should not be surprised that our early universe had a low entropy. Now we are not going to succeed at doing that once and for all. We're going to discuss the different possibilities. One of these possibilities might end up being correct, but the truth is we currently don't know why the early universe had such a low entropy. It's one of the research questions that modern cosmologists are trying to figure out.

In this lecture, we are going to tease you a little bit. We're going to talk about a very particular idea that didn't work out, an idea that goes all the way back to Boltzmann himself and maybe even a little bit earlier than that, trying to explain why the early universe had such a low entropy. The basis for this explanation or at least an inspiration for thinking about it can be thought of as coming from a quote from Carl Sagan. Sagan once said, "In order to make an apple pie you must first invent the universe." Now I think what he was talking about was the fact that to make an apple pie it's not only true that the apple pie sits in the universe, it's also true that the apple pie came from apples and flour and so forth, which means that to make that apple pie you needed apple orchards, you needed wheat fields, that means you needed a biosphere and earth. You needed a sun and you needed the whole galaxy and you needed the whole universe.

Sagan is simply pointing out that every little piece of our universe inherits what it is from the larger story in which it was invented. However, a modern cosmologist would say that's not necessarily so. We can invent models of the universe in which the easiest way to make an apple pie is to just make the apple pie out of the surrounding primordial chaos, not to make a universe at all, not to make a solar system, a galaxy, much less the hundred billion galaxies we see in our current universe. That's the essence of the scenario that was proposed by Boltzmann to try to understand the past hypothesis.

Another way of motivating this idea was to go back to Friedrich Nietzsche in the 1880s. Now, Nietzsche was not a physicist, but it is interesting that he was writing in the 1880s when there were debates going back and forth among the physicists and the mathematician and the chemists about what is entropy and why is there an arrow of time and so forth. Nietzsche asked us to imagine a thought experiment and his thought experiment was to say, imagine that a demon comes down, they loved talking about the demons in the 19th century, the demon comes down and tells you that you live in a world which is not just the only occurrence of this world, but that the world will repeat itself over and over and over again, that the life you are leading now you will lead again and again and again and you have led it in the past an infinite number of times.

The reason why Nietzsche wants you to think about this thought experiment is not for physics reasons. You are not trying to demonstrate something about the nature of time, he is thinking about morality. He wants you to say I would like to live a life that I would be happy to live an infinite number of times. That's a moral piece of guidance. The question is, is that a realistic cosmology? Does the demon have a chance of being right? Is it plausible to imagine that the universe just repeats itself many, many, many times?

Well it wasn't long after Nietzsche did his little story that a mathematician and physicist Henri Poincare proved a theorem that we now call the

Recurrence Theorem. What the Recurrence Theorem says is that if you have a bunch of particles, you have some system, for which the particles can only take on a finite number of states or if they can only move within a bounded region. What he was thinking of was the planets moving around the Sun in the solar system, they don't fly off to infinity, the planets just orbit around and Poincare proved that if that happens, if that's your system, it's somehow bounded, it can't roll off to infinity and it lasts forever, then whatever state you find that system in at one point in time will repeat itself an infinite number of times toward the future. That's the Poincare recurrence theorem.

One interesting thing is that Poincare did not try to prove this theorem. He stumbled on it in the attempt to prove something else. He was trying to understand that the solar system was stable. He wanted to prove that the solar system would not go flying off to infinity and he in fact did prove it. He submitted a paper to the journal, he won a prize for doing this wonderful proof, and when they were proofreading his paper, when they were checking every step, Poincare realized he had made a mistake. He had made an assumption which isn't always true. He went back to fix the mistake and he ended up inventing what we now call Chaos theory.

Poincare ended up showing that if you include the gravitational effect of the planets on each other, not just the planets pulled by the Sun, but the planets pushing on each other, it is very likely that one of the planets in the solar system will zoom off to infinity. It is very hard to control things when they can go infinitely far away. In the process of doing that, he proved that if you can't go infinitely far away then you are doomed to recurrence. Things that happen will happen again. As close are you are to any one particular configuration, you will come that close again an infinite number of times.

This doesn't sound very surprising to us. If you have a finite number of things you can do, and you have an infinite amount of time to do them, then yes, you will keep doing the same thing over and over again. But, when you think about the implications of these ideas for the real world it becomes a little bit disconcerting. For example, we put a glass on the table. It has warm water in it, we put an ice cube in it, the ice cube will melt. Entropy will increase. But, if we leave that glass of water there forever, for an arbitrarily long period of time, Poincare's Recurrence Theorem says that that ice cube will reform. The random motions of the molecules in the glass of water will convert themselves back into an ice cube, and then it will melt again, and then it will reform and melt an infinite number of times.

This is very, very surprising to our intuition, which says that entropy increases. Poincare is saying that if you wait long enough the entropy will spontaneously decrease. Now we already talked a little bit about the fact that entropy is allowed to decrease in Boltzmann's way of thinking about it. If you have a box of gas or a glass of water that is in equilibrium, it will have fluctuations up and down and occasionally just through the random motions of the molecules you will move into a lower entropy configuration. But, Poincare's theorem is stronger than that. He does not say there is some small probability that this will happen. He says the probability is one, it will certainly eventually happen, all you have to do is wait long enough.

That is a little bit more shaking to our conviction that entropy is supposed to increase. But, fundamentally it is simply a consequence of reversibility. If you can go there you can come back. If you wait long enough you will come back. This theorem of Poincare's was turned against Boltzmann by a man named Ernst Zermelo. We now know Zermelo in the mathematics community as one of the founders of set theory, but at the time in the late 19th century he was a young guy. He was basically what we would now call a post doc. He had just gotten his PhD. He had studied under Max Planck, the famous German physicist, and he was a mathematically inclined guy so he seized on Poincare's Recurrence Theorem to argue against Boltzmann's understanding of entropy.

Zermelo's argument was very, very easy. He says if motion is periodic, if the motion of the molecules in a box of gas or a glass of water does something and then will do it again an infinite number of times anything that goes up will eventually go back down and then go up again. Zermelo says, how can you possibly prove that entropy will increase if I can prove that the entropy will eventually decrease and then increase again and going on forever? That doesn't seem to be the world in which we live. This is now called Zermelo's Recurrence objection. It's a close cousin of Lohschmidt's reversibility objection.

Lohschmidt said yes there are trajectories where entropy is increasing, but there is an equal number where they are decreasing, how do you say that one is more probable than the other. In some sense, Zermelo's objection is a little bit stronger. He says, if entropy is increasing now Poincare's theorem says it is certain that some time in the future it will decrease. That is saying that you haven't shown, at all, there is a tendency for entropy to go one way or the other. You haven't really established what we would truly call an arrow of time at all.

Well there is a simple answer to Zermelo's objection. The question is whether or not you are willing to accept it. The simple answer is that in the visible world, in the observable part of reality the entropy was in fact very low in the past, in the observable past. That is what we would now call the past hypothesis. The point is that there's a period of time in the universe that we have access to. We now know it is about 13.7 billion years, but that period of time is much, much shorter than the timescale over which anything interesting would recur according to Poincare's Recurrence Theorem. If you have a glass of water with say 10^{25} molecules in it, the amount of time it would take for that ice cube to reform is something like 10 to the 10 to the 25 years or second. It doesn't really matter what units you use, 10 to the 10 to the 25 is such an enormously mind-bogglingly large number that it doesn't matter.

Boltzmann and other people said who cares about the Recurrence theorem. It's something that is so far in the future I don't need to worry about it. Maybe it would happen, but maybe the laws of physics don't let the universe last that long. All we have to explain is the recent past, so he actually focused his mind on what we now call the problem of the past hypothesis. What Boltzmann had, and other people didn't at the time, was this firm conviction that entropy is fundamentally statistical; that it doesn't need to go up. What he didn't have that we now have is any idea about quantum mechanics or general relativity or the Big Bang. Boltzmann worked in a world that was governed by Newtonian physics, so he thought of space and time as absolute and also eternal. They were there forever.

When Boltzmann had to ask himself the question, what should the universe look like? What is the natural configuration for the universe? He took the

Newtonian space time, space and time are there forever, and you filled it with particles so there's an infinitely big space, there are infinitely many particles scattered throughout the universe and most of the time they're in thermal equilibrium. That is what the particles would like to do. After all they've been there forever. In Newtonian physics time does not begin or end, you have an infinitely long time the particles would come to thermal equilibrium. But, Boltzmann also knows that thermal equilibrium doesn't last forever, it fluctuates. Most of the time any particular set of particles remains at the high entropy equilibrium state, but there will be occasional fluctuations to lower entropy states.

Now, these are very rare as we have discussed, but they are inevitable. In fact, arbitrarily large fluctuations to lower entropy configurations will eventually happen. If you wait long enough, the air in this room will go all the way into one tiny little corner and then come back. It will take a very long time, but in this Newtonian eternal universe, you have forever to wait. At this point, Boltzmann makes a very unusual move for the physicist of his time. This is something that, these days we would take very straightforwardly, but in the 19th century no one made this argument mainly the anthropic principle. The idea that if you have a big universe, if you have a universe where there are many different conditions in many different places, then we human beings don't get to see typical parts of the universe. If typical parts of the universe are inhospitable to the existence of life then we are going to find ourselves only in those regions that are hospitable.

The anthropic principle says our job is not to construct a theory of cosmology where the entire universe is hospitable to us. Our job is to construct a theory where those regions that are hospitable look like the universe in which we actually live. Boltzmann's strategy was to take this eternal Newtonian universe with fluctuating entropy all over the place and apply to it the anthropic principle. Now the anthropic principle, even today, has a somewhat shady reputation. It can seem empty. You can phrase it as simply saying we live where we can live. That is obviously true, but it's irrelevant unless we live in a universe where we can live some places and we cannot live others.

For example, here in the solar system there is a lot more volume in the space in between the planets than on the surface of any of the one planets so you might wonder, should we be surprised that life arose on the surface of the Earth rather than in between the planets? But, that is a dumb thing to wonder, no one wonders that. Life would just not survive very long in the interplanetary space of the solar system. We live in the place in the solar system that is hospitable to life so we are not surprised to find ourselves there. There are also arguments against the anthropic principle saying that it is a move of desperation or that somehow it is spiritual and spooky and it implies that the universe was designed for us. None of that is necessarily true. I don't want to say that the anthropic principle is the right way to think about the universe, or not, but it is certainly an important part of cosmology. If we live in a universe that is big enough to have very different regions, to have regions that look like the universe we see in some places, but other regions that look completely different. If that's our universe then the anthropic principle tells us why we live here rather than somewhere else.

Boltzmann knew about the fact we discovered previously that life itself depends on entropy being low and increasing so he said that in this large universe of his, most of which is in thermal equilibrium, most of the places of that universe cannot support life. If the universe is in equilibrium, life cannot exist there. But, there will be fluctuations, there will be regions of space in which the molecules just randomly find themselves in the low entropy configuration and then they bounce back. If you wait long enough, there will be large fluctuations. If you wait really, really, really long there will be a fluctuation large enough to form an entire galaxy. Now remember, in the late 19th century they didn't know about the Big Bang, about different galaxies. They thought that our galaxy was the whole universe, what we now call the Milky Way. They thought there might be millions of stars in it. We now know there are 100 billion or more stars in the Milky Way.

What Boltzmann suggested was that there could be random, smoothly distributed gas and dust in the universe, but if you wait for eternity it will simply happen to the random motions of that stuff that you collect in something that looks like our galaxy. Outside that region it will not look like our galaxy, but here it will. There is essentially what we would now call a multiverse. There're different regions in this infinitely big and eternally persisting universe that look hospitable to life and many other regions that don't. What Boltzmann is suggesting is maybe our universe is just something that happens from time to time. Maybe our galaxy just is randomly formed on occasion and then dissolves back. Maybe that is how we find ourselves living in this eternal, mostly equilibrium, Newtonian universe.

Now it's interesting to think about the fact that, in Boltzmann's scenario, the entropy is not increasing forever. It is like a box of gas that is usually in thermal equilibrium. The entropy can decrease locally because of some fluctuation and then it will increase again. The overall evolution of the entropy is actually symmetric. It goes up and down equally often, so you might worry, isn't Boltzmann going to predict that entropy increases just as much as it decreases? That is not the right way to think about it. The point is that we define the past to be the direction of time in which the entropy was lower. As we argued, the direction of time in which entropy was lower is the direction we remember. The region of time in which the entropy is higher is the direction to which we can make choices. It doesn't matter what coordinate you use, whether you have your clock running clockwise or counterclockwise, what matters is we would always define the past to be the direction in which the entropy was lower.

If entropy goes up and goes down and then comes up again there're two possible places you can live, during while it's going down or during while it is going up. Both people, both sets of people would define the past as the minimum point of the entropy. This scenario is potentially compatible with the data. You can argue that Boltzmann was not the first one to think about this scenario. You can go all the way back to 50 BC to the ancient Roman poet Lucretius. Lucretius, like Boltzmann, was a believer in atoms. He was a follower of Epicurus and Democritus, the ancient Greek atomists, and he wrote a poem called *On the Nature of Things* where he tried to explain how the universe works in terms of the atomic theory. He was faced with the same problem that Boltzmann was faced with. He didn't know about entropy or so forth, but he knew that we lived in a universe that was complex and orderly and there was a whole bunch of structure in our universe that did not look like the mindless interaction of atoms.

Lucretius proposed that if you wait long enough those mindless interactions of atoms would eventually arrange themselves in the form of a universe. That is exactly the same scenario that Boltzmann had, but he just didn't have the math, but it is a compelling idea that the random evolution of chaos ultimately creates the universe around us. I want to say that it's a good idea because it is a well-formed scientific hypothesis by which I mean it makes predictions. Sadly those predictions can be tested and the hypothesis completely fails, so don't get me wrong. I am talking about the scenario in detail, but the final verdict on the story is that it is false.

This was pointed out by Sir Arthur Eddington, an astrophysicist in England, in the 1930s because he pointed out that, unlike Lucretius, Boltzmann has equations, Boltzmann can talk about the exact likelihood of different fluctuations in entropy. In particular, Boltzmann should know that small fluctuations in entropy are much, much more likely than large fluctuations. Basically any fluctuation in entropy is unlikely, and a large fluctuation is like many small fluctuations right on top of each other. So, you are much more likely to get a tiny fluctuation from thermal equilibrium to just outside thermal equilibrium than you are to get a large fluctuation from thermal equilibrium into a whole giant galaxy with millions or billions of stars.

Boltzmann's scenario makes a prediction, namely, you tell me what you want to make, you want to make a planet or a galaxy or a star or whatever. You want to make it in this universe that lasts forever, just like Newton said, and it has all these fluctuations in it. Then Boltzmann's scenario predicts that you make that thing you want to make via the tiniest possible fluctuation that could make it. For example, if you want to make the galaxy then you can make the galaxy. That's fine; it is possible. But, if you want to make a star with some planets, you would make the star and the planets, you would not make the millions or billions of other stars. You wouldn't need to make the entire galaxy.

If you wanted to make just the Earth you wouldn't need to make the other planets in the solar system. If you wanted to make a person you wouldn't need to make the Earth. You could just fluctuate the random molecules around you into the shape of a person and you might say that sounds very unlikely, and it is very unlikely, but it is much more likely than fluctuating into the entire earth with many, many people on it. So, Carl Sagan was wrong. If you are in this Boltzmannian scenario where you have an eternal Newtonian universe with random fluctuations, the easiest way to make an apple pie is just to fluctuate the apple pie. It is much harder to fluctuate the galaxy and then the Sun and then the Earth than the apple orchards, etcetera, just so you can get that pie. In Boltzmann's universe, the pie will appear all by itself spontaneously out of the chaotic surrounding universe.

What does this have to say about you and the past hypothesis? Well you think that you have memories, that you remember your 10th birthday, you remember purchasing these lectures and so forth, but in the Boltzmann/Lucretius scenario, in this idea where the universe lasts forever and things like you and me fluctuate into existence, it is immensely more probable that you fluctuated into existence with those memories randomly out of the surrounding chaos rather than actually having those memories happen. The actual appearance of you with the photograph of John Lennon is much more likely to randomly fluctuate into existence than to represent a real event in the past. In fact, there is no reason in this scenario for you to believe any of the memories you think you have.

Another way of putting it is that you don't even need you. That is to say if you consider this scenario and ask, what are the most likely ways for conscious observers to arise in this universe? If you think that a conscious observer, a thinking person, is defined by their brain and their central nervous system, then you don't need to fluctuate a whole body into existence, you can just fluctuate a brain. It is whatever memories or thoughts or impressions that you want this to have. It will always be easier to just fluctuate that brain than to fluctuate a whole universe and have that brain learn in the conventional way. This scenario or this horrible implication of this scenario, is called Boltzmann Brains.

The idea is that if we live in chaos, if the universe simply fluctuates, almost all observers in that universe would be disembodied brains. We are not disembodied brains, therefore we do not live in that universe. We need to do better than that. But, when faced with this argument you should be asking yourself, maybe you are asking yourself, how do I now that I'm not a Boltzmann Brain? How do I know that I did not fluctuate into existence yesterday? There are two arguments against that possibility. They are sort of a sloppy argument that convinces most people and there's a better argument that really is completely convincing, but might leave a sour taste in your mouth. The simple and short argument is simply that if we lived in a universe that had randomly fluctuated into existence, let's say that you believe you and the room that you're in right now randomly fluctuated into existence, then the Boltzmann scenario predicts that as soon as you leave that room it should be thermal equilibrium.

If you believe that you are in some city and outside of your current location the city exists, Boltzmann's scenario says the impression of that city is much easier to put into your brain by itself than actually creating the city. So, once you leave where you are there shouldn't be any city there. The problem is that even when you do leave where you are you see a city, it is still much easier to fluctuate that person into existence than to fluctuate the rest of the universe. The real reason that Boltzmann wasn't right is something called cognitive instability.

The real problem with Boltzmann's scenario was that even if you do fluctuate exactly like you exist now into existence randomly, all of your impressions about how physics works, how math works, how logic works, how philosophy works, they are most like to also have fluctuated randomly into your brain. Even if this scenario is true, you can never have any rational justification for believing that the scenario is true. You can never have any reason to believe the thoughts you think you have if you live in this universe. That is why it's called cognitive instability. Once you believe it, you realize you have no rational reason for believing it.

Even if we do live in the universe described by Boltzmann and Lucretius and their friends, we have no reason for thinking that we need to do better. Let's emphasize that's what we are here to do, to do better. It's interesting to think that here was a fully sensible, well-developed theory of physics that predicted that we randomly fluctuated into existence, but that theory made predictions and the predictions are not right. We're not advocating this way of thinking of it, we are saying that that was worth considering and we need to move on.

This is not just a historical curiosity. It turns out that many of our best theories of cosmology today, as we will talk about, make the prediction that the overwhelming majority of conscious observers in those cosmologies are Boltzmann Brains. Or if you like, the overwhelming majority of situations like we find ourselves in right now, people like you and me with the memories we think we have, have randomly fluctuated into existence. We don't want that to be true so one of the biggest challenges for modern cosmologies, cosmologies that try to account for the past hypothesis is to make our impressions of the past be reliable, to make sure that if you and I exist, if apple pies exist, it is because there were universes in which we are embedded that gave rise to us, not that we just fluctuated into being here.

Complexity and Life
Lecture 14

In the last lecture, we talked about a cosmological scenario that attempted to account for the arrow of time just by imagining that the universe was infinitely big and lasted infinitely long and there were random fluctuations that could give rise to people like us. We found that it didn't work. If we lived in that universe, individual complex creatures could be made through random fluctuations, not through billions of years of evolution. In the next few lectures, we'll return to the real world and see how what we know about the actual features of life on Earth—including how human beings think and perceive time—matches up with what we think about the arrow of time.

Defining Complexity

- In the 1960s, the Soviet mathematician Andrey Kolmogorov defined the complexity of something as the length of the shortest description that captures everything relevant about that thing.

- When we talk about the complexity of some living being or nonliving object in the universe, we coarse grain. We don't list the position and momentum of every atom; instead, we describe a certain living being as having a certain number of proteins, molecules, and so forth. The complexity of an organism is Kolmogorov's complexity applied to that coarse-grain description.

- It's important to understand that low entropy or high entropy doesn't necessarily correspond to simple or complex. The example of milk stirred into coffee shows us that complexity can arise when entropy is in between very low and very high.

Neither low entropy nor high entropy corresponds to simple or complex; complexity can arise when entropy is in between very low and very high.

- What's true for the cup of coffee is also true for the universe. Basically, the universe began 13.7 billion years ago in a low-entropy state and a simple state. It was homogeneous. It was smooth everywhere, and everything was densely packed.
  - The far, far future of our universe is also very simple. Everything will scatter to the winds, and we will have empty space once again.

  - The middle universe that has medium entropy is complex. It is today that we have galaxies and stars and planets and life on those planets.

- Complexity depends on entropy; it relies on the fact that entropy is increasing. We don't have to worry about how complexity can arise in a universe evolving toward a heat death. The simple fact that entropy is increasing is what makes life possible.

Erwin Schrödinger's Definition of Life
- A careful construction of the second law of thermodynamics says that in a closed system—an isolated system that is not interacting with the rest of the world—entropy either always increases or remains constant. But a living being is not a closed system; it interacts with its environment constantly. The second law does not, however, rule out the existence of a living organism. In fact, it allows for life to exist.

- One of the best ways of thinking about this was put forward by Erwin Schrödinger, the pioneer of quantum mechanics. Schrödinger came up with a wonderful definition of what life is: It is something that goes on doing something much longer than you would expect it to.
  - To understand what Schrödinger had in mind, imagine a glass of water with an ice cube in it. Over the course of a few minutes, the ice cube melts, and the water in the ice cube comes to equilibrium with the water around it. Once that equilibrium is reached, everything stops.

- Now imagine a glass of water with a goldfish in it. The goldfish does not come to equilibrium with the water around it. It maintains its integrity even though, by mass, most of the goldfish is also water. If you give it food, you're allowing that goldfish to take advantage of energy in a low-entropy form, and it can last a long time. This is what Schrödinger had in mind: A living being can put off the approach to thermal equilibrium for a long time.

Schrödinger tells us that life is something that goes on doing something much longer than we would expect it to.

- What's going on here is actually very similar to the action of Maxwell's demon in maintaining a low-entropy situation without violating the second law. The demon is making use of the fact that he lives in a low-entropy world to keep his box of gas in a low-entropy situation. By itself, the box would equilibrate; both sides would come to the same temperature. The demon increases the entropy of the universe by writing things down in his notebook and later erasing them.

- If the external world were already in equilibrium, the demon could not do that. The entropy of a system that is in equilibrium cannot be increased.

• This is a general paradigm for thinking about how life persists. Schrödinger tells us that the complexity of a living being can last longer than you would expect it to. It does not come to thermal equilibrium because there is plenty of room for entropy to grow.

Low Entropy in the Biosphere

- We typically say that the Sun is the source of energy for living beings on Earth. But it's not the energy we get from the Sun that matters; it's the fact that the energy is in low-entropy form. The Sun is a hot spot in an otherwise cold sky. If the whole sky were the temperature of the Sun, we would get much more energy than we get now, but we would come to equilibrium with the sky very quickly, and all life would cease.

- Likewise, if the whole day and night sky were the temperature of the real night sky—completely dark and cold—the Earth would quickly come to the temperature of the night sky, and life would cease. The reason we have life is that the whole sky is not at the same temperature, and we do not quickly come to thermal equilibrium.

- We give back to the universe just as much energy as we get from the Sun. But if we don't get a net gain of energy from the Sun, what is it that drives life on Earth? The answer can be found in the fact that the energy we get from the Sun is in the form of visible light.
    - For every 1 photon of visible light we get from the Sun, we give back 20 photons to the universe, each with about 1/20 of the energy. We radiate infrared light to the universe, which means that we increase entropy by a factor of 20.

    - Our biosphere is far from a closed system. Energy comes in, we increase its entropy by a factor of 20, and then we give it back to the universe in the form of infrared radiation.

- Can we account for the increase of entropy quantitatively when we compare the amount that we have increased the entropy of the universe through infrared radiation to the amount that we have decreased the entropy of the universe by inventing life here on Earth?
    - The answer is yes, we can do the necessary calculations, and when we do, we find that the Earth has increased the entropy of

the universe over the course of the history of life by 4 billion times the amount that the biosphere has decreased the entropy of the universe.

- o The bottom line here is this: The origin of life on Earth and the existence of life on Earth do not violate the second law.

The Origin of Life

- Before we talk about the origin of life, it's important to note that the process of evolution is not working toward the goal of greater complexity—or any goal at all. The way evolution works is that individual species and the genes and DNA they carry flourish or don't in whatever environment they are in, and when they flourish, they reproduce and create more of themselves. Sometimes flourishing means being complex, but sometimes it doesn't.

- There are two schools of thought about the ways in which life could have started on Earth, one of which is called "replication first." In this view, the most important feature of life is that a living being can reproduce itself. Some biologists argue that RNA molecules could have come first and reproduced themselves. Later, these reproducing molecules formed cells around themselves and became the first single-celled organisms. We can think of this as software before hardware.

- The competing school of thought is called "metabolism first." This school argues that the use of complex chemical reactions to take advantage of the low-entropy environment on the early Earth came first. This is hardware before software.

- From one point of view, it makes sense to have hardware without software. The metabolism first school says that the first life was just a chemical reaction, something like fire, taking advantage of the fuel around it to get going.
  - o The early Earth most likely had an atmosphere that was rich in hydrogen and carbon dioxide. The interesting thing about this atmosphere is that it was low entropy. To increase entropy—

which is what the world wants to do—that hydrogen and carbon dioxide would be converted into methane and water.

- ○ The problem is that to go from the original situation of low entropy to the situation of higher entropy requires a chain of reactions that is very difficult to get started. The hypothesis is that those complicated series of reactions started in very specific geological formations.

- ○ If this story is true, the reason life started is that it was trying to increase the entropy of the primitive atmosphere on Earth. There were many ways to rearrange the fundamental structure of the things in the Earth's early atmosphere that would have had a higher entropy, but to get there required something complicated, like life.

Suggested Reading

Carroll, *From Eternity to Here*, chapter 9.

Gell-Mann, *The Quark and the Jaguar*.

Schrödinger, *What Is Life?*

Questions to Consider

1. What is the best definition of "life"? Is a virus alive? A computer program? A forest fire? A galaxy?

2. What are the different ways in which living organisms increase the entropy of the universe around them? Could life as we know it exist in a high-entropy (equilibrium) environment?

Complexity and Life
Lecture 14—Transcript

That was fun. In the last lecture, we did a little thought experimenting. We talked about a cosmological scenario that attempted to account for the arrow of time just by imagining that the universe was infinitely big, lasted infinitely long and there were random fluctuations that could give rise to people like us. We found that it didn't work. If you lived in that universe, you would make individual complex creatures just through random fluctuations, not through billions of years of evolution both cosmological and biological.

In the next few lectures, we are going to return to the real world and we are going to say alright let's take what we think is actually true about the universe. We did start with a low entropy condition at about 13.7 billion years ago. We've been evolving ever since. Let's try to put the actual features of life here on Earth, including how human beings think and perceive time into this context. Let's see how that matches up with what we think about the arrow of time.

One problem with this project, obviously, is that the idea of life is not something we understand very well. We know that living creatures are complex, but our understanding of what life is is not very well developed. Thinking about anything living is always going to be more difficult than thinking about physics. That is why physicists talk about boxes of gas with partitions and so forth, once you start talking about biology or psychology or neuroscience, you cannot simplify things to boxes of gas. We don't even have a good definition of what life is. We think that living things are complex in some sense. We often talk about the fact that they reproduce, they create other living beings; they metabolize, they take energy in from their environment and put it to good use. Sometimes we talk about information processing, that something we might think of as alive can think about the world or represent it.

But, we don't know which exactly is important and which is not. Is a computer alive, is a flame alive, is a galaxy alive? These are questions to which we don't know the answer probably because they are not good questions to ask. We know that these things exist. We should not worry about whether or not they fit into the definition of life or not. Likewise what we're going to concentrate on now is life as we know it. We are not going to try to come with an absolutely clear definition that could tell us whether something is going on on another planet is alive. Rather, given how life works here on Earth, how could it have arisen in the evolution of the universe?

Let's start by thinking about the fact that life is complex, living beings are complicated. They have a lot of different moving parts. Your first step might be to examine this notion of complexity. What does it mean for something to be complex? It is one of those things where we know it when we see it, but that doesn't mean that it's easy to come up with a definition. Immediately we are going to worry about the fact that the whole focus of our course is the Second Law of Thermodynamics, the fact that entropy is increasing and the universe is becoming more disorderly.

One way of thinking about that is that the universe is tending toward what is called a heat death. Once the universe reaches equilibrium, there won't be anything going on anymore. Equilibrium means static, nothing happens. The temperature will be the same everywhere. There will be no more living beings. Why is it, we should be asking ourselves, that this tendency toward heat death has given rise to what we think of as life?

Let's first start by trying to give some definition to the notion of complexity. Like life we don't have a single definition. There're different kinds of complexity out there, but there is a definition that is most common and it will be good enough for us. It was proposed by Soviet mathematician Andrey Kolmogorov back in the 1960s. He was thinking in the 1960s they were very interested in information theory and the new science of computer science. They were thinking about strings of numbers and wanted to talk about well when is a string of numbers or for that matter some physical object complex versus noncomplex.

Kolmogorov says well imagine you are providing me with the information I need to reconstruct that number or that physical situation. Imagine you have some language, some fixed way of talking. In the back of his mind obviously he had a computer language, something that could be very, very formal and rigorous, but we can think about talking in English or in Russian if you would like, in some specific language come up with the briefest possible description of the thing you are trying to describe, the shortest possible string of words or symbols that would let somebody else reconstruct the thing you're talking about. What Kolmogorov said was that we should define the complexity of something as the length of the shortest description that captures everything relevant about that thing.

For example, imagine a hundred digit number, a random hundred digit number, so just a list of a hundred different digits from 0 to 9, that's complex by Kolmogorov's definition because in order to describe it I need to give you a hundred digits. There is no shorter way to capture what is in there if it's a truly random hundred digit number. On the other hand, if it's a nonrandom number, if it is some specific number there might be much more concise ways of describing it. If I tell you 10^{100} power, that is a hundred digit number, a 1 followed by a hundred zeros, it's a 101 digit number, but it requires far fewer than a hundred characters to describe. I just tell you 10^{100} power.

Pi is a number that requires an infinite number of digits to describe, but not to tell you what it is. I just tell you it's Pi or I tell you it is the ratio of the circumference of a circle to its diameter. These numbers are less complex by Kolmogorov's definition because we can describe them more concisely. Notice that when we're using this definition of complexity what matters to us is that we have chosen a coarse-graining. Just like when we talk about entropy, when Boltzmann said there's every possible configuration of a system, but we cannot observe which microstate we're in, we only observe certain macroscopic features. The same thing happens in the real world when we talk about the complexity of something.

When we talk about the complexity of some living being or nonliving being, some object in the universe, it's not that we have to list the position and momentum of every atom, in that case everything that has the same number of atoms would have an equal complexity. Instead we coarse-grain, we tell you what we care about. We say that a certain living being has a certain number of moving parts, a certain number of proteins and molecules and so forth, and that is the way that we describe it. That's the coarse-grain description that we give.

By that definition, the complexity of an organism is Kolmogorov's complexity applied to that coarse-grain description. What is the shortest possible way I can convey to you which coarse-grained features we are talking about. As a simple example let's go back to one of our favorite ways of describing entropy increasing, mixing milk into your cup of coffee.

We know that if the milk is separate from the coffee; that's a low entropy configuration. If you mix them together that's a high entropy configuration. Even if you didn't do the counting of the atoms you would know that was right because it is easy to mix the milk into the coffee, it's hard to get it to unmix.

Let's think about this process from the perspective of complexity. Imagine an image of a cup of coffee with milk put on top, very, very carefully you put the milk on top of the coffee so they are separate from each other. That's a low entropy configuration. The coffee is on the bottom, the milk is on the top. It is also a very simple configuration. I don't need to give you a lot of information to describe it, I just did. Coffees on the bottom, milk is on the top. Now imagine I take that cup of coffee and I mix it together, I mix it up all the way. I go to a high entropy configuration where all the coffee and all the milk molecules are mixed together. However, it is still a low complexity configuration. It's a very simple configuration. All of the milk and coffee are mixed together, that is a very short description that tells you everything you need to know.

Low entropy can be simple and high entropy can also be simple. There is no necessary relationship between a simplicity or complexity of something and its entropy. They are two different ways of characterizing some physical situation. Now go back to the middle period of our process of mixing the milk with the coffee. Before we were totally mixed we began to mix the milk with the coffee and in your mind's eye you can picture what is happening. The tendrils of the milk begin to mix in with the coffee. If you took a picture of the half-way process, in between the low entropy segregated milk and coffee and the high entropy mixed milk and coffee, what you would see is complex because in between when some of the milk has been mixed, but not all of it, there is no simple description that lets you precisely reconstruct what the milk and coffee configuration look like.

If you took a picture of it you would have to describe at every part of the picture well there's so much milk here, so much coffee there, and so forth. In between the low entropy state and the high entropy state was a medium entropy state that was very complex compared to the extremes of high entropy and low entropy. This is an absolutely crucial insight into the relationship between entropy and complexity. It's not that low entropy or high entropy corresponds to simple or complex, complexity can arise when the entropy is in between very low and very high.

This goes for a cup of coffee, it also goes for the universe. If we think about the evolution of our real universe, we will talk about this in great detail later, but basically the universe began 13.7 billion years ago in a low entropy state and also a very simple state. It was homogeneous. It was smooth everywhere and everything was densely packed, that was the universe. The far, far future of our universe is also very simple. Everything will scatter to the three winds, four winds, whatever it is, we will have empty space once again. The early universe is simple, the late universe is simple. The early universe has low entropy, the late universe has high entropy. The middle universe that has medium entropy, that is on a journey from low entropy to high, is complex. It is today that we have galaxies and stars and planets and life on those planets.

The way to think about the relationship between complexity and entropy is that complexity depends on entropy, it relies on the fact that entropy is increasing. If you like it is a side effect of the fact that entropy is increasing. The people who worried about this question—how in the world can we get complex things like life in a world where the Second Law of Thermodynamics says that everything is evolving toward a heat death where everything is smooth?—needn't have worried. In fact, it is not despite the Second Law of Thermodynamics that complexity can exist, it is because of the Second Law of Thermodynamics that complexity can exist.

That is the right way to think about why there is life on Earth. It doesn't guarantee that there should be life on Earth. This way of thinking doesn't tell you whether there is life on Europa, one of the moons of Jupiter, or around some other planet around some other star, but it tells you that the simple fact that entropy is increasing does not in any way make it surprising that we have complex structures like life. The fact that entropy is increasing is what makes life possible.

You will often hear people say well it doesn't matter that there is a Second Law of Thermodynamics to try to explain why there is life because life is an open system. If you think about the absolutely careful definition of the Second Law, it says that in a closed system, in an isolated system that is not interacting with the rest of the world entropy, always either goes up or remains constant. A living being is not a closed system by any stretch of the imagination, a living organism is constantly interacting with its environment. Strictly speaking, the Second Law simply says nothing about any individual living being. You certainly can't say that the Second Law rules out the existence of a living organism.

We can say more than that. We can say that the Second Law allows for life to exist. Let's think about how this actually works out. One of the best ways of thinking about it was put forward by Erwin Schrödinger, the physicist who is a pioneer of quantum mechanics. Later in life, after he had helped develop quantum mechanics, he moved to Ireland and became fascinated by biology. Schrödinger wrote a tiny little book based on some lectures that he had given called *What is Life*. It's a very interesting book to read. It predates our discovery of DNA, but Schrödinger's little book was actually inspirational for Francis Crick, one of the men who discovered DNA, it helped Crick decide to leave physics and go into molecular biology and understand the way that DNA actually works.

Schrödinger was thinking about biology like a physicist would, so he was trying to understand how living organisms can persist, how they work. He came up with a great definition of what life is. It is not the final definition by any means, it's not even very rigorous, but it gets you on the right track. Schrödinger says that life is something that goes on doing something much longer than you would expect it to. Now what is that supposed to mean? What do we mean how long we expect it to? How long do we expect something to go on doing something?

What Schrödinger has in mind is imagine once again our favorite example of a glass of water and we put an ice cube in it. Over the course of a few minutes the ice cube melts, the water and the ice cube comes to equilibrium with the water around it. Even if it is not a water ice cube, there are these plastic ice cubes that are made of a different substance, but you put them in your refrigerator, put them in the water, and it will cool it off, but it will still equilibrate. The temperature of a plastic ice cube comes to be the same as the temperature of its environment, just like Clausius would've said, and then once you reach equilibrium everything stops.

On the other hand, imagine putting a goldfish in a glass of water. The goldfish does not come to equilibrium with the water around it. It maintains its integrity even though, by mass, most of the goldfish is also water. If you give it food, if you feed it, what you're doing is allowing that goldfish to take advantage of energy in a low entropy form, food is fuel, it can be used to do useful work and then that goldfish can last a very long time. If you don't feed it, it will eventually die. If you feed it, the goldfish can last a long time.

This is what Schrödinger had in mind, a living being even in an environment can put off the approach to thermal equilibrium for a long time. What is going on? What's going on is actually very similar to Maxwell's demon. Remember we talked about this thought experiment from James Clark Maxwell that tried to ask whether a little demon could maintain a low entropy situation without violating the Second Law of Thermodynamics. The demon sits on top of a box and the demon starts the box in equilibrium, but by cleverly letting some molecules go one way and some molecules go the other way, the demon constructs a lower entropy situation in the box and then by constantly keeping track of the molecules and updating his little record book, the demon can maintain that low entropy situation as long as he wants.

What is going on is the demon is making use of the fact that he lives in a low entropy world in order to keep his particular favorite box in a low entropy situation. By itself, the box would equilibrate, both sides would come to the same temperature. The demon remember increases the entropy of the universe by writing things down in his notebook and later erasing them. If the external world were already in equilibrium, the demon could not do that. You cannot increase the entropy of a system that is in equilibrium. So, the demon makes use of the fact that he lives in a low entropy world in order to maintain the complexity of his box, the difference between the temperature on one side and the temperature on the other side.

That is a very general paradigm for thinking about how life persists, how Schrödinger says a living being lasts longer than it should. What he's saying is the complexity of a living being can last longer than you would expect it to. It does not come to thermal equilibrium because there is plenty of room for the entropy to grow. Again it's not just that the entropy is one value or another, but living things, complex organisms in general can maintain their integrity by taking advantage of the fact that they are increasing the entropy of the rest of the world.

Let's think about how this works in the real world example of the biosphere here on Earth. We live here on Earth. We have the Sun in the sky that is clearly a source of energy for living beings here on Earth. It's almost all of the energy that we make use of. Now, that's a typical way of talking about how life on Earth makes use of the Sun. We get radiation from the Sun; that gives us energy; we use the energy to do things. But of course, if you have sat through all these lectures you know better than that. It's not the energy that we get from the Sun that matters. It's the fact that the energy is in a low entropy form. In particular, the Sun is a hot spot in an otherwise cold sky. If the whole sky were the temperature of the Sun, we would get a lot more energy than we get now. However, we would come to equilibrium with that sky very, very quickly. We would become the temperature of the Sun. We would equilibrate and all life would cease. Even though we have a lot more energy we can't be alive because we would be quickly brought to thermal equilibrium.

Likewise if the whole night sky were dark, if the whole day and night sky were the temperature of the real night sky completely dark and cold the Earth would very quickly come to the temperature of the night sky, it would equilibrate and life would cease. The reason there can be life is because the whole sky is not at the same temperature and we do not quickly come to thermal equilibrium. What actually happens is that the energy we get from the Sun is in a low entropy form, we give back to the universe just as much energy as we get from the Sun. We get radiation during the day, we give it back, we radiate into the universe almost exactly the same amount of energy that we get. It's not exactly the same because of global warming, but to a very, very good approximation the energy we get equals the energy we give back.

You ask, what is it that drives life on Earth if we don't get a net gain of energy from the Sun? The answer is the energy we get from the Sun is in the form of visible light. We can see it. That's most of the energy, most of the photons that we get from the Sun. But, then we process that energy for every one photon of visible light we get from the Sun, we give back 20 photons to the universe each with about 1/20 of the energy. What that means is that we radiate infrared light to the universe and what that means is that we have increased the entropy by a factor of 20. Our ecosystem, our biosphere, is very, very far from a closed system. Energy comes in, we increase its entropy by multiplying it by 20 and then we give it back to the universe in the form of infrared radiation.

You can actually do the calculation to see by how much the entropy of the biosphere has increased, the entropy of the universe has increased for that matter by the function of the Earth. People sometimes wonder, as I mentioned, how it can be that there's all this complex life here on Earth just because entropy keeps increasing. Why would you have complexity increase in the biosphere? Well you can ask can you account for the increase of entropy quantitatively when you compare the amount that we have increased the entropy of the universe through our infrared radiation to the amount that we have decreased the entropy of the universe by inventing life here on Earth.

One way of doing that is to imagine that the biosphere, all of the mass in living things here on Earth was once in its highest entropy form and is now in its lowest entropy form. That's obviously not right, it's somewhere in between on both ends, but it's a conservative estimate. The maximum amount by which life on Earth could have gone down on entropy is what you would get if you imagined it started at the highest possible entropy and ended with the lowest possible entropy. The biosphere has about 10^{15} kg and we know the surface temperature of the Earth. That means we can calculate what the entropy of the biosphere would have been if it were in thermal equilibrium. It never was in thermal equilibrium precisely, but that's the maximum entropy it could have. You work out the numbers; it's about 10^{44}.

To go from the maximum entropy that the biosphere could have to the minimum entropy it could have all you need to do is decrease the entropy of the biosphere by 10^{44}. In order for that to not violate the Second Law of Thermodynamics as the entropy of the biosphere goes down as random muck organizes itself into living organisms the entropy of the universe needs to go up by more than 10^{44}. You can again do the calculation, you can say, at what rate does the Earth increase the entropy of the universe by taking visible light from the Sun and radiating it back to the universe as infrared radiation? The answer is it takes about one year for the Earth to increase the entropy of the universe by 10^{44}. In other words, turning it around, the Earth has increased the entropy of the universe over the course of the history of life by 4 billion times the amount that the biosphere has decreased the entropy of the universe.

The Second Law is in very good shape. The origin of life on Earth and the existence of life on Earth do not violate the Second Law, it's completely compatible with what we know about the increase of entropy. Life on Earth is a tiny little part of ashen [phonetic 0:23:05] compared to the huge amount by which we have increased the entropy of the universe. Now there is one thing sort of as an aside that I want to get clear here. The process of evolution, the going from simple single celled organisms at early times to the very complex organized biosphere that we have today is not teleological. It is not right to think of evolution as working toward the goal of greater complexity. Evolution doesn't work toward any goal at all. The way that evolution works is that individual species and the genes and the DNA they carry flourish or don't in whatever environment there in and when they flourish they reproduce and create more of themselves.

Sometimes to flourish means to be complex, to have a complex brain or very accurate vision, sometimes not. Very often in the history of evolution you will see species that have lost abilities that they used to have, a mammal climbs underground and it loses over the course of generations the ability to see. It's not that evolution wants life to be more complex, it's that there are many ways to be complex, sometimes those ways are very, very useful, therefore life adopts them. But, the underlying laws, the underlying processes don't have any goal in mind, they just take advantage of whatever circumstances they're in.

The reason I want to emphasize that is because I am about to tell you a story that suggests that in some sense life is taking advantage of a function that complexity has. Complexity does actually help us do certain things and that is one explanation for why there is life at all, not just how life has evolved over the course of the last several billion years, but why life appeared. The science of the origin of life is very, very undeveloped. We don't know how life starts even here on Earth, much less how life might start in some very, very different environment. But as physicists, we can choose not to sweat the details about the actual geology and atmospheric science of the Earth at early times and ask, on the basis, of physics does it make sense to us that life started at all?

Fortunately, the answer is yes. Remember we argued that complexity is compatible with the Second Law, in fact, it is parasitic on the growth of entropy. When entropy increases complexity will often follow until entropy starts increasing close to equilibrium and the complexity is going to have to go away. Think about the early earth, the Earth soon after it formed, it didn't take that long, I mean maybe hundreds of millions of years, but still not billions and billions of years, for life to appear on the young earth. Again, we don't know how it started, but there are two schools of thought about the different ways in which it could have started.

One school of thought is called the replication first. This is the idea that the most important feature of life is that a living being can reproduce itself, not just continue to exist or create one other organism, but create more than one other organism like itself. To do that you need some molecule that conveys information. Biologists talk about an RNA world where RNA molecules, which are the partners of DNA molecules, the RNA molecules could have come first; they could have stored genetic information and reproduced themselves and only later do these little reproducing molecules form around themselves cells and make the first single celled organism and then those organisms could reproduce and evolve and come to the marvelous complexity of life that we see today. That's one school of thought.

The competing school of thought which is a little bit less popular is called metabolism first. It is a little bit less popular, but it's still a respectable way of thinking. Metabolism first says that it's not the information and the replication that came first, it's the use of complex chemical reactions to take advantage of the low entropy environment in the early earth that came first. One way of thinking about it is hardware before software. The software is the information in the RNA. The hardware is the actual cell and its machines inside that make use of the fuel that it can get from its environment.

From one point of view, it makes sense to have hardware without software. It seems difficult to have software without hardware. One of the advocates of this point of view is a scientist named Mike Russell who says I could take the computer out of my Prius and it would still go, but if I take the engine out of my Prius it doesn't do anything at all. The metabolism first school of thought says that the first life was just a chemical reaction, it was just like fire, it was taking advantage of the fuel around it to get going.

In particular, we can think about what the chemistry was like on the early earth, in the early days. Again, we don't know very, very well. we have theories and we have data and we're trying to do our best, there's a very good understanding that we think is accurate that says that the early earth had an atmosphere that was rich in hydrogen, two hydrogen atoms making a molecule and carbon dioxide, one carbon atom and two oxygen atoms. The interesting thing about such an atmosphere full of hydrogen and carbon dioxide is that it is low entropy, which those exact molecules hydrogen, carbon, and oxygen, there's a much higher entropy configuration namely methane and water. Methane is one carbon and four hydrogens, water is one oxygen and two hydrogens. So, to increase the entropy, which is what the world wants to do you would convert that hydrogen and carbon dioxide into methane and water.

The problem is that to get there, to go from the original situation of low entropy to the situation of higher entropy, there is no one simple reaction. To get there you need to go through formaldehyde, which is one carbon, one oxygen, and two hydrogens, and that requires a decrease in entropy. The first step on this chain of reactions that wants to happen is very difficult to get started. It doesn't just go, it is not like a piece of wood where you can spark it and it will go to flame. You need a complicated series of many reactions to get the entropy to increase. The hypothesis is that those complicated series of reactions started in certain very specific geological formations. There were little membranes that separated mostly positively charge atoms from negatively charged atoms. This differentiated structure allowed a series of complicated reactions to start that allowed the low entropy configuration to increase to the high entropy configuration.

If this story is true, the reason why life started is that it was trying to increase the entropy of the primitive atmosphere on Earth. I am going to use the word trying, don't take me too seriously, there was no goal in mind, but there are many ways to rearrange the fundamental structure of the things in the Earth's early atmosphere which would have had a higher entropy, but to get there you needed something complicated like life. Given enough time that complication would happen and here we are as a consequence of it. It is very interesting to contemplate the thought that life is not only compatible with, but actually an outgrowth of the Second Law of Thermodynamics and the arrow of time.

The Perception of Time
Lecture 15

In this lecture, we'll try to answer a question that many of us have when we're talking about the mysteries of time: Why am I always late? Of course, we can't come up with a specific answer, but we can talk about how our brains and bodies measure the passage of time and how we perceive that passage. The reason we feel the passage of time is that our bodies have clocks in them, such as our heartbeats and our breathing. But as we'll see, biological clocks are not very reliable compared to mechanical or electronic clocks because our bodies are affected by many things that are outside of our control.

Biological Rhythms

- One way of thinking about an advanced organism, such as a mammal, is as a network that includes the brain and the different systems in the body, such as the nervous system, circulatory system, and so on. On the basis of network theory, we can make predictions of how the rhythms cascade through the body.

- Biological networks move faster in smaller animals, and smaller animals have faster heartbeats than larger animals. They also have shorter life-spans. It's an interesting feature of our biology that somehow a shrew and an elephant have similar "blueprints," although these take reality in different forms—a tiny form for the shrew and a larger one for the elephant. This kind of scaling law goes all the way down to the cellular level.

- Unlike shrews and elephants, human beings are affected by another variable, the culture in which we find ourselves. We all know that different cultures approach time differently. In fact, various studies have been done to try to measure the pace of life in different cultures.

- Not surprisingly, people in higher-population areas have been found to walk faster than they do in smaller-population areas.

- Researchers comparing the pace of life in cities over time have found that over the last 20 years, the pace of life has increased by 10 percent. In areas that are rapidly industrializing, the increase in the pace of life is even greater.

The areas of the world that are developing, industrializing, and gaining high technology the most quickly also see their pace of life increasing the most quickly.

Perceiving the Passage of Time

- We might think that the human brain is similar to a computer because it clearly computes in some sense, but the way the brain came together is very different than the way a computer program is written. The brain evolved over billions of years by an incremental process in which random possibilities were tested. A computer program, in contrast, is usually written to address a task from the top down.

- In the brain, information for any task you might want to do is probably shared among many different structures. In particular, keeping time is a highly distributed feature in the brain.
  - Neuroscientists have discovered that they can remove the cerebral cortex (the advanced parts of the brain) in rats, and the rats are still able to tell time. This means that whatever we are doing when we are measuring duration and time, it's not a matter of our conscious brain; there are unconscious things going on.

- o Another experiment showed that mice with intact brains could keep track of at least three different rhythms.

- Inside our brains is more than one timekeeping device. In fact, we can sort of roughly categorize the different kinds of timekeeping. One part of the brain keeps track of what time of day it is, another part keeps track of how much time has passed during certain tasks, and yet other parts are more or less like alarm clocks. They keep track of time before some relevant future event.

- Neuroscientists have been able to isolate at least three different things that affect our perception of the passage of time: (1) pulses in the brain, (2) sensory input and focus, and (3) the accumulation of memories.

Pulses in the Brain
- Different neurons in the brain do work via pulses, and together, multiple levels of pulses help us perceive the passage of time.

- These pulses can be affected by stimulants, such as caffeine, and depressants, such as alcohol. When you drink caffeine, your internal clock seems to speed up compared to the outside world. Stimulants or depressants are believed to affect the neurotransmitters that send signals from neurons to other cells in the brain.

- The neurons send these neurotransmitters, such as dopamine, in the form of pulses. Caffeine, alcohol, and other drugs can make it easier or harder for these neurotransmitters to be sent, and that's what speeds up or slows down our internal clocks.

Sensory Input and Focus
- When you are focused on a task, you don't pay as much attention to the outside world, and in some sense, you also don't pay as much attention to your internal clock. Your internal timekeeping device seems to slow down while the outside world speeds up.

- In contrast, if you're bored or are not focused on any one task, the opposite effect happens. Your internal clock seems to go faster while the outside world slows down.

Accumulation of Memories
- Often, in high-stress situations, time seems to slow down, by which we mean that your internal clock speeds up, but the rest of the world seems to slow down.

- Researchers have found that conditions that create stress and speed up the internal clock do not help us perceive the outside world. Still, subjects in high-stress experiments report that time slowed down for them in recollecting a stressful event.

- One theory explaining this phenomenon is that the more memories we accumulate, the more time seems to have passed. When you are in a high-stress situation, your brain does its best to record absolutely everything. It accumulates a huge amount of data, even though it does not perceive things any more quickly than it would otherwise. When you think about the situation afterward, you have more memories—more data to leaf through—and, therefore, it seems as if more time has passed.

- This hypothesis gets some support from the fact that time seems to pass more quickly as we age. For example, when you were a child, summer seemed to last forever, but when you get older, it seems to rush by.
    - It may be that when you were young in the summertime, such activities as going to the beach were new to you, but when you get older, you've been to the beach and you don't take in as much new information about the experience as a child would. Thus, time seems to pass more quickly for you compared to when you were a child.

    - Experiments in which 20-year-olds and 60-year-olds are asked to estimate the passage of time show that younger subjects are much more accurate in their estimations than older ones. For older people, it takes longer for the same amount of subjective time to pass.

- Another hypothesis tries to quantify this experience of the passage of time by saying that the amount of time we experience grows logarithmically with our age.
    o If you've ever been on a boring plane ride, it seems to last forever while you are experiencing it, but when you recall it after the fact, it seems to go by quickly. You might remember that you were bored, but you don't have some elaborate memory of every single event because none of the events was interesting.

    o Because you were not making new memories, your subjective time after the fact seems to make the trip quite short.

Experiencing the Present Moment
- We all think that there is a moment called "now" that we are perceiving. But if you think about it, what we call the present moment isn't the present moment. It takes time for the information we are gathering to get to us and for our brains to process that information.

- If you touch your nose and your toes at the same time, you will notice that you feel the touch simultaneously, although you shouldn't. The amount of time it takes the nerve signal to travel from the nose to the brain is much less than the amount of time for the signal to travel from the foot to the brain. The brain takes into account that our feet are farther away than our faces.

- It turns out that what we consider to be the correct moment right now is actually about 80 milliseconds in the past. You can measure this by watching a person dribbling a basketball and moving away from you. You see and hear the sound at the same time in your brain until the person gets so far away that it takes more than 80 milliseconds for the sound to reach you. At that point, there is a mismatch between the sound and the sight.

The Stanford Marshmallow Experiment

- The Stanford marshmallow experiment shows that our orientation toward time is a crucial component of how we live the rest of our lives.

- In this experiment, children are given one marshmallow and offered another one if they can wait a few minutes before eating the first. Psychologists claim that the difference between eating the first marshmallow immediately and waiting to get two marshmallows later shows something about a person's orientation toward time.

- There are some people who, in some sense, dwell in the past; what they care about most is what happened in the past. Others live in the present; they want that marshmallow right now. Still others are future oriented; they will sacrifice the opportunity to eat the marshmallow in front of them to get the marshmallow in the future. This ability to take the future just as seriously as we take the present is a good predictor for how we approach many situations in our lives.

Suggested Reading

Levine, *A Geography of Time*.

Zimbardo and Boyd, *The Time Paradox*.

Questions to Consider

1. Can you think of reasons why the pace of life is faster in dense, developed areas? How would you test these hypotheses?

2. When do you perceive time to be moving faster and slower? How would you interpret that perception in terms of your biological rhythms and the accumulation of memories?

3. How would you describe your personal orientation toward time?

The Perception of Time
Lecture 15—Transcript

You have been very good, if you are following the lectures in order you are more than half-way through the course and I thought that as a reward in this lecture we would answer the question that everyone has when they're first talking about the mysteries of time, namely why am I always late. It is not a question to which I can give a specific answer, it has more to do with your circumstances than with any scientific principles. But, what we can do is talk about how human beings perceive the passage of time. How we, in our brains and in our bodies, measure time as it passes. We already talked about in earlier lectures the idea of using a clock to measure time. A clock is something that does the same thing over and over again in a repetitive and predictable way compared to other clocks.

The reason why we human beings feel the passage of time is because our bodies have clocks in them. We have things that happen in a rhythm over and over again, you have your heartbeat, your breathing, the pulses in your central nervous system. Even Galileo, when he was in his cathedral in Pisa, and he first figured out that a pendulum rocking back and forth took the same amount of time. The way that he did that was to compare the oscillation of the pendulum to his heartbeat, to the best biological clock that he had available.

However, biological clocks are not very reliable compared to a good mechanical or electronic clock. That is because our bodies are affected by many things that are outside our control, whether we are tired or exhausted, whether we have adrenalin rushing through our system. Also our mental state, are we focused on something or are we distracted by the world around us. Finally, the way that we get memories and keep them affect how we perceive the passage of time. We do feel time passing. In many ways it is pretty accurate, but there is always something that makes us not completely perfect compared to let's say a good atomic clock.

To set the stage for this discussion, let's just talk about biological rhythms more generally. In any organism, there are things that happen over and over again, that's what a rhythm is, something that repeats itself over and over. The heartbeat is probably the single most fundamental rhythm that an advanced living organism has. We can compare how these rhythms work in different kinds of animals. Just to keep everything else as constant as possible we can compare how mammals have different rhythms inside their bodies.

One way of thinking about an advanced organism like a mammal is it's a network. You have the brain and you have all the different things coming out of the brain, the nervous system, your circulatory system, and so forth. You can actually make predictions on the basis of network theory for how the rhythms cascade through your body and then compare those predictions to data. Geoffrey West at the Santa Fe Institute who studies complexity, he was a physicist at first, but he went into complexity studies as that field came to life and became more exciting, has asked the question about how the networks in different mammals are affected by their size.

You find two of the most basic relationships are that if you are a smaller animal your networks move faster, in particular your heart beats faster the smaller you are, the less weight you have the faster your heartbeat is. On the other hand, he found that smaller animals also live shorter lives. The lifespan, the typical lifespan of a tiny little shrew is much shorter than that of a giant elephant or a blue whale. If you think about this a smaller animal has faster heartbeats and also doesn't live as long, a larger animal has a slower heartbeat and lives longer, you might wonder, do these affects cancel out? The answer is yes. As a very rough rule of thumb every mammalian species, on average, has the same number of heartbeats in its lifespan. That number turns out to be 1-1/2 billion. Roughly speaking every mammal gets 1-1/2 billion heart beats in its life.

Now, don't take this too seriously. When I tell this to people they sometimes worry that this is a strict rule, that no matter what you do it's almost like predicting the future, you will live for 1-1/2 billion heartbeats. They wonder should I exercise more, should I exercise less to get more or less heartbeats. That's not how it works. It is an average. Of course, many animals live a little bit shorter, a little bit longer, it is an interesting feature of our biology that somehow a shrew and an elephant have similar blueprints just taking reality in different forms, in a really tiny form for the shrew and a bigger one for the elephant. This kind of scaling law goes far beyond organisms. It goes all the way down to the cellular level.

West and his collaborators were able to use this kind of scaling relationship to make predictions for the rate at which individual cells behave in certain way if they are in an organism or outside an organism. They made predictions which hadn't yet been tested. After they made the prediction they were tested. The idea that a human being is a complex network of smaller functions talking to each other is a very good fit to how real animals live.

Of course, unlike shrews and elephants and blue whales, we human beings have another variable, which is the culture in which we find ourselves, the environment around us, the human environment in which we are embedded. We all know that different cultures approach time differently. The psychologist Robert Levine, who studies how human beings interact with time is based in Fresno, California, but he sometimes spends time in other countries giving lectures and so forth. He tells a wonderful story of his first semester in Brazil when he was going to give a lecture. He was scheduled to go from 10 a.m. to noon, it was a regular class he was teaching. He was walking down the street and he asked someone what time it was and they said it was 9:05. He didn't have a watch himself. He says, oh I have plenty of time to get to my lecture at 10 o'clock.

About half an hour passes and he asks someone else what time it is and they say it's 20 minutes after 10. He panics, he rushes to the lecture room and there are almost no students there yet. He says what is going on and so he asks the students what time it is. One says it's 9:45, they're looking at their watches, one says it's 9:55, one says it's 5 minutes past 10. What Levine realized was that no one's clock in Brazil reads the same time, and nobody cares. The students sort of came in very, very gradually to the 10 a.m. class. By 11 a.m. the room had more or less filled up, but it wasn't that the students were lazy or didn't want to go to the lecture, when noon came along he said that in California when the end of the lecture was scheduled you would know, every single student would be rustling in their chair looking at the clock ready to go. In Brazil, none of the students moved. Who cares it was 12 o'clock, the end of the lecture, they were still interested. They had questions, they kept discussing. It became 12:30, eventually he had to escape.

The point is that different cultures approach time differently. Levine also spent time in Japan and he says, of course, compared to what you would do in California, the Japanese found that his natural timekeeping was incredibly lax and kept telling him to hurry up. In Brazil, they kept telling him to calm down. It is not these anecdotes that are really important. What is important is data, can we actually do a scientific experiment to justifying these stories about how different cultures perceive time. And the answer is yes. A bunch of different studies have been done to basically try to measure the pace of life in different cities, different cultures, different environments.

For example, you can take how long it goes if you walk up to a post office and you went to order some stamps. Psychologists have sent people to order stamps in different post offices in different cities throughout the world and found consistent differences in how long it takes to get those stamps to get to you. There is also a study where you can just measure how fast people are walking. You sit there in the café, you watch people walk by, you measure how fast they're moving, and once again you find systematic differences in different cultures and in different cities.

For example, you will not be surprised to learn that in higher population areas people walk faster, people literally will walk faster down the streets of New York City than they will down the streets of a small town in Massachusetts. It is not because everyone else is walking faster, it's just because that is what the environment does to you. Another obvious effect is technology and industrialization. Psychologist Richard Wiseman has compared the pace of life in cities over time. He does this walking experiment, he sees how fast people are walking down the street, but rather than comparing one city to another one, he compares one city 20 years ago to the same city today.

What he found was that over the last 20 years the pace of life has increased by 10 percent. That doesn't sound like a lot, 10 percent, but it is only in 20 years. How quickly will it go over the next century? Wiseman also found that the effects are different from place to place. Singapore was up 30 percent, Guangzhou, a city in China, was up 20 percent. The areas of the world that are developing and industrializing and gaining high technology the most quickly, also see their pace of life increasing the most quickly.

Nevertheless, your stereotypical guesses aren't always right. It turns out that New York City is far and away not the fastest walking city in the world. In fact, it's behind cities like Dublin and Madrid and Copenhagen. There are many complex features that go into how quickly life is lived in different environments. Technology and density are very important, but we are very far away from having a complete theory of how this works.

Another place where we are very far away from having a complete theory is how individual people perceive the passage of time. There is one lesson that we have certainly learned, which is that the human brain is not a computer program. We think about the human brain as very similar to computers because the human brain clearly computes in some sense. We can mimic some things the brain does in a computer program, but the way the brain came together is very different than the way you would write a computer program. The brain developed over literally billions of years of evolution of life here on Earth by an incremental change through random possibilities being tested and seeing what works.

A computer program, on the other hand, is usually written from the start. You have a task in mind and you are going to design it from the top down in the most efficient way you can. Therefore, in a computer program if you want to store data for example, well you have an array that you can put the data in and that's that. In the brain, any task that you might want to do is probably shared among many different pieces because that's what was useful at the time. In particular, when it comes to keeping time to measuring duration there are different parts of the brain that kick in. It is a highly distributed feature. It's not like a simple clock that you have in your computer, there's some chip that has a rate that you can buy a faster and faster chip, the brain has many, many pieces each one of which contributes to our understanding of time.

As one experiment to demonstrate this that is very vivid, neurologists did an experiment with rats. What you keep in mind is that neurologist neuroscientists hate rats. They are constantly torturing these poor little rats. The rats have a brain much like we do. They are mammals. As far as all of life is concerned, rats are pretty close biologically to human beings. They have a complex brain. They have a cerebral cortex, which we have also. It is the gray matter in our brain. The cerebral cortex is the outside part of our brain, it's where we do our higher information processing. The life of consciousness is happening within our cerebral cortex. There are also other organs inside the cerebral cortex which are more primitive. We think of them as giving us subconscious brain functions.

You can actually take a rat and remove its cerebral cortex, remove the advanced parts of its brain, but the rat will still be alive. It won't be able to do mazes and other things that we teach rats to do, but what the neurologist neuroscientists have found, is that rats can still tell time even with their cerebral cortex removed. They give them a little task like pushing on a lever. If they push on the lever one every 40 seconds they get a food reward. The rats without a cerebral cortex can measure the time interval of 40 seconds. That means that whatever we are doing when we are measuring duration and time, it's not a matter of our conscious brain there are unconscious things going on.

There's more than one thing going on. Another experiment with mice was able to show that mice with their whole brain intact could keep track of at least three different rhythms. These mice had three different paddles that they would push to get food and to get the food one paddle needed to be pushed every 10 seconds, one had to be pushed once every 30 seconds, and one had to be pushed every 90 seconds. The mice were able to separately keep track of the time intervals necessary to get the rhythms right in all three paddles simultaneously. You might think, well, 30 seconds is 3 × 10 and 90 second is 3 × 30 so you just press this one three times and this one once, etcetera. But, you could actually turn on and off the rhythms, you could displace the rhythms so that the 10 second one was out of phase with the 30 second one and the 90 second one, the mice could still keep track.

Inside our brains is more than one timekeeping device. In fact, you can sort of roughly divide it up into different kinds of timekeeping. As you are listening to this lecture right now there is a part of your brain that is keeping track of what time of day it is. This is your circadian rhythm that tells you when you wake up, when you go to sleep. There is another part of your brain that is keeping track of how much time has passed since the lecture started or since you started listening to it. There are yet other parts of your brain that are more or less like alarm clocks. They keep the amount of time before some relevant future event. If you want to go to dinner or go to bed or something is going to happen, how much time do you have before that happens, your brain is keeping track of that.

Now, this is a complex set of operations. We don't have a grand unified theory of everything. This is why brain science and biology is much more complex than physics is. But, neuroscientists have been able to isolate at least three different kinds of time perception, three different ways in which we increase or decrease the rate at which time seems to pass for us. The three ways are number one, pulses, this is sort of the basic way that you would keep track of the time in a clock or with a pendulum going back and forth. There are pulses in our brains and we simply count them. Another thing that affects the passage of time is our sensory input and focus, what are we paying attention to? And finally and sort of most intriguingly, the way in which we accumulate memories affects our notion of how much time has passed. The more memories we accumulate, the more time we attribute to what happened.

Let's consider these three aspects in order. When we say counting pulses, it makes you think that there is a little part of your brain that is almost like a pendulum going back and forth and that is not true or at least we don't know of any single part of the brain that acts exactly like a clock, like a chip in your computer. Rather there are multiple levels of pulses, the different neurons in your brain do work via pulses. It is not that the neuron is constantly sending signals, it is that there is a signal or there is not a signal so neurons turn on and off. There are multiple levels of pulses due to all the neurons in our brain and together they help us perceive the passage of time.

Again, you will not be surprised to learn that we can affect how quickly those pulses go. For example, we can affect them through drugs, stimulants make our pulses beat faster. If you drink caffeine, the pulses in your brain that keep track of time go a little bit faster, depressants will slow them down. If you have alcohol or other depressants, even something like marijuana, the pulses decrease in their rate and it takes us longer to accumulate a certain perceived amount of time.

Now, whenever we talk about this kind of phenomenon we have to be very, very careful because people say well you drink caffeine and time speeds up, but in fact, as we talked about way back when in the early lectures, time is always moving at 1 second per second. When you want to say that time speeds up what you mean is that one clock has sped up compared to some other clock. When you drink caffeine your clock speeds up, but what that means is that when you compare it to the world the world has slowed down so your clock has sped up compared to the outside world. You have to be sure to get that right when we're talking about affects on our internal clocks. These affects from stimulants or depressants, we think, are affects on the neurotransmitters that send signals from our neurons to other cells in our brain.

Neurotransmitters like dopamine and other chemicals are what the neurons send and they do it in the form of pulses. Caffeine or alcohol or other drugs can make it easier or harder for these neurotransmitters to be sent that speeds up or slows down our internal clocks. But, that is usually not what makes you late. If you want to know why you are always late, it might not have anything to do with caffeine or booze or anything like that. It's probably due to the second aspect which is the sensory input and focus. When you are focused on a task, when you have a really hot first date or when you are in a really engrossing project for work or at home that you are really in love with, you don't pay as much attention to the outside world and in some sense, which psychologists still don't understand, you also don't pay as much attention to your internal clocks.

Your affected internal timekeeping device slows down. The outside world speeds up. If you're having a really interesting date or conversation with someone you think that a half an hour has passed, but in fact it's been two hours. Your focus on one task makes it harder for you to tell time. Usually if you are late, it's because you've been focusing on something else. Contrary wise if you are bored, if you are on a plane ride and there's nothing going on, if you have a tedious aspect to your job, your attention is constantly flitting around. You are not focused on any one thing and the opposite affect happens. Your internal clock seems to go faster, the outside world slows down. It seems to take forever to get that plane ride across the country simply because you are bored and your attention keeps wondering.

The final aspect is the rate at which we form new memories. This is a fascinating thing that neuroscientists are just beginning to understand. You might have, as an example, this idea that if you're in a high stress situation time seems to slow down, but which we mean your clock speeds up, but the rest of the world seems to slow down. This was noticed by a neuroscientist named David Eagleman long before he was a working neuroscientist. When he was a child he once fell out of a tree and it was not so long that he was really hurt, long enough that he was very scared as he was falling out of the tree and he remembered, as he grew up and became a working neuroscientist, that the world seemed to slow down. If you have ever been in an accident or a very, very stressful sudden situation, everything around you seems to move more slowly.

Eagleman wanted to know is it really true that in these high stress situations your internal clock beats faster therefore making the rest of the world seem to slow down. The problem with this question, as a working scientist, is that to answer it you need to scare people and that's not really good procedure when it comes to human subjects. But what Eagleman did was to figure out a way to scare people in a way that wasn't really dangerous. He threw them off of a tall building, but he threw them off of a tall building onto a trampoline so it was actually perfectly safe, and while they were falling off the building, the theory was they would get scared, there would be adrenalin going through them even though they knew there was a trampoline down there. They weren't surprised by the trampoline. He had them actually do little recognition tests. He game them little pieces of equipment that would flash numbers on a screen and it flashed them either fast or slow if they were so fast they would be flashing so fast that you couldn't recognize the numbers. If they were going slow then you had the time to see what the numbers were.

What Eagleman did was to compare how good you were at recognizing the numbers. When your adrenalin was running and you were falling out of the building versus when you were just sitting at a desk calmly. The answer is there was no difference. Despite the fact that there was adrenalin going and your pulses were racing, you were not any better at recognizing the quick flashing numbers than you would have been sitting calmly at your desk. Therefore, in that sense, the speeding up clock inside didn't help you perceive the outside world. Nevertheless if you talked to the subjects afterward, they said that the outside world slowed down. They had a perception that time had slowed down for them in recollection after the affect.

There's a theory about what is going on. The theory is and it's just a theory, it's not something that we can test very accurately that the more memories you accumulate, the more time seems to have past. When you are in a high stressed scary situation, your brain does its best to record absolutely everything that has happened. It's scared. It's looking around. It's accumulating a huge amount of data even though it's not perceiving things any more quickly than it would otherwise. When you think about that event afterward you have more memories, you have more data to leaf through and therefore it seems to us like more time has passed.

This hypothesis gets a little bit of support from a related fact, which again everyone knows even if you haven't been in a scary situation you know that time seems to pass more quickly as we age. When you are older the summer seems to rush by. When you were a kid the summer seemed to last forever. This is not just an anecdote. This is not just an idea that people have, this is something that you can test. The theory that Eagleman has, and other people had, is simply that when you are young in the summertime you are going to the beach and whatever, it's all new to you. Everything around you is a new experience. When you are older you've been there before, it's a little bit more blasé so it seems to pass more quickly for you compared to what it did when you were a child.

This is something that again we can try to test. The way that neuroscientists have tried to test it is to give 20-year-olds a test where they're simply told, "Starting now, tell me after 3 minutes have passed." And then they give the same test to 60-year-olds. The answer is that 20-year-olds are pretty good at measuring about how long 3 minutes are so they have no clocks around them. They're just sitting there quietly in an empty room and they say 3 minutes have passed and on average the actual amount of time that has passed is about 3 minutes and 3 seconds, very, very accurate. But you do exactly the same test to 60-year-old people and when they say 3 minutes have passed, the actual elapsed time is more like 3 minutes and 20 or 40 seconds.

So, 20-year-olds are better at estimating time than 60-year-olds. For an older person it takes longer for the same amount of subjective time to pass. The theory is that that is because they don't create as many new memories, but the fact is that it does take longer. There is even a hypothesis that tries to make it quantitative that says the amount of time we experience grows logarithmically with our age. It will be very hard to put exact data about that, but it makes sense to us. Think back to that boring plane ride. If you are on a boring plane ride it seems to last forever, but if you recall it after the fact it seems to go by very quickly. You might remember that you were bored, but you don't have some elaborate memory of every single event because none of the events were interesting. You were not focusing so your subjective time at the time seemed to last forever, but you were also not making new memories, so your subjective time after the fact seems to make the trip actually quite short.

Another aspect that is very interesting when it comes to how human beings perceive time is a simple statement that we live in the past, so forgetting about measuring time using our internal clocks, what about when we actually just perceive the moment now? We all think, whether we are presentists or eternalists, that there is a moment called now. We are perceiving it, we're looking around, we're getting data, sensed data from outside of ourselves and we experience what we call the present moment. It turns out, of course, and if you thought about it a little bit it would make sense that what we call the present moment isn't the present moment. That's because it takes time, number one, for the information to get to us and, number two, for our brain to process that information.

The most obvious example of this is if you see lightning very, very far away, you see it, and we all know that if lightning is far away it takes time for the sound to get to us. You can even count how many seconds it takes and figure out how far away the lightning storm is. That is because sound moves more slowly than light. But it's not just the external world that matters, it is the internal world that matters as well. For example, you can do this at home, I won't demonstrate, but you can touch your nose and you can also touch your toes. Do it at the same time, put your finger to your nose and to your toes and what you will notice is that you feel the touch simultaneously, if you perform the touch simultaneously you will feel the touch simultaneously. That seems very natural to us.

But, it shouldn't be natural because the amount of time it takes the nerve signal to travel from our nose to our brain is much less than the amount of time it takes the nerve signal to travel from our feet to our brain. Now the nerves are moving pretty quickly, but we as our brains are very, very good at measuring tiny differences in time, more than good enough to be able to measure the difference between a signal coming from our feet and a signal coming from our face. Nevertheless we don't perceive that difference. We perceive the simultaneous touches as simultaneous events.

Why is that? It is because our brain knows that our feet are further away and takes that into consideration. It turns out that what we consider to be the correct moment right now is actually about 80 milliseconds in the past. The way that you can get that number, one simple way to do it is to watch a person dribbling a basketball. As they're dribbling the basketball you both see it and hear it just like the lightning bolt and your brain corrects for the fact that it takes the sound longer to get to you. As the person dribbling the basketball moves away, you see and you hear the sound at the same time in your brain until they get so far away that it takes more than 80 milliseconds for the sound to get to you. At that point, suddenly there's a mismatch between the sound that gets to you and the sight. That is because you get the vision of it and you put together that in the conscious now before the sound can get to you.

We live about 80 milliseconds in the past. Really the primary lesson to learn from this is that consciousness is kind of a messy thing. It is very difficult to understand what is going on in the brain because it's a very elaborate mechanism. This is why physics is the right thing, right field of study to go into if you have a short attention span. Speaking of which, people have different attitudes toward time and this has nothing to do with culture. This is just within any one culture. Individual human beings approach time differently. It turns out psychologists have shown that our orientation toward time is a crucial component of how we live the rest of our lives.

There is a great experiment that illustrates this called the Stanford marshmallow experiment. It was first done by Walter Mischel, a professor at Stanford in 1972, and it is a simple set up. You offer a child a marshmallow, you fool the child first. Every psychology experiment involves fooling the subject into thinking that it's an experiment about something else. You make the kid do something and as a reward you say here's a marshmallow for being a good test subject, but then you say I have another marshmallow, you first check if the kid likes marshmallows, I can give you a second marshmallow if, I need to go out of the room for a few minutes, if you wait and you don't eat that first marshmallow I will give you a second one when I come back.

Roughly speaking when you do this experiment on 3- or 4-year-old kids, only about a half or a third of them can actually wait for you to come back. The kids eat the marshmallow, some of them, others are very, very patient, they use elaborate strategies of self-deception, looking around, trying their best not to look at the marshmallow. The claim of modern psychologists is that the difference between eating the first marshmallow immediately and waiting to get two marshmallows later shows you something about your orientation toward time. There are people who really, in some sense, dwell in the past. What they care about most, what they are talking about the time are what happened in the past, what happened they were growing up, high school, college, whatever. There are other people who live in the present, they want something right now, they want that marshmallow, there it is and they're going to eat it. There are other people who are future oriented. They know that there's some benefit coming to them in the future, they will sacrifice the marshmallow right in front of them to get that marshmallow in the future. They are future oriented.

This is not a matter of being rational or irrational. Economists will tell you it makes perfect sense to discount something that is a reward that you won't get until the future. If someone offers you $10 now or $11 ten years from now, you should just take the $10 now because our brain says how do we know that they will be around ten years from now to give us the $11? That is a perfectly rational thing to do. Nevertheless, the ability to take the future just as seriously as we take the present turns out to be a good predictor to how we approach many things in our lives.

The psychologist Philip Zimbardo, also at Stanford, has done a follow-up study on Mischel's marshmallow study. He is studying now the kids who were three and four years old, they're now in high school and in college. What he found was that children who waited, who got the second marshmallow by being good, scored higher on their SATs, had better behavior in school, and more scholastic achievement. It turns out that our attitude, as human beings, toward time is a crucial component in making us who were are.

Memory and Consciousness
Lecture 16

Both remembering the past and predicting the future are crucial for human consciousness, for how we process information and make sense of our lives. In this lecture, we'll return to psychology and neuroscience, and we'll see even more clearly that the human brain is a complicated place—much harder to understand than the physical universe. The single lesson we will learn over and over again is that the brain does things in a very distributed way, which is exactly as we would imagine given evolutionary theory about how the brain came to be.

The Hippocampus and Place Memory

- Probably the single most important part of the brain related to memory is the hippocampus. The hippocampus is part of the limbic system, which is interior to the cerebral cortex. Many specific kinds of memory functioning are served by the hippocampus, such as place memory—the memory we have for different places in the world.

- To study various functions of the brain, neuroscientists look at the firing of individual neurons when certain mental procedures are taking place. From such research, the hippocampus has been found to have place cells that give us place memories. These cells always fire when we are trying to recognize places we are in or places in photographs or other depictions. It's fascinating that memory of places is confined to a different part of the brain than, say, memory of a person or a song.

- Of course, it's not true that the place cells in the hippocampus have one cell per location. There is a pattern of cells in the brain that lights up when you're in one place and a different pattern that lights up when you are somewhere else. These location memories are some of the strongest memories we have.

- In complicated memories, it's sometimes true that a single neuron will always fire. One researcher discovered that whenever he showed any individual patient a picture of Jennifer Aniston or a picture of the TV show *Friends*, the same neuron would light up. It is not that the "Jennifer Aniston neuron" contains everything you know about Jennifer Aniston, but that the concept of Jennifer Aniston gets that neuron to light up.

Other Brain Regions and Memory

- The hippocampus is located inside the medial temporal lobe, and the combination of those two structures helps with memory formation, with consolidation of memories, and with explicit memories—facts, events, and the order of different sequences in time.

- The cerebellum, which is in the back of the brain, next to the cerebral cortex, helps with motor learning and procedural memory. The mammillary bodies are associated with recognition memory and unconscious memory.

- The amygdala is the region of the brain that gives rise to emotions. It fires strongly whenever you have a strong emotion; in particular, fear is associated with the amygdala. In the phenomenon known as memory enhancement, if you are experiencing strong emotions and perceiving something for the first time, your memory of that experience will be vivid. This phenomenon is not seen when the amygdala is damaged.

Unreliable Memory

- People who suffer from anterograde amnesia cannot form new memories. This condition can be caused by drugs or brain injuries, in particular, damage to the hippocampus.
    - A person who can't form new memories seems to have lost something that is central to what we think of as being a person. He or she can't learn anything new.

- - However, those who suffer from this condition inevitably score happier on psychological tests than people with fully functioning memories.

- Even for people whose brains are fully functioning, we all know that memories are not perfectly reliable. Not only do we misremember things, but we can have dramatically detailed false memories that are as vivid and true to our perception as any real memory is.

- Misperception, misinterpretation, and false similarity between different things can cause you to misremember what happened.
    - In one famous experiment, subjects were presented with a list of words that had some similarities, such as sugar, candy, honey, and so on.
    - When asked to remember the words later, not surprisingly, they added related words, such as chocolate, to the list and remembered them as being on the list just as vividly as the words that were actually on the list.

- True memories in our brains can be degraded by interference with the process of remembering. Researchers have shown that suggesting a false aspect of a scene (a stop sign in place of a yield sign) caused subjects to incorrectly remember other things about the scene.

- Wholly new memories can also be put into the brain. In the "Lost in the Mall" study, subjects were told five stories about their childhoods, knowing that one of the stories was false. They were then asked to fill in details about the stories—anything they could remember—or say that they didn't remember the event. About 25% of the subjects "remembered" the false story and filled in the details with no problem.

- In addition to false memories, people can also experience false forgetting. Research has shown that the hippocampus has suppressed activity and the frontal cortex has increased activity in

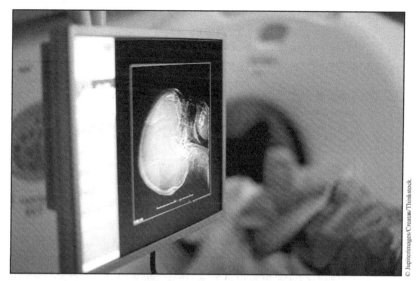

Researchers use functional magnetic resonance imaging (fMRI) to study the brain while it is thinking.

people who are told to suppress thoughts of certain words. It's as if the conscious brain is beating up on the unconscious brain, trying to get it not to remember certain words.

A Theory of Memory

- We don't have a full theory of memory yet, but we do have clues about some aspects of memory. One such clue is that remembering the past seems to be a similar function in the brain as imagining the future.

- This clue has yielded a theory that the brain stores data from which it can reconstruct or re-create images and details. This helps us understand why false memories are just as vivid as true memories. If false memories require only a "script" in the brain—not a fully detailed set of pictures—it is much easier to imagine how they could be introduced and how we could be convinced that they are true.

Consciousness

- Imagining the future is perhaps an even more important part of being human than remembering the past. Thinking about the future is the key to what makes us conscious. Consciousness is not well understood, but it has certain key features: symbolic thinking; abstract thinking; the ability to contemplate alternatives; and the ability to determine what is real, what might be real, and what might happen in the future.

- Like memory, all these functions are highly distributed in the brain and very complex. There are probably many stepping stones along the way in evolution to get from a single-celled organism to a conscious being, and it's impossible to say at which step actual consciousness arose. It's likely that different steps contributed to consciousness in different ways.

- Linguists have argued that a crucial step in the development of consciousness was the development of grammar, in particular, the ability to use language in the subjunctive mood, to say, if you do this, I will do that. The ability to use language in that way enables people to make agreements and develop more elaborate societies based on thinking about the future. This is a relatively late step in the history of consciousness.

- An earlier step in the development of consciousness is the ability to make decisions. Malcolm MacIver, a biomedical engineer at Northwestern University, has an interesting theory in this regard.
  - MacIver pointed out that the visual field of ocean-dwelling organisms is limited. Thus, such organisms must react quickly to what they see.

  - Once organisms crawled onto land, they could see for much longer distances and had more time to consider threats or opportunities. Evolution would favor organisms whose brains allowed them the ability to contemplate alternatives and the future.

- Still, the actual process by which we consider alternative futures and make decisions about them is still mysterious. It seems to be the case that human consciousness is more like Congress—operating by subcommittee—than like a dictatorship, with one totalitarian self giving instructions from the top down. Some researchers have shown that the brain may make decisions before we even know about it.

- Such studies raise numerous questions: How conscious are we really? How much free will do we have? If my brain decides to do something before I even know about it, who is it that is doing the deciding? Science can't give us definitive answers; it can only explore how the brain works. It may be best to think of our selves as the emerging phenomenon that results from all the different parts of the brain coming together to make us who we are.

Suggested Reading

Carroll, *From Eternity to Here*, chapter 9.

Schacter, *The Seven Sins of Memory*.

Questions to Consider

1. Do you think your own memory has improved or deteriorated over time? What kinds of memory are you best at?

2. Do you think it's possible that any memories you feel are completely true are actually false? Have you ever been confronted with evidence that something you remember vividly didn't actually happen that way?

Memory and Consciousness
Lecture 16—Transcript

Probably the most direct and immediate way in which human beings relate to time is by projecting ourselves into other moments of time, mental time travel. In other lectures later on, we are going to talk about physical time travel, hopping in a spaceship and moving around through a curved space time and ending up in the past, but that is probably against the laws of physics as they will eventually be understood. Mental time travel on the other hand is something we do all the time. We either have memory, putting ourselves in the past trying to remember what was going on at some previous time. We also have projection into the future, trying to imagine something that hasn't happened yet.

Both of these procedures, remember the past, predicting or projecting into the future, turn out to be crucial for human consciousness, not just for remembering and projecting other times, but for how we process information and make sense of our lives. To investigate this just a little bit obviously requires going back to psychology and neuroscience. Even more than in the last lecture what we are going to learn is that the human brain is a complicated place. It's a much harder thing to understand than the physical universe. The single lesson we will get over and over again is that the brain does things in a very distributed way.

Think about a memory in particular. We sometimes think about memories and how they work in our real human brains in the same way that we think about a photograph or a file on the hard drive of our computer. It is some data stored in some place reflecting an image or some description of a past event. The human brain is not like a computer and human memory is not like a hard drive. If you think about it a little bit this makes perfect sense. For one thing, evolution which gave us the brains we have, didn't drive towards building a computer it built itself up very gradually through accidents and taking advantage of random changes in our DNA. Even more subtly than that it just would not make sense to imagine that a part of the human brain was left empty until we filled it with some memory.

On a computer hard drive you have the empty memory and then you put a file here, you put a file there, you put a file there until you filled up your hard drive. The human brain doesn't work like that. It doesn't have one part of the brain which is physically connected with one memory, another part which is connected with another memory and so forth. It's a much more subtle and complex process exactly as you would imagine reflecting on how the human brain came to be according to the theory of evolution. Given that sort of obvious understanding about how memory works, we can nevertheless look at particular parts of the brain and see what role they play in memory. It is not going to be that there is a memory organ, there is going to be different parts of the brain that do different things and nevertheless we can identify some of the things that we do.

Probably the single most important part of the brain when it comes to memory is the hippocampus. Remember we talked about the cerebral cortex, the gray matter that is the outside part of our brain where we do our higher brain functionings. The hippocampus is part of the limbic system which is inside connected to, but interior to the cerebral cortex and the hippocampus is the most important part when it comes to memory. There are many specific kinds of memory functioning served by the hippocampus, one simple one is place memory. Place memory means exactly what it says, it is the memory we have for different places in the world. You look around and you recognize that you are at home or your at work or you are in your favorite restaurant or you are in your car. We can ask the question, how does that happen? What is the actual process that goes on in your brain when you look around and recognize that you are in this place rather than some other place. This is something that is just at the beginning of the kind of thing that neuroscientists are trying to understand.

What we can do right now are experiments where we look at the firing of individual neurons when a certain mental procedure is taking place. For example, you can be asked to remember a certain memory or just look around and recognize a certain place and the neuroscientists can look at specific neurons in your hippocampus or elsewhere and they can figure out what fires. It turns out that the hippocampus has particular cells called place cells, which give you place memories. These cells always fire when you are trying to recognize what place you are in or trying to recognize other places in photographs or depictions. This is a fascinating fact that memory of places, knowing where you are or where a certain picture is of, is confined to a certain part of the brain, a different part of the brain than let's say remembering a person or remembering a song. There are literally cells that help you remember what place you're in.

But, it is not that the place cells in the hippocampus have one cell per location. It's not like, here is the cell that fires when I am at home, here is the cell that fires when I'm in my car or anything like that. Again, you would not expect that to be true because there would be a whole bunch of unused cells depending on how many places you had in your brain. Rather what happens is that there're patterns of place cells that fire. If you are in a certain location than about 50 percent of the place cells in your hippocampus will be firing. If you are in a different location, again about 50 percent of those cells will be active, but it'll be a different 50 percent. There'll be overlap. It's not like there is this 50 percent and this 50 percent, therefore you can only remember two places. There are patterns that are forming in your hippocampus and those patterns repeat themselves every time you recognize the same place.

There is some pattern of cells in your brain that lights up when you look around and notice that you are at home. There is a different pattern of cells that lights up when you look around and notice that you are in your work or some other building. These location memories, as it turns out, are some of the strongest memories we have. We all know this sort of intuitively if you drive to work, or if you commute even by walking, or by taking the train. We all know that at some point the repetition of this, the doing it over and over again transfers that memory to a deep part of our brain and you actually don't know need to be paying attention to where you are going. You can do the commute even if you're driving your car, even if you are literally moving the steering wheel and pushing on the accelerator. You don't really need to be paying close attention to what you're doing. That part of the memory you have of where you are going and how to get there is utterly subconscious and very, very strong.

This fact that memory of places is so strongly embedded in our brains actually helps you in a memory trick. There are people who get into the *Guinness Book of World Records* and so forth for memorizing long lists of random numbers or long lists of words. It turns out that these people don't have genetically superior memories. There are techniques for remembering complicated, unrelated lists.

One of the best techniques is to think of a complicated place, to think of some castle with many, many rooms in it for example, and it doesn't matter that the castle isn't real, you can invent a place, you can create in your brain an elaborate place with many, many different sub-places. Your brain can remember the map very, very well and then when someone gives you a list of random words or random numbers, the well-trained rememberers will put every word or every number in a different room in the place that they have constructed in their minds. When it goes back to remembering the list, they don't remember the list itself, they know oh what was the number that was in room number one, what was the number that was in the living room, what was the number that was in the dining room etcetera. They are taking advantage of the fact that remembering places and maps is much better and stronger in our brains than remembering random lists of numbers.

Another intriguing thing that has been learned by the neuroscientists is that, even in complicated memories, you can have a single neuron which will always fire. For example, there is something called the Jennifer Aniston neuron. This is very surprising because 100 years ago there was no such thing as Jennifer Aniston, but a psychologist doing an experiment noticed that when he showed a list of pictures, many, many different pictures to test subjects he was looking at a certain set of neurons and you can't map one neuron and one brain to one neuron in another brain, but in any individual patient whenever he showed either a picture of Jennifer Aniston or a picture of the TV show *Friends* or the words Jennifer Aniston, the same neuron would light up. This is again not because that one neuron contains everything you know about Jennifer Aniston, it is because there are many other neurons that the neuroscientist is not keeping track of, but the concept of Jennifer Aniston gets that neuron to light up. It's not the picture, it's not the words in the name, it is the very concept. This is just again another illustration of how complex memory really is, about how the many different ways we remember things play out in our brains.

One of the things that we have learned is that different types of memory are associated with different parts of the brain. We said the hippocampus is number one, the primary memory keeping part of the brain and in particular its associated with place memories, with where we are. The hippocampus is located inside something called the medial temporal lobe and the combination of those two things help with memory formation, with consolidation of memories, your memories get more accurate the more you think about something, and with what we call explicit memories, facts, events, the order of different sequences in time.

There is also the cerebellum which is in the back of our brain, next to the cerebral cortex, which helps with motor learning and procedural memory. When you're driving that car, you don't need to actually think exactly about where do I put my key? When do I turn the steering wheel? and so forth, your brain takes care of that and that is stored in the cerebellum. There are other organs: There are the basal ganglia; there are the mammillary bodies for recognition memory, smell, unconscious memory. The point is—this is not a quiz, you don't need to remember all the different parts of the brain— but the point is that different parts of the brain, which presumably evolved biologically over many, many millions of years for different purposes all play a role in what we call memory.

There is also a part of the brain called the amygdala. The amygdala is actually not a place where memories are stored. It is actually a place where emotions come to be. The amygdala starts firing really strong whenever you have a strong emotion, and a particular fear is something that is associated with the amygdala. In patients where the amygdala has been damaged or removed, it is found that they don't feel fear. That is obviously important. You can ask, what does it have to do with memory? There is a phenomenon known as memory enhancement. If you are perceiving something for the first time, if you are seeing something, hearing something, learning about something, at a time of strong emotions, if your emotions are strong the memory that you have of that experience will be better. You will be able to better recall specific details later on. They will be more vivid to you a long time later. It turns out that this phenomenon of memory enhancement goes away when the amygdala is damaged.

Not only do different parts of the brain literally store the memories, but other parts of the brain where the memories are not stored come into play when it comes to storing them more or less efficiently. That's just how complex the brain is. How do we know all of this?

Sometimes it's because we do studies or we put someone in an fMRI machine and sort of try to map out activity in the brain. Other times it's when we literally open up the brain in a little rat or in a mental patient, but sometimes it is because there are people who have damage to different parts of their brain. We talked about people whose amygdalas are damaged and do not feel fear response in the same way. There are other parts of the brain that can be damaged that lead directly to problems with memory. If you have ever seen the movie *Memento*, it was about a guy who could not form new memories. This is called anterograde amnesia as opposed to retrograde amnesia. Retrograde means going backwards. Retrograde amnesia means you forget the things you used to know. Anterograde amnesia means you cannot form new memories. It is caused by drugs, by brain injuries, in particular damage to the hippocampus. The hippocampus is crucial for forming new memories.

The subject in *Memento*, in the movie, had an accident and after that moment could not remember anything more than a few minutes. As you can imagine this creates very interesting cinematic situations, but it's a very real syndrome. There are patients who cannot form new memories. As you might imagine, a human being who can't form a new memory has lost something that is very central to what we think of as being a person. You meet a person, you give them your name, you wait five minutes into the future and they can't remember meeting you or what your name is. They literally live in the present plus whatever memories they formed before the event that gave them the anterograde amnesia, but they cannot learn anything new.

The interesting thing is that when you give people with anterograde amnesia tests, psychological tests to see how well they are doing, patients with this kind of amnesia inevitably score happier than regular fully functioning human beings. I don't have any deep moral lessons to draw about this, it is certainly a tragedy when a person has this form of brain damage, but it's interesting to contemplate the relationship between memory formation, happiness, and the struggles that we have going through life.

Even for people whose brains are fully functioning, we all know that memories are not perfectly reliable. We sometimes misremember things; that is not a surprising thing. What is surprising is precisely how unreliable memories can be. Not only can you misremember things, but you can have dramatically detailed false memories that are as vivid and true to your perception as any real memory is. Scientists and neuroscientists in particular are fascinated by the process by which memories are formed, both true ones and the false ones. There are factors of misperception, misinterpretation, false similarity between different things that can cause you to misremember what happened. In one famous experiment subjects were presented with a list of words, but the words were not random. They had similarities between them. For example, your list of words would say sugar, candy, honey, other sweet kind of things. Then later on some time would pass and you're asked to remember the list of words.

Now you will not be surprised to hear that sometimes you misremember, you added some words that weren't there. You will not be surprised to learn that the kinds of words that are typically added were not random. They were in the same kind of category as the list. So, if your list was sugar, candy, honey, you would remember words like sweet or chocolate or other related words. The reason why this isn't surprising to us is we think well the brain is trying to remember, trying its best to come up with words that were on the list. It remembers that the list was full of words involving candy or sweetness so we will put some words in there that are related to candy or sweetness.

However, that is not the whole thing that is going on. It turns out that once the patient recalled this incorrect word like chocolate for the list, the psychologist could come back and say actually that word was not on the list. The patient would say yes it was. Their memory of that word once they invented the fact that it was there is just as vivid as the real words that were actually on the list. This is true even if the subjects who were brought in for the experiment were told ahead of time that the purpose of the experiment is to test false memories. It is absolutely fascinating that the falseness of a memory has nothing to do with how vivid it is or how confident we are that it is true.

Another example of how memories that are false can get into our brains is interference. You have a true memory that is in your brain and psychologists can degrade that true memory, make it less vivid to you, by interfering with how you remember it. There is another test in which subjects were shown a picture of a Datsun car, so you know when this was from, the '70s and the '80s. There was a Datsun stopped at a yield sign so the patients were asked to look at this picture. They were later asked, did another car pass the Datsun at the stop sign? Even though the actual picture had a yield sign in it, they were asked as if the picture had a stop sign in it. What the result was was that they were less likely to get right the answer was well there was another car in the image even though the stop sign and yield sign had nothing to do with the other car. The fact that the experimenter put this false thing into their brain made them less likely to correctly remember other things about the image.

Finally of course, there are wholly new memories that can be put into the brain. The psychologist Elizabeth Loftus did a study called now the *Lost in the Mall* study. Subjects were told five stories and they were told that these stories came from their close relatives who told them cute little anecdotes about their childhood. Four of the stories were true stories, one was a false story just as elaborate, filled in with plausible details about a time they got lost in the mall and had to be rescued by some passerby and were eventually returned to their parents. What the subjects were asked to do was simply fill in the details, to say anything they could remember about those stories or to say simply they didn't remember that event at all. One of the five stories was false, it turns out that about 25 percent of the subjects remembered that story, the false one, and filled in the details no problem. Even when they were told that one of the five stories was false, 25 percent of the subjects could not figure out which story was false. To them the false story was just as true as the true ones.

In fact, I have sometimes served as a science consultant on Hollywood movies, because as an expert in time travel science, extra dimensions, as you can imagine there is plenty of movies that want to know how that might happen. It was interesting and amusing to me when I talked to a friend of mine who was a psychologist, she was outraged by the movie *Inception*. I saw the movie, I thought it was great, I didn't have any scientific complaints, but if you know the plot of the movie *Inception*, it was about an elaborate scheme to try to implant a false memory in someone's brain. To a working psychologist, that is a silly premise for a movie because implanting false memories is the easiest thing in the world.

It's not just false memories, there is also something called false forgetting. If you tell a subject to not thing of a word, don't think of the word snorkel, now at first you might think that this is hard to do. If I tell you don't think of a pink elephant, you will inevitably start thinking of a pink elephant. It turns out that you can really not think of a word, you can suppress your thoughts about it, you can think about other things to distract yourself. The psychologist Michael Anderson took patients and put them in a functional MRI machine that could map out which parts of their brains were active while they were doing certain tasks and he asked them to not think of certain words.

Interestingly the hippocampus, which remember is what we use to form memories, had suppressed activity when they were trying not to think of this word, but the frontal cortex, the part of our central cortex, one of the conscious higher order parts of our brains, was highly active. It was as if the conscious brain was beating up on the unconscious brain trying to get it to not remember certain words and it works very well. When you're asked later, the subjects were not able to remember what the word was they were told to not think about.

What is going on? What is happening in the brain where all these complicated are occurring? Again, we don't know. We don't have the full theory of memory yet, but we do have certain individual clues. One clue is that remembering the past seems to be a very similar function in the brain to imagining the future. Daniel Schacter at Harvard, Kathleen McDermott at Washington University, and other neuroscientists have done studies where, again, they put you in an fMRI machine so they can look at what parts of your brain are active, where the blood is pumping in your brain, and they ask you to do two things. They give you a cue word like puppy and they say remember a time when you got a puppy or you saw a puppy, so you conjure up a memory in your brain. Then they say now imagine a time in a future when you get a puppy or you see a puppy or something similar to the memory.

The interesting thing was that when you are thinking about a future event that hasn't happened or when you are thinking about a past event that has happened, the same parts of your brain become engaged. Remembering the past is a very similar brain activity to predicting the future. Once again, this brings us to a theory, that is the data. The theory says that memory is not like bringing up an image of a photograph from your computer hard drive. It is not like when you have a memory you store every bit of the videotape someplace in your brain. Rather, when you have a memory you don't store the videotape, you store the script, the screenplay. You have some data from which the brain can reconstruct the image and the details. You have a set in your brain that says this is what my living room was like when I was 10 years old, you have a script somewhere else that said, oh the time when the puppy came to my living room.

When you remember that event, so this theory goes, you take the script you plug in the sets, you bring out some actors and you recreate the event in your brain in precisely the same way that you create the event in the first place if you're imagining the future. This can help us understand why false memories are just as vivid as true memories. If false memories only require a script in our brain, not a fully detailed set of pictures, it is much easier to imagine how they could get in there, and how we could be convinced that they are true.

Memory aside, there is also the future. We talked about how memory is an important part of being a human being. Imagining the future, which is very similar to remembering the past, is perhaps an even more important part of being a human being. In a very real sense, thinking about the future, contemplating the future, is the key to what makes us conscious. Now I certainly cannot tell you what consciousness is, if I couldn't tell you what life was before, you shouldn't rely on me to tell you what consciousness is. It is, again, something that is not well understood by anybody. We do agree that consciousness has certain key features, symbolic thinking, the ability to think abstractly, the ability to represent ourselves as well as the outside world, the ability to contemplate alternatives, to say not only what is real, but what might be real or what might happen in the future.

Like memory, all of these functions are highly distributed through the brain and very complex. There are probably many, many stepping stones along the way in evolution to get from a single celled organism to a conscious being like ourselves and we can debate at what step along the way actual consciousness arose. The more likely story is that there were different steps, all of which can contribute. For example, the ability to imagine different futures is absolutely important. If you can't say, well this could happen or that could happen, what should I do to make the preferred outcome happen, you cannot count yourself as conscious.

Linguists have argued that a crucial step in the development of consciousness was the development of grammar. In particular, the ability to use language, not just to point at things and say that is a rock, that is a cave, that is a tree, but the ability to speak in the subjunctive mood, to say if you do this, I will do that. Once you have the ability to use language in that way you can make agreements with other people, not just convince them to do something by threatening them. You can actually make up a contract, you can develop more elaborate societies based on thinking about the future. That is a relatively late step in the history of consciousness.

There is also an earlier step that also refers to how we think about the future, which is the idea of making decisions. Malcolm MacIver at Northwestern University has a theory that one of the most important steps in the development of consciousness, happened 350 million years ago when the first fish crawled out from the sea onto the land. He is not a neuroscientist or a philosopher, he is an engineer, and he studies fish and land animals. He points out that in the sea you can only literally see in front of you a few meters. When you ire inside water, when you are in the ocean, your visual field doesn't extend that far. When you climb out onto land suddenly you can see for hundreds and thousands of meters so MacIver's theory is that if you are an ocean dwelling, sea dwelling animal the only thing you have time to do is to see something and react. You see something, it's either food or it's an enemy, make a quick choice.

Once we climb out into the air, now we can see for a much longer distance. That means that we can see both threats and opportunities coming, and have much more time to take advantage of them. Once you can see a much larger distance our brains are now selected by evolution for the ability to contemplate the future, to keep in your mind alternative possibilities. That was not a useful skill, MacIver says, when you are a fish or an eel or an octopus. It is a useful skill once you climbed out on land. This is a step toward consciousness, the ability to think about different things at once.

Having said all that, we will admit that the actual process by which we do consider alternative futures and make decisions about them is still very mysterious. It seems to be the case that human consciousness is more like congress than like a dictatorship. It's more like a parliament where there are many little subcommittees arguing with each other than like a top down structure where there is one totalitarian self that gives instructions to everything else.

There is a classic study done by Benjamin Libet from UC San Francisco in the 1970s that basically proves that you make decisions, at least in some context, before you even know it. What Libet did was to have you watch a moving dot moving around in a circle, it was moving quickly so it was basically like watching a clock. Libet asked you to make a decision, just randomly sort of decide when you're going to push a button and when you made that decision mentally note where the dot was. Essentially you are saying at what time you made the decision to push the button, and then of course you push it. He found to nobody's surprise, that you decided to push the button 2/10 of a second before you push it. It takes some time for the signal to go from your brain to the finger that actually does the button pushing.

At the same time, Libet hooked people up to an EEG machine, an electroencephalogram, and he recorded activity in the brain. What he found was that there is a particular part of the secondary motor cortex that always fired 1/2 of a second, 5/10 of a second before you actually pushed the button. In other words, the brain does something before you consciously make the decision to act, 3/10 of a second before you know that you are going to push the button some part of your brain knows that you are going to push the button. In subsequent tests, this was shown that you could even stretch that out to longer than 3/10 of a second and a great deal of debate has gone back and forth.

The question that studies like this raise is how conscious are we really? How much free will do we have? If my brain decides to do something before I even know about it who is it that is doing the deciding. This may or may not be a good question. All that science is going to do is tell us how it works. We can decide what to make of it ourselves. How it works is that the brain is a complicated thing. There are many little sub-processes going on. The way to think about ourselves is not as a little dictator telling our brain what to do. Ourselves are the emerging phenomenon that we get from all the different parts of the brain coming together to make us who we are.

Time and Relativity
Lecture 17

In earlier lectures, we talked about the arrow of time and the problem of entropy; in this last part of the course, we turn to the question of why the early universe had such low entropy. Cosmologists don't know the answer to this question, but we will look at some plausible explanations. This means that we need to understand cosmology and the force that is most important for cosmology: gravity. Of course, when we're talking about gravity, we also need to understand relativity, Albert Einstein's theory of space and time. This lecture compares Newton's and Einstein's views of spacetime and gives us a new way to picture the idea that time is not universal but personal.

A Four-Dimensional Viewpoint
- A four-dimensional understanding of the universe is optional in a Newtonian viewpoint but becomes absolutely necessary with relativity.
  - In three dimensions, we have particular points in space, but in four dimensions—spacetime—instead of points, we have events. An event is neither a place nor a time; it is both. It is a location in spacetime.

  - Spacetime, the four-dimensional reality of our universe, is a collection of an infinite number of events, just as space is a collection of an infinite number of points indexed by the three dimensions of space.

- Both the Newtonian view of the universe and the Einsteinian view encompass three dimensions of space and one dimension of time; the difference is that the two views divide up spacetime into space and time in very different ways.

- Newton thought that space and time were separate and absolute. If we think of an event as a point in space at a moment in time, in this view, for any one event, there is something called the present moment. This is a three-dimensional slice through reality, and that moment is the same time as the event. It is all of space at one moment in time. Any one moment has a present; everything before it is its past, and everything in front of it is its future.

A Robot Army

- Let's think about how we could use clocks to construct Newtonian spacetime, in which any one event has a three-dimensional slice associated with it—the whole universe at one moment in time.

- Imagine that we have a large robot army, and each robot has a clock and a spaceship. We synchronize the clocks on our robots and send them throughout the universe to every point in space.

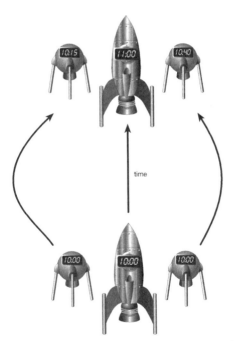

If we send two robots with synchronized clocks into space on different paths, their clocks will not be synchronized when they return.

- We now have a label at every event in spacetime. We know where each robot is in space and we know when it is in time. What we mean by the universe at one moment is simply the set of events where all the robot clocks read the same time. When they all say 6:00 p.m. on November 5, that is a moment in the history of the universe.

- With Einstein's relativity, there is no such thing as one moment of time all throughout the universe that everyone agrees on. To Einstein, space and time are combined into spacetime, but space by itself and time by itself are not absolute; they are relative. What we call space and what we call time are different for different observers.

- The way we send our robots equipped with clocks throughout the universe affects the answer we get in our thought experiment. Different observers with different sets of robots that were sent throughout the universe using different spaceships would get different notions of simultaneity. They would divide the universe up into space and time in different ways.

- If we send out robots following different paths through the universe and bring them back, their clocks will have measured different amounts of elapsed time. It's not that there is anything wrong with our robots; it is that the amount of time elapsed on a particular trajectory through the universe depends on how the robot moves through the universe. If the robot is moving on a curved path at high velocity, it will always measure less elapsed time when it comes back.

A Muon Clock
- There is no such thing as one moment spread throughout the universe that everyone can agree on. What time it is at one point depends on who does the measuring. One of the most dramatic and immediate experimental verifications of this fact comes from an elementary particle called a muon, a heavier cousin of the electron.

- Pions decay into muons, which then decay into electrons and other particles. Muons decay very quickly, in about 2 microseconds. They can be used, in some sense, as a clock.

- Traveling through interstellar space in the form of cosmic rays are protons that smash into atoms in Earth's upper atmosphere. They create pions and other elementary particles, and those quickly decay into muons. The upper atmosphere where the muons are created is about 15–20 km above the Earth, and the muons should decay after traveling about 1 km, yet they're able to reach the ground on Earth.

- Why do muons reach us before decaying? The answer is that their clocks—their lifetimes—are not ticking in the same way as ours. Because muons are moving so fast with respect to us here on Earth, they "feel" less elapsed time. To them, it takes less than 2 microseconds to go from the Earth's upper atmosphere to the Earth itself.

Notions of Time
- At a fundamental level, there are different notions of time, which sometimes coincide and sometimes do not.
  - We can define one idea of time as the time measured by an observer, the time that is elapsed by a clock.

  - Another notion of time that is a bit more abstract is time as a label to measure what moment of the universe we're talking about.

  - To Newton, these two notions of time were the same. But Einstein says these notions are not the same. Time is personal; how much time passes depends on how you move through the universe. The time that you feel depends on your trajectory through the world.

- Consider the universe as a football field. In this universe, one way of measuring how much distance the running back has traveled is to say what yard line he reaches. That is like the Newtonian idea of absolute time. But the actual distance the running back travels involves running back and forth to avoid tacklers. To get from the 10 yard line to the 20 yard line, the actual amount of distance the running back has to go is usually more than 10 yards.

- In space, this makes sense to us, and Einstein says that the same thing is true in time, but the amount of time that has elapsed for us, measured by our clocks, depends on how we connect two different events in the universe by our motions through spacetime. What Einstein is doing is replacing the idea of space at any one moment in time with a much more subtle way of thinking about spacetime.

A Light Cone
- According to relativity, everyone measures the same speed of light: about 300,000 km per second. With that in mind, think about an experiment in which you flick a light bulb on and off quickly. As you do so, photons move away from the light bulb, so there is a sphere of space that is the distance reached by all the photons. Let's say you flick the light bulb on and off very quickly, so that just one small shell of photons will leave it.

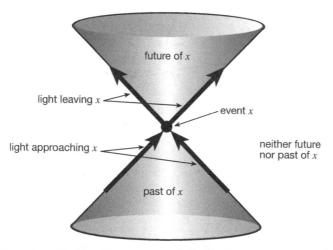

Relativity replaces "space at one moment of time" with light cones—the paths of all possible light rays in spacetime to and from a fixed event.

- In space at any one moment of time, there would then be a spherical region that is where the photons from the light bulb have traveled given a certain amount of time.

- Now imagine that we took a picture of every three-dimensional moment in space to track the expansion of all those photons and we stacked those moments on top of each other, like the pages of a book.

- At the moment you flicked the light bulb on and off, the photons were all very close; they were at the point where the light bulb is. A second later, all those photons were 300,000 km away; 2 seconds later, they were 600,000 km away. As

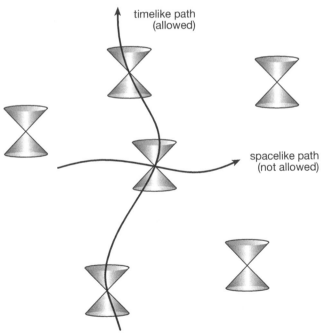

In a Newtonian point of view, we must march forward in time; with Einstein, we must stay inside our light cones.

we stack up these different pictures of space, we see that the photons going from a point out in all directions form a cone in spacetime, a light cone.

- The fact that the speed of light is the same to everybody means that different observers construct the same light cones. If somebody else moving by you at a speed of some hundreds of thousands of km per second in a spaceship also flicked on a light bulb at the same event as you did, those photons would do the same thing as yours did; they would form the same light cone in spacetime.

- Einstein tells us that instead of dividing the universe into space and time, we need to divide the universe into the light cones that we can imagine making.

• The light cones give us a structure on spacetime because we cannot travel faster than light. If I flick the light bulb on and off very quickly right now, I will get a bunch of photons leaving, and I can never catch up to them. They will always be receding from me at the speed of light.
- If I draw that light cone in spacetime, what I am drawing is my personal future. My future is only the points inside that light cone. An event outside that light cone is inaccessible to me. I can reach a particular point in space outside the light cone but not a particular point in spacetime.

- Newton would have divided spacetime into past, present, and future. Einstein divides spacetime into the past of my event, that is, the interior of the light cone that is approaching me at this point in spacetime; the future of my event, the interior of the light cone that spreads out from me in spacetime; and the unreachable parts of the universe, the parts that I would have to travel faster than the speed of light in order to reach.

• The lesson is that in special relativity, time is not universal; it is personal. We can slice the universe up into equal moments of time, but there is no one right way to do that. The division of spacetime into space and time is not absolute anymore; it's not a Newtonian universe.

Suggested Reading

Carroll, *From Eternity to Here*, chapter 4.

Greene, *The Fabric of the Cosmos*.

Thorne, *Black Holes and Time Warps*.

Questions to Consider

1. Why is it that relativity seems so counterintuitive to us? Can you imagine a world in which the speed of light was much slower—say, 100 miles an hour or just 5 miles an hour—but everything else was unchanged?

2. Can you think of other experiments (feasible or otherwise) that would test Einstein's description of spacetime?

Time and Relativity
Lecture 17—Transcript

By now we are two-thirds of the way through the lectures and the three sections actually fall into a certain kind of pattern almost as if I had planned it that way from the start. The first third was setting the stage. We talked about what time it is, how the arrow time is the most important mystery, how certain things do not explain the arrow time. In the second third, we sort of took very seriously the problem of entropy. Entropy is why there is an arrow of time. Boltzmann explained to us what entropy is. And, we still had puzzles about why the entropy was so low in the past, but we also made the case that if we were to understand that we would understand lots of other things, why remember the past and not the future, why we can make choices about the future and not the past.

Now, in the last third of the lectures, we are going to really try to explain why the early universe had such a low entropy. The bad news is that we're not going to succeed. Cosmologists don't actually know why the early universe had a low entropy, but we will think about what the plausible explanations are. What that means is we need to understand cosmology, the study of the whole universe and when we need to understand cosmology that means we need to understand gravity. Gravity is the force in the universe that is the most important and influential one when it comes to cosmology.

When we need to understand gravity in the modern world, that means we need to understand relativity. Relativity is Albert Einstein's theory of space and time and especially gravity. Once we dig into that, we can understand not only cosmology and the expansion of the universe and the Big Bang, but also interesting questions like how time began. Can you travel backwards in time? This makes sense because Einstein's theory is a theory of space and times. Even if we weren't interested in the past hypothesis and the low entropy of the early universe, we would still want to talk about relativity anyway.

To really understand relativity the secret is to take a truly four-dimensional viewpoint. We can say that the universe is made of four dimensions. There are three dimensions of space and one dimension of time. What we mean by that is that to locate something you need to give me those four pieces of information, where you are in three-dimensional space, and when you are in one-dimensional time. If you want to meet somebody at 5 p.m. at a certain location, you need to say where that location is and give the time. Nevertheless we tend to think in terms that say that time and space are very different. We tend to think of the universe as a three-dimensional universe with stuff in it. We tend to think that reality is a three-dimensional thing and that three-dimensional thing happens over and over again as a function of time. As time passes three-dimensional reality evolves.

The four-dimensional point of view, which is optional in a Newtonian viewpoint, becomes absolutely necessary when relativity comes along. Let's think about what that means. In three dimensions, we have points. We say here is a particular point in space. It's different than all these other points. In four dimensions, in what we call spacetime, instead of points we have events. If you say this particular point at 6 p.m. on October 25, that one moment of time is an event. An event is not a place, nor is an event a specific time, it's both. It's a specific point at a specific time. It is crucial that we mean not that there's an important event, an event is not a happening, either a glamorous event or a non-glamorous event, it's not something that occurs, it's just a location in spacetime. Just like we have locations in space, if you have a location in space at a specific time that's what an event is.

Then spacetime, the four-dimensional reality of our universe is just a collection of every event, just like space is a collection of an infinite number of points indexed by the three dimensions of space. Spacetime is a collection of an infinite number of events. What we want to do is to understand those events and internalize it so well that we can throw away the division of spacetime into space and time, to treat it as one truly four-dimensional thing. This is a very difficult trick to pull off intuitively. We're not used to thinking four-dimensionally, but we can train ourselves to do better and better.

The real difference here is the difference between the Newtonian view of the universe and the Einsteinian view of the universe. They both have a four-dimensional spacetime. In both Newton and Einstein, you need three dimensions of space, one dimension of time. The difference is that they divide spacetime up into space and time in very, very different ways. If you are Isaac Newton or any of Newton's followers, you thought that space and time are separate and they are both absolute. When we say that space and time are absolute, we mean that they are given, we know what they are. There are no choices to be made. It's not up to us. Different people don't have different kinds of space or different kinds of time.

In a Newtonian point of view you would say here is an event, here is a point in space at a moment in time and for any one event there's something called the present moment. That is to say to any one event there is a whole three-dimensional slice through reality and that moment is the same time as this event. It is all of space at one moment in time. It is a notion of simultaneity. What set of events happened at the same time as this particular one event? The picture in your mind is of a four-dimensional universe and time slices the universe into moments of time. In a Newtonian point of view, both time and space are there given to us. Any one moment has a present, everything before it is its past, everything in front of it is its future.

Now whenever we get to these confusing issues of space and time, it is very helpful to ground ourselves by thinking operationally, in other words by thinking about clocks. One way of thinking about time is the time is what clocks measure, so when we say in a Newtonian spacetime any one event has this three-dimensional slice associated with it, the whole universe at one moment of time. What you should be asking yourself is how would I construct that using clocks? What would I actually do to measure this concept? Rather than taking this God's eye view of the whole universe, what would I build and what would I write down in my notebook.

There is a great book by Peter Galison who is a historian of science called *Einstein's Clocks & Poincaré's Maps*. Einstein, of course, you know of. He is the founder of relativity. Poincaré you have heard of in the context of the recurrence theorem. Poincaré is the French mathematician and physicist who invented chaos theory, showed that if you have a finite number of things that can happen, things would eventually recur. He was one of Europe's leading thinkers at the turn of the century around the year 1900. He did a lot of work that was prior to Einstein basically laying the groundwork for what we currently call relativity.

Poincaré was very interested in the symmetries that underlie the theory of relativity. The symmetries that say no matter where I am physics is the same, no matter how fast I am moving physics is the same, and no matter what direction I am pointing in physics is the same. It was Poincaré who put all those symmetries together. In Galison's book, *Einstein's Clocks & Poincaré's Maps* he makes the point that we think about Einstein and Poincaré as these very abstract thinkers, people who are concerned with the very essence of space and time and reality far removed from our everyday lives.

In fact, Einstein worked in the Swiss Patent office and Poincaré worked as the President of the French Bureau of Longitude. Remember the problem of longitude, one of the leading issues in the nature of time tried to make people build better clocks to figure out where we are on the surface of the Earth. Poincaré played a big role in helping people try to figure out what longitude they were on how to draw maps. Einstein, in his patent office, he wasn't just looking at any old gizmo. At the time in the early years of the 1900s, one of the most important gizmos to patent were clocks. The point is that that was the time when transportation between cities became easier and easier.

It used to be that every different city had its own time zone. You would just have the actual time that was associated with where the Sun is above you in the sky. But, once transportation and communication became easier, it became interesting to sort of relate the times in one city to other locations. Even within one town, it was important to synchronize your clocks. So, many of the patents that Einstein actually looked at were for better and better clocks and methods for synchronizing the clocks. Galison's argument is that despite the reputations of Einstein and Poincaré as very, very abstract deep thinkers, they were both heavily influenced by their down-to-earth day jobs by the need to understand map, pictures of space, and clocks, measurements of time. When you want to think about spacetime, you should think about what the maps do and what the clocks do.

To make sense of Newtonian spacetime, imagine actually constructing that slice through the universe that we call the present moment. What you might imagine doing is building a large robot army, each robot has a little clock on it and they all have spaceships. Even though we are being practical this is certainly a thought experiment, but that's okay, imagine you have a lot of robots and you can send them anywhere in the universe. Your trick to constructing what you mean by a single moment at one moment in time throughout the universe is to synchronize the clocks. You bring your robots together, you set their clocks to read the same amount, and then you send them throughout the universe.

Imagine we could put a robot with a clock at every point throughout space. We know that the clocks read the same thing because when they were here next to us we synchronized them. Now you basically have a label at every point, at every event in spacetime. You know where the robot is so you know where it is in space and it has a clock so you know when it is in time. Basically through this thought experiment you can try to make real Newton's absolute space and time. You can say what I mean by the universe at one moment is simply the set of events where all of my robot clocks read the same time. When they all say 6 p.m. November 5th, that is a moment in the history of the universe.

The problem is that Einstein figured out that this technique doesn't work, that even as a thought experiment you cannot take robots synchronize their clocks and send them all throughout the universe. In relativity, in Einstein's point of view, there is no such thing as one moment of time all throughout the universe that everybody agrees on. To Einstein, space and time are combined into spacetime, but space by itself and time by itself are not absolute. They are relative, that's why that word is attached to this theory. What you call space and what you call time will be different for different observers.

Why is this true? What happens to our robot army? It is not that we have another army that fights our robot army. It is that the way that we send our robots equipped with clocks throughout the universe affects the answer we get. Different observers with different sets of robots that were sent through the universe using different rocket ships, would get different notions of simultaneity. They would divide the universe up into space and time in different ways. One very, very simple way of seeing this is just take one robot, here I synchronize it with my clock, I send it off so now I know that when it says 3 p.m. on my watch it says 3 p.m. on the robot's watch, but I worry that that robot has had a hard time, I bring it back. Its spaceship brings it back here and I notice that when the robot comes back it is no longer synchronized. The clock doesn't read the same anymore. That is not a fault of my robot, Einstein says, that is a fault of spacetime itself.

This is the best way, I think, to conceptualize the difference between relativity, Einstein's spacetime, and Newtonian spacetime. If you send out clocks following different paths through the universe and bring them back they will have measured different amounts of elapsed time. If you are on a spaceship yourself and you are not firing your rockets or anything, you are just floating through space on some constant trajectory, either standing still or moving, Einstein says it doesn't matter whether you are standing still or moving. There's no absolute way you measure that, but you send out little probes, you send out little satellite robots, satellite spaceships, they both have clocks on them, they started synchronized, they go out exploring, they come back, Einstein says they will always be out of sync.

It's not optional. It's not that you're robots aren't good enough. It's not that there's a technology failure, it is that the amount of time elapsed on a different trajectory through the universe depends on how you move through the universe. It works in a very consistent way, namely any time that you're moving on a crazy path through the universe, a curved path where you're firing the rockets, going at high velocities, and then coming back, when you come back you will always have measured less elapsed time. The way Einstein says to maximize the amount of time spent in between two events in spacetime, is to do your best not to move or to equivalently move at a constant velocity, acceleration, curved paths through spacetime always decrease the amount of time that your clocks measure.

This is not just abstract nonsense. This is not just some theoretical speculation. It's a deep profound idea. We're saying that there is no such thing as one moment spread throughout the universe that everyone can agree on. We're saying that what time it is at one point depends on who does the measuring. That's a deep fact. But, it's also experimentally verified. One of the most dramatic and immediate experimental verifications comes from the idea of a muon. A muon is an elementary particle. It is like a heavier cousin of the electron. The electron, of course, plays a crucial role in atoms. Muons don't play a crucial role in anything because they're heavier than electrons, they decay very, very quickly. The lifetime of a muon is about 2 microseconds, 2 millionths of a second.

If you had a muon right here it would decay. Pions decay into muons, which then decay into electrons and other particles. Muons, in some sense, are a clock. They are not a very good clock, but if you have a large collection of muons, over half of them decay every 2 millionths of a second. By looking at how your collection of muons diminishes over time, you are essentially measuring how much time has passed. From measurements in the laboratory, we know that the lifetime is about 2 millionths of a second. Nevertheless, we find that we can discover muons here on Earth that came from cosmic rays in the sky. In fact, this was the way that muons were first discovered by Carl Anderson at Caltech. He was measuring cosmic rays and he measured the existence of this particle that looked like an electron, but was much heavier.

All that sounds fine until you plug in the numbers. How do you make these muons from cosmic rays? It's not that muons are traveling through interstellar space. If they were they would decay very quickly. What travels through interstellar space in the form of cosmic rays are protons and sometimes heavier atomic nuclei, but those protons smash into atoms in our upper atmosphere here on Earth. They create pions and other elementary particles and those quickly decay into muons so you might want to ask yourself do the muons have enough time to reach the ground and be detected. Well their lifetime is 2 microseconds, you know how fast they are moving, it's almost at the speed of light, so you plug in and you figure out that the muons should decay after travelling about 1 km.

The problem is the upper atmosphere where the muons were created is 15–20 km in the air. It is a surprise if we didn't know relativity, which of course we did. It would be a surprise to realize that muons are able to reach the ground. If their clocks were ticking at the same rate that our clocks are ticking here on Earth, they would have decayed long before they reached our detectors here at sea level. They don't. Why do the muons reach us before decaying? Because their clocks are not ticking in the same way ours are. They don't have a literal wristwatch, but the lifetime of a muon is a kind of a clock. It is a predictable thing. Because they are moving so fast with respect to us down here on Earth, essentially they feel less elapsed time. To them it is less than

2 microseconds to go from the Earth's upper atmosphere to us down here so they are able to make it before they decay.

This sounds a little bit crazy or at least a little bit hard to swallow, that how much time passes for you depends on how you are moving through the universe. What's going on is that as we mentioned way back when there are different notions of time when we talk about what time is at a fundamental level. Sometimes these notions coincide, they are the same, but sometimes they don't. In particular, we can define one idea of time as the time that is measured by an observer. The time that is elapsed by a clock. The nice thing about a clock is it's a small thing. It's not spread all throughout the universe, it's a localized object, a mechanical or electronic machine and it measures a certain amount of time.

There's another notion of time, which is a little bit more abstract. Time uses a label to measure what moment of the universe we're talking about. If we are Newton or even if we are Einstein we can slice up the universe into moments of time and then we say it's 6 p.m. here and everywhere throughout the universe. The reason we thought that everyone would agree when we did that is that to Isaac Newton these two notions of time were the same. Time was absolute. The time that you measure in the universe passing everywhere and for everybody was the same as the time that a clock would measure, but Einstein says they are not. Einstein says that time is personal, that how much time passes depends on how you move through the universe. The time that you feel depends on your trajectory through the world.

Again, this sounds a little weird, but in fact, once you really internalize this idea that time is like space, then it begins to make sense. Think of a football game or a football field with a running back trying to avoid being tackled. A football field is nice because there're markers that tell you what yard line you're on. You are on the 10 yard line, the 20 yard line, the 30 yard line, and what you want to do as the running back is you want to gain yardage. You want to go from the 10 yard line to the 20 yard line. One way of measuring how much distance the running back has traveled is to say what yard line they get to. That is like the Newtonian idea of absolute time. It measures where you are in the universe, now the universe is this football field, but the actual distance the running back travels involves a lot of going back and forth and of trying to avoid tacklers. To get from the 10 yard line to the 20 yard line, the actual amount of distance the running back has to go is usually much more than 10 yards. This is very, very natural. Just to get from one place to another we don't always move in straight lines. We often travel a lot more distance than the actual distance between two different places.

In space, this is no surprise to us, this just sounds perfectly obvious. Einstein is saying that the same thing is true in time, but the amount of time that has elapsed for us measured by our clocks, depends on how we connect two different events in the universe by our motions through spacetime. What Einstein is doing is replacing the idea of space at any one moment in time with a much more subtle way of thinking about spacetime. The short answer to how Einstein divides up spacetime comes in the concept of a light cone. This is a somewhat difficult to visualize idea, but once you get it, it is extremely helpful in thinking about relativity so we are going to spend just a little time trying to understand what light cones are.

A light cone is not something that exists in space. It is not a cone you can draw somewhere or construct with clay. It is something that exists purely in spacetime, and the reason why light is special is because relativity says that everyone measures the same speed of light. It is also the fastest speed that there is, but more importantly it is invariant. It is constant. If I sit here and I measure the speed of a light beam going by, I get a certain answer, 300,000 km per second. If someone else goes by here at 200,000 km per second and measures this same speed of light, you might think that the velocities should add together, but they don't. That person passing by also measures 300,000 km per second. The speed of light is the same to everybody.

With that in mind, think about sitting where you are right now and flicking a light bulb on and off very, very quickly so you have a bunch of light rays or photons depending on how you want to put it. They move away from the light bulb so there is a sphere in space which is the distance to which all the photons got. If you really imagine that you flicked on and off very, very quickly just one little shell of photons is leaving that light bulb. In space, what we would have at any one moment of time is just a spherical region, which is where the photons have gotten from the light bulb to any point given the amount of time they had to go. It is the speed of light times that time.

What about in spacetime? Even though the light forms a sphere around us in three-dimensional space, when we portray it, we usually portray it two-dimensionally so every moment has a circle of light. Imagine we took a picture of every three-dimensional moment in space and we tracked the expansion of all these photons and then we stacked those moments on top of each other just like we stack the pages of a book or the frames from a film on top of each other to make a picture of spacetime. At the moment you flick the light bulb on and off, the photons were all very, very close, they were at the point where the light bulb is. A second later all of the photons are 300,000 km away, 2 seconds later all the photons are 600,000 km away. So, as you stack up these different pictures of space, what you get is the photons going from a point out in all directions forming a cone in spacetime. That is the light cone. This is as fast as you can possibly go in the universe.

The fact that the speed of light is the same to everybody means that different observers construct the same light cones, so if somebody else moving by you at a speed of some large number of hundreds of thousands of km per second in their spaceship also flicked on a light bulb at the same event as you did, their photons do the same thing as yours did, they would form the same light cone in spacetime. What Einstein is really saying is that instead of dividing the universe into space and time, we divide the universe into the light cones that we can imagine making. We don't need the actual light bulb. We don't need to actually flick anything on and off. There doesn't need to be any actual light. This is part of the structure of spacetime itself. You could starting at any event draw the light cone coming out of it or draw the light cone going into it from all the hypothetical light rays that would reach that point rather than leave it.

Usually when people give the popular introduction to relativity, this is not the language they use. They talk the language of time dilation and length contraction. Time dilation is the idea that if a clock is moving very quickly by you, you would see it tick more slowly. Just like the muons going by you seem to take more time as they measure it to get down to here on Earth. Length contraction is the idea that if something moves by you, it is squeezed in the direction. It is contracted along the direction in which it is moving. As my favorite example, the large Hadron collider in Geneva, this giant particle accelerator, moves protons very, very close to the speed of light and then smashes them together. You think of a proton as a spherical collection of quarks and gluons, but when they actually smash together they're more like pancakes. The contraction in the direction in which they're moving smashes these two pancake-like structures together.

Now it's a perfectly legitimate way to talk about relativity in terms of time dilation and length contraction. But to me I find it confusing because you are trying to measure something at different points in space at the same time, which is exactly what Einstein tells you not to do. You are trying to measure the length of something, different points in space, at one moment in time. I think it is actually much more helpful to think of physics absolutely locally, to say what can I measure here at this point, forgetting about the changes in lengths and time as things go by me, think about what their clocks would read when they return to where I am. With that philosophy in mind, what are the light cones getting us? The light cones are giving us a structure on spacetime because we cannot travel faster than light. If I flick my light bulb on very, very quickly right now, I have a bunch of photons leaving and the point is that I can never catch up to them. I can never go faster than they are so they will always be receding from me at the speed of light no matter how hard I try, no matter how powerful my rocket ship is.

If I draw that light cone in spacetime, what I am drawing is my personal future. My future is only the points inside that light cone, but there is some event, some point in space and time that is outside the light cone, then in order to get there I would need to travel faster than the speed of light. That means that that point is simply in accessible to me. I can get to that point in space, but not to that point in spacetime. I cannot get to that point in space at that particular time. What Newton would have done is to divide spacetime into past, present, and future. What Einstein does is to divide spacetime into the past of my event, the interior of the light cone that is approaching me at this point time, the future of my event, the interior of the light cone that spreads out from in spacetime, and the unreachable parts of the universe, the parts that I have to travel faster than the speed of light to get to.

The lesson is that in special relativity time is not universal, it is personal. It is measured along an observer's world line. Now don't get me wrong, you can slice the universe up into equal moments of time. I can talk about what time is it right now at Alpha Centauri or the Andromeda galaxy. The problem is not that you can't do it, the problem is that there is no one right way to do it. There are infinite number of ways to slice up spacetime into space and time and none of them is right and none of them is wrong. The division of spacetime into space and time is not absolute anymore, it's not a Newtonian universe.

As a result of that the rules for moving through spacetime are a little bit different and this is going to become important because soon we are going to talk about time travel. The Newtonian rule is that there is a past, present, and future and you must keep going toward the future, whatever time you are at right now, one second from now you will be one second into the future. What Einstein says is that you must stay inside your light cone. The reason why that is going to become important is because in special relativity, what we are talking about right now, the light cones are fixed, they're rigid, they're there once and for all. They are the absolute elements of space and time, but in general relativity you can move the light cones around so the fact that all we need to do is stay inside our light cone is going to become crucially important.

One of the ramifications of this, for us in this course, is that we care about Laplace. Remember Laplace said if I knew the universe at one moment of time, I could predict its future and retrodict its past. Einstein says there is no such thing as the universe at one moment of time once and for all. Don't worry about that, the point is that Laplace still works. It still remains true that if we know everything about the universe at any one moment of time, we can predict the future and retrodict the past, the only difference is that there is no unique way to say what you mean by any one moment of time. The fact that it's non-unique doesn't matter. That means there are many, many ways to give the data of everything in the universe going on at any one moment of time. The important thing is you can work forwards and backwards. So, special relativity doesn't help us with the arrow of time, it's still a puzzle, we need to move on to general relativity to understand space and time a little bit better.

Curved Spacetime and Black Holes
Lecture 18

In the last lecture, we talked about special relativity, the theory that replaces Newtonian absolute space and absolute time with a single four-dimensional spacetime. But that spacetime is still rigid and fixed. In this lecture, we'll talk about general relativity, Einstein's theory of gravity in which spacetime can bend and twist. This is what gives rise to the force we call gravity. As we'll see in a future lecture, the fact that spacetime can evolve will be important for our understanding of the past hypothesis and the event known as the Big Bang.

The Principle of Equivalence
- Special relativity is not a theory of physics but a backdrop, a meta-theory. It is a stage on which other theories, such as electromagnetism or the strong and weak nuclear force, play out. But Einstein was unable to make a theory of gravity that would fit with special relativity.

- Einstein realized that gravity is different than all the other forces of nature because it is universal. Electromagnetism, for example, is not universal; it acts on different particles in different ways. But in a gravitational field, those particles will all fall down and will all fall at the same rate. The principle of equivalence is Einstein's way of formalizing this idea. The equivalence is between gravity and acceleration.

- Imagine that you are in a sealed room, and you decide to do some physics experiments, such as dropping different objects and measuring the rates at which they fall or measuring the frequency of radiation from an atom.
    - In the room, you are in a gravitational field; however, Einstein says that if you imagine doing those same experiments in a zero-gravitational field, in a rocket ship that is accelerating, you will get exactly the same answers.

- An accelerating rocket ship pushes you in the direction in which it's accelerating; you are pulled to the ground, just as you are when there's a gravitational field beneath you. According to Einstein, this is not just a similarity but an equivalence. You can do all those experiments in the rocket ship that is accelerating, and you will get exactly the same answers as if you did them in a gravitational field.

- Gravity is impossible to detect in a local region of spacetime. There is no experiment you can do to determine whether or not you are in the field of gravity or in a rocket. This is completely different than electromagnetism.

The Geometry of Spacetime

- Einstein tells us that if gravity is universal, then perhaps gravity isn't a field on top of spacetime; perhaps it is a feature of spacetime itself, in particular, the curvature of spacetime. Matter and energy can bend spacetime, and that bending is what we call gravity.

- Euclidean geometry is a kind of flat geometry; it's the geometry of ordinary three-dimensional space. But Einstein thought that spacetime might not obey the rules of Euclid. This way of thinking about gravity is usually illustrated by the image of a rubber sheet with a bowling ball in the middle. The bowling ball distorts the shape of the rubber; the effect of this bending of the rubber sheet is analogous to the effect of gravity.

- In general relativity, just as in special relativity, there are light cones at every point, but because spacetime is curved, those light cones can bend and twist. Further, the curvature can be great in some places and small in others, positive in some places and negative in others. The light cones that form the structure of spacetime in special relativity still do that in general relativity, but now, they can move around.

- It's that distortion of light cones in the solar system caused by the gravitational field of the Sun that makes the Earth orbit rather than moving in a straight line.
  - One way to think about general relativity is to think that particles, including the Earth, are doing their best to move in straight lines. But because spacetime itself is curved, there are no straight lines. That is what gives rise to what we call gravity.

  - This is why the Moon goes around the Earth and the Earth goes around the Sun. They're trying their best to move in straight lines, but spacetime itself is curved.

  - If the gravitational field becomes very strong, the light cones can tilt so much—spacetime can be affected so strongly by gravity—that black holes are formed, regions of the universe where gravity is so strong that light itself cannot escape.

Entering a Black Hole
- As we said, light cones tell us where light can go and, because we cannot travel faster than the speed of light, where we can go. At any one point in your lifetime, there is a light cone moving forward into spacetime away from you, and according to the rules of relativity, you need to stay inside that light cone; that is your future.

- It's also true that gravity can be so strong that spacetime becomes curved in such a way that the light cones tilt and form black holes. Moving slower than the speed of light, the only thing you can do is to move into a black hole, and you can never come out.

- Inside the black hole is a place where the curvature of spacetime becomes infinite. This is what we call the singularity. The gravitational field is so strong that everything shrinks to an infinite density.

- Note that the singularity is not a place but a moment in time. It is in the future.

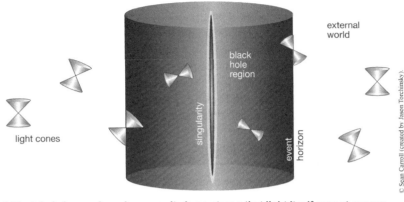

A black hole is a region where gravity is so strong that light itself cannot escape.

- When you enter the black hole, you are forced to hit the singularity because everything inside your light cone becomes focused onto that singularity.

- In fact, if you're inside a black hole and you try to fire your rocket engines to move away from the singularity, you will hit it even faster because less time will elapse for you as you age forward in time. You don't travel toward the singularity; you age into it.

- At the singularity, time comes to an end. Past that, we can't say what happens. In general relativity, all we can say is that it is a boundary, known as the event horizon. It is the edge of spacetime itself.

- If you were to fall into a black hole, you wouldn't notice as you passed by the event horizon. If it is a big enough black hole, you wouldn't even be aware that you are doomed, that you are going to hit the singularity and you can never get out.
  - If the black hole has the same mass as the Sun, the time it takes you to go from the event horizon to the singularity is about 1 millionth of a second.

- If you fall into a black hole feet first, your feet will be pulled apart from your head by the stronger gravitational field they feel. The technical term for this is spaghettification.

Varieties of Black Holes

- Although we can't see them, we believe that we have strong evidence for real black holes in the universe. A black hole affects the gravitational field around it so strongly that other forms of matter near the black hole react in very noticeable ways. Gas and dust fall into the black hole, heat up, and give off x-rays, gamma rays, and other forms of radiation. We can detect that radiation and infer that a black hole is present.

- Black holes seem to come from different places and, therefore, have slightly different properties. The most well-known kind of black hole comes from the collapse of a massive star.
    - A star is actually held up by the heat energy inside it. At the center of the Sun, nuclear fusion is taking place that emits radiation. This radiation heats up the gas and plasma around it, and the pressure of that plasma is what keeps the Sun in its shape. Someday, that fuel will be exhausted, and the Sun will collapse.

    - A star typically collapses to a white dwarf, but if it's heavy enough, it will become a neutron star. A neutron star that is, say, 20 times the mass of the Sun will collapse and make a black hole that is about 3 times the mass of the Sun.

- There is a supermassive black hole at the center of our galaxy. Cosmologists believe that most galaxies in the universe have giant black holes at the center. These black holes may be a million or a billion times the mass of the Sun. The black hole in the Milky Way has 4.1 million times the mass of the Sun and is spinning extremely rapidly.

- Middle-weight black holes may be thousands or hundreds of thousands of times the mass of the Sun. There are also primordial black holes from the very early universe that perhaps weigh only grams. Some physicists have speculated that we could even make black holes ourselves in the Large Hadron Collider.

The Expansion of the Universe

- A separate consequence of the curvature of space and time is that space itself can expand; the size of the universe can change over time.

- In trying to model the universe as a uniform distribution of matter using general relativity, Einstein discovered that the universe must be either expanding or contracting; it could not be static. To account for this apparent contradiction with the data, he posited a cosmological constant that balanced the force of matter and yielded a static solution to cosmology.

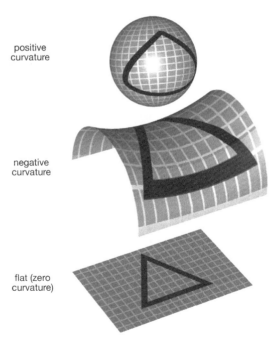

Uniform curvature in space may be manifested in different ways.

- In the 1920s, Edwin Hubble made the prediction that Einstein should've made: The universe is expanding. Given that fact, we can run the movie forward and the universe will get increasingly dilute, but we can also run the movie backward. In the past, things were closer together and, therefore, the gravitational fields were stronger.
  - Winding the movie backward in time is similar to following the formation of a black hole forward in time. Inside the black hole is a singularity in the future because all the matter collects in one place.
  - In the universe, there is a singularity in our past, which is what we call the Big Bang.

Suggested Reading

Carroll, *From Eternity to Here*, chapter 5.

Susskind, *The Black Hole War*.

Thorne, *Black Holes and Time Warps*.

Questions to Consider

1. A "white hole" is just a time-reversed black hole. How would you describe a white hole without referring to the concept of a black hole?

2. Imagine you're falling into a black hole, all the way into the singularity. What do you see as you look "outward" along your trip?

Curved Spacetime and Black Holes
Lecture 18—Transcript

In the last lecture, we talked about special relativity, the theory that replaces Newtonian absolute space and absolute time with a single four-dimensional spacetime, but still that spacetime is there. It's rigid, it's not changing even though it's not absolute space and absolute time in some sense it is a fixed spacetime. In this lecture, we're going to talk about general relativity. This is Einstein's theory of gravity where spacetime can actually change. It can move, it can bend and twist, and that is what gives rise to the force that we call gravity.

It is interesting to think about the different ways in which special relativity and general relativity came about. Special relativity was completed in 1905 and it's certainly Einstein who gets the credit for putting the capstone on special relativity, for putting it in its place, to understand how it really worked. It was not a single person's effort. It was actually the culmination of a group effort that took many, many years. People like Poincaré, Fitzgerald, Lorentz, Minkowski, many different physicists put their ideas together to invent what we now call special relativity.

On the other hand, general relativity, the theory of gravity is really Einstein's alone. General relativity is the reason that Einstein is Einstein. That was his tremendous achievement that was more or less singular. Historians debate whether or not we would even have general relativity if Einstein hadn't come along. We probably would, but it might have taken a lot longer to get there. Unlike special relativity, general relativity is actually a specific theory. It's a little bit tricky to get the names right because special relativity allows for many, many different theoretical ways of thinking. General relativity is a particular implementation of how gravity works. Special relativity is basically not a theory of physics itself, it's a backdrop, it's a meta-theory. Special relativity is a way of constructing, in particular, physical theories.

For example, quantum field theory, which is the way that we think nature works in a fundamental level, is based on special relativity. Quantum field theory is what you get when you take quantum mechanics and try to marry it to special relativity. Special relativity is this stage on which other theories play whether it is electromagnetism or the strong and weak nuclear force or whatever new theories you would like to invent. General relativity, on the other hand, comes from the fact that there is a force of nature that we all know about called gravity and certainly whatever framework you have for talking about your other theories gravity better fit into it.

Einstein knew this. He was actually inspired by electromagnetism when he invented special relativity because the symmetries of electromagnetism are the symmetries that make special relativity go. After he completed that, he was fascinated by gravity, he knew that gravity existed. He knew that the thing to do in special relativity is to take a theory and make it special relativistic to take a theory that already exists and make it compatible with the ground rules of special relativity. He tried to do that with gravity. He tried to make a theory of gravity that would fit in with special relativity.

But, it didn't work. What he was that you could write things down, but none of them fit what we know gravity to do. We have a very successful theory of gravity in Newtonian gravity. To try to recapture that success, in the context of special relativity, turned out to be very, very difficult. That's what ultimately led Einstein to a dramatic departure from special relativity. The idea of general relativity, the theory of gravity that Einstein completed in 1916, is that gravity is not a force sitting on top of spacetime. In electromagnetism, which was the paradigm for how to work in physics at the time, there are fields. There is the electric field, the magnetic field, and they move through spacetime. It was Einstein who figured out that gravity should be different, that gravity would be the stage itself, not just an actor on the stage.

Einstein's point of view changed how we think about spacetime. Even though special relativity replaced Newtonian absolute space and absolute time, spacetime itself was still fixed. General relativity, according to Einstein, says that spacetime itself can evolve, it can change. That's going to be extremely important for our purposes where we want to understand the past hypothesis. We want to understand gravity and cosmology, what was going on near the Big Bang. If spacetime itself can change, that will be crucial to understanding why the Big Bang was the way it was.

The crucial insight that made Einstein know that Newtonian gravity was not supposed to be built into special relativity, but we needed something completely different, is something called the Principle of Equivalence. This is one of Einstein's famous thought experiments. Without going outside, Einstein is able to come up with some great truth of nature. What Einstein realized was that gravity was different than all the other forces of nature because it is universal. If you have electromagnetism, for example, if I have an electron in an electric field that gets pushed in some direction. If I have a proton in the same magnetic field, the electron has a negative charge, the proton has a positive charge, the proton gets pushed in the other direction. If I have a neutron that doesn't have charge at all, it doesn't get pushed at all by the electromagnetic field.

Electromagnetism is not universal. It acts on different particles in different ways. But gravity is universal. Put that electron, that proton, that neutron in a gravitational field and they will all fall down and they will all fall down at the same rate. That was shown by Galileo many, many years ago. The Principle of Equivalence is Einstein's way of formalizing this idea. The equivalence is between gravity and acceleration. Imagine that you are in a sealed room where you cannot see outside, you feel the gravitational force holding you to the ground and you do some physics experiments.

You will notice that if you drop different objects they fall at different rates. You can also do non-gravitational physics experiments. You can measure the frequency of some radiation from an atom or you could measure the charge on the electron. What Einstein says is that if you also imagine doing those same experiments in zero gravitational field, but in a rocket ship that is accelerating, you will get exactly the same answers. Think about an accelerating rocket ship, it pushes you in the direction in which it's accelerating. You are pulled to the ground, effectively, just like you are when there's a gravitational field beneath you. Einstein says that it's not just a similarity, it is an equivalence. You can do all those experiments in the rocket ship that is accelerating; you will get exactly the same answers as if you did them in a gravitational field.

Gravity is impossible to detect in a local region of spacetime. There is no experiment you can do to say whether or not you are in the field of gravity or whether or not you are in a rocket. This is completely different than electromagnetism. Just put an electron and a proton in there and you can immediately tell whether there is an electrical field. What does this mean? It's one thing to do the thought experiment. Anyone can do that. It takes Einstein to figure out what the consequences of this thought experiment really are. What are the implications of this idea?

What Einstein says is that if gravity is universal, if you can't really do an experiment to tell that there is gravity then maybe gravity isn't a field on top of spacetime, maybe it is a feature of spacetime itself. That's one little leap that he made. The next leap was yes, it is a feature of spacetime, in particular, it is the curvature of spacetime. Einstein imagined that spacetime, just like space, can have a geometry. It doesn't need to be flat, it doesn't need to be Euclidian, it doesn't need to be the case that you have a triangle and add up the angles in between, all of the angles inside the triangle add up to 180 degrees. That doesn't need to be the case. Einstein says that matter and energy can bend spacetime and that bending is what we call gravity.

Think about regular geometry. This is Euclidian geometry. It's the same geometry that we are taught in high school and was taught by the ancient Greeks. There are various principles of Euclidian geometry. For example, if you have parallel lines you start them out parallel, you follow them for an infinite distance, they will always stay precisely the same distance apart. That is what it means to be parallel. But, as mathematicians thought about nature of geometry over the years, they realized that that was actually not sort of an unavoidable part of geometry. This idea that parallel lines stay the same distance apart is an assumption and we can make other assumptions. It is not even that dramatic, we can do it a very down to earth way.

Think about our lives here on the surface of the Earth. We live on a sphere; a sphere is not a plane. Here on a sphere if you start two parallel lines they will eventually come together. That's a different kind of geometry. It's not violating the rules, it's just a different set of rules. There can also be geometries that are shaped like a saddle where if you start parallel lines they will move apart from each other. There are actually different ways to do geometry. There is Euclid, which is kind of flat geometry, it's the geometry of a tabletop or ordinary three-dimensional space. There is also curved geometry.

Einstein said maybe spacetime doesn't obey the rules of Euclid, and to be honest I have no idea what inspired him to say that. It's just one of those brilliant moments in the history of science where someone comes up with a really killer idea. Einstein's saying maybe the geometry of spacetime is curved is one of those killer moments. Einstein's way of thinking about gravity was usually illustrated by a rubber sheet. Imagine you have a rubber sheet that you put a bowling ball in the middle of, so the bowling ball distorts the shape of the rubber, it pulls it down. Now, I have to say because people always complain about this, this is just an analogy, this is not how gravity really works because of course the bowling ball is pulled because it is sitting in a gravitational field.

The analogy is trying to tell you that the effect of this bending of the rubber sheet is analogous to the effect of gravity. Once you put that bowling ball in the rubber sheet it bends it in a certain way and now if you imagine rolling a marble on the sheet without the bowling ball the marble would just roll in a straight line because the sheet would be flat. It would be Euclidian geometry. With the bowling ball, now the marble goes around the ball. It is distorted by the distorting twist in the rubber sheet. The fact is that the geometry of spacetime can give rise to what we call gravity. In general relativity, just like in special relativity there are light cones at every point.

If you take an event in spacetime and you follow the paths of all hypothetical light rays that could leave that event or could come into that event from the past, that still forms a light cone. But, because spacetime is curved, because as a nontrivial geometry those light cones can bend and twist. It is not quite as simple as the examples that are often given of parallel lines moving apart or coming together because those are examples where the curvature is constant throughout the world. On the surface of the Earth, in the idealization where the Earth is a perfect sphere, there is just one kind of curvature everywhere on the surface of the globe. But, in general relativity the curvature can be big some places, small other places, it can be positive some places, negative other places. It could even be positive in some directions, negative in some directions so initially parallel lines can do all sorts of crazy things.

The point is that these light cones that formed the structure of spacetime and special relativity still do that in general relativity, but now they can move around. They can change, they can twist, they can bend. It's literally that distortion of the light cones here in the solar system caused by the gravitational field of the Sun that makes the Earth orbit around rather than just moving on a straight line. One way of thinking about general relativity is to think that particles and that includes the Earth, anything that is not affected by forces other than gravity are doing their best to move on straight lines. That is what objects want to do. They want to move on straight lines, but because spacetime itself is curved there are no straight lines in the curved geometry of the world. That is what gives rise to what we call gravity. That's why the moon goes around the Earth, the Earth goes around the Sun. They're trying their best to move on straight lines, but spacetime itself is curved.

This curvature can obviously get very dramatic. If what we call the gravitational field becomes very, very strong the light cones can tilt so much, spacetime can be affected so much by gravity that you make what we call black holes. Black holes, I am sure you have heard of, they're regions of the universe where gravity is so strong that light itself cannot escape. But now that we know what a light cone is, we know a little bit about gravity and general relativity we can understand that a little bit better. The conventional way of thinking would be well just like the gravitational field is stronger on Jupiter or the Sun than it is here on Earth because they're more massive, if you have a lot of mass the force that is pulling on you becomes so strong that light cannot escape.

That's kind of true. It is good enough to give you a perfectly acceptable way of thinking about black holes, but you can do better if you think about the light cones. Remember the light cones are telling you where light can go and because you cannot travel faster than the speed of light the light cones are telling you where you can go. At any one point in your lifetime, you are in a vent. There's a light cone moving forward into spacetime away from you and the rules of relativity say that you need to stay inside that light cone. That is your future, so the way of thinking about a black hole is that the gravity is so strong that spacetime has become curved in such a way that the light cones tilt over and the only thing that you can do while moving slower than the speed of light is to move into the black hole.

The light cones are tilted toward the center of the black hole, you can't escape from the black hole without actually moving faster than light and since you can't do that the black hole is a place you can go in, but you can never come out. Inside the black hole, of course, is a place where the curvature of spacetime becomes infinite. This is what we call the singularity. You have probably heard about a singularity as well. The gravitational field is so strong that everything shrinks to an infinite density, the curvature of spacetime is infinite, our equations blow up and we call that a singularity.

There is one subtle feature that isn't usually gotten across which is that the singularity is not a place. We talk about it and we draw pictures of it as if the singularity is at the center of a black hole, but really the singularity is a moment in time. It is in the future. When you enter the black hole, when you are staying inside your light cones, you are forced to hit the singularity because everything inside your light cone becomes focused onto that singularity. You can no more avoid hitting the singularity than you can avoid hitting tomorrow.

In fact, remember we talked about the fact that the more you move in fast trajectories through the world the less time elapses on your clock. If you're inside a black hole and you try to escape, if you try to fire your rocket engines to move away from the singularity, guess what, you will hit the singularity faster because less time will elapse for you as you age forward in time. You don't travel toward the singularity, you age into it. At that singularity, at that moment of time, time comes to an end. Past that, we can't say what happens. In general relativity all we can say is that it is a boundary. It is the edge of spacetime itself. General relativity might not be the final word when you are inside a black hole, but right now that's the best we can do. A singularity is a place that time ends.

The singularities are really the parts of general relativity that are the most dramatic. It's general relativity predicting its own downfall. The theory says there is a place you can go where the theory breaks down. But, when you fall into the black hole and you cross the boundary between outside the black hole and inside, general relativity works perfectly well. That boundary is called the event horizon. It's called a horizon because it's the place where once you go in you can't come out. You can never escape back to the rest of the world. The light cones won't let you. They're all forcing you into the singularity.

However, the horizon itself is just space and time. There's nothing there, there's no wall. It's not like the surface of the Earth or the surface of another planet where there is a boundary that says here is outside the planet, there is inside the planet. A black hole is just a region of space. It is not a solid object. If you fall into the black hole, if you pass by the event horizon you don't even notice. If it is a big enough black hole you wouldn't even be aware that your future is doomed that you are going to hit that singularity and you can never get out. There's no sign post at the event horizon saying that it is a one-way trip.

Depending on how big the black hole is the actual events of your voyage can be different. If the black hole has the same mass of the Sun, if it's a solar mass black hole the time it takes you to go from the event horizon to the singularity is about 1 millionth of a second so you won't get a lot of time to enjoy the ride. Along the way, remember you are entering a region of very strong gravitational fields and those gravitational fields pull you in some directions and push you in other directions. As you are falling in, if you fall feet first your feet are a little bit closer to the black hole than your head it so your feet feel a stronger gravitational field.

The transverse directions, your sideways size is actually squeezed together so your feet are pulled apart from your head, you are made thinner and thinner in the other direction. The technical terms for this is called spaghettification. That's what happens if you fall into a black hole. It's that you are ripped to pieces by the tidal forces of gravity according to the rules of spaghettification. It's not a pretty sight, no one should fall into a black hole. In particular, don't believe what *Star Trek* and other science fiction movies tell you that you go into the black hole and you come out somewhere else in the universe. That is not a black hole, that's a worm hole. We will be talking about the possibility of a worm hole. A worm hole, maybe you can travel through it. In a black hole you will just cease to exist so it's not a good idea.

These black holes are not completely hypothetical. We believe that we have extremely strong evidence for real black holes in the real universe. You can't see the black hole itself. It's more or less black, but the black hole affects the gravitational field around it so strongly that other forms of matter near the black hole react in very noticeable ways. Gas and dust falls into the black hole, it heats up because it keeps banging into other particles and that hot gas and dust gives off X-rays and gamma rays and other forms of radiation. We can detect that and infer that there is a black hole there.

It turns out that in the real world there seems to be several different kinds of black holes, not kinds intrinsically, but coming from different places and therefore having slightly different properties. The most famous kind of black hole comes from the collapse of a massive star. A star is actually held up by the heat energy inside. At the center of the Sun, there is nuclear fusion going on that is emitting radiation. This radiation heats up the gas and plasma around it and the pressure of that plasma is what keeps the Sun in its shape. The Earth is kept in a shape because the atoms and molecules of the Earth take up space, but that's not true in a star. In a star it is just the atoms are moving around, but bumping into each other. Someday that fuel will be exhausted, you will use up all the nuclear energy you had and then the star will collapse.

It will typically collapse to a white dwarf where again the individual electrons and protons and neutrons take up space, but if it's heavy enough, the collapse is past that, the electrons and protons join together to make neutrons and you're left with nothing but neutrons. Now you have a neutron star. We think that if you have an even more massive situation, if you have a very, very heavy neutron star it will continue to collapse all the way to a black hole. If you have a star that let's say has 20 times the mass of the Sun it will collapse and make a black hole that is about three times the mass of the Sun. That's because in those processes of collapsing it ejects a whole bunch of matter into the surrounding space. We think that we have seen these massive star black holes because they often have binary companions. You have the black hole, you have another star next to it and that other star can leak onto the black hole, give off X-rays, and we can detect it.

Even better is the super massive black hole at the center of our galaxy. Cosmologists believe that most galaxies in the universe have giant black holes at the center. Whereas a stellar black hole might be three or five or ten times the mass of the Sun, a super massive black hole is a million times the mass of the Sun or more, all the way to a billion times the mass of the Sun. We think that the early form of these black holes are what power quasars in the distant universe and we think that our Milky Way galaxy, the center of which is in the constellation of Sagittarius, has a giant black hole in it. This is not just speculation. We can see stars move around an invisible object in the center of the galaxy. You can take movies of them over the courses of several years, you discover that there must be something incredibly massive and incredibly tiny relatively speaking to make the stars move in these very tight, fast orbits.

You can infer that our black hole in the Milky Way has 4.1 million times the mass of the Sun and it's spinning extremely rapidly. There are other kinds of black holes that you can get. There are middle weight black holes that might be thousands or hundreds of thousands of times the mass of the Sun. There are primordial black holes that maybe weigh grams from the very, very early universe. You can even imagine making black holes yourself at the large Hadron collider. There are physicists who have speculated that if you smash protons together at these very, very high energies gravity might be stronger than we think and you could actually make a little black hole.

If you did the world would not end, the universe or the Earth would not be destroyed. In fact, what would happen is that black hole would evaporate very, very quickly. The kinds of collisions that you make at the large Hadron collider are made all the time in outer space so we are not doing anything the universe hasn't done already. It's a very speculative idea. The odds are small, but it's actually true, but it's fascinating to think that we could actually make a black hole in the lab.

The completely separate consequence of general relativity because space and time are curved, space itself can expand. What you call the size of the universe can change over time. Einstein realized this in the very early days of general relativity when his mind turned to cosmology and he was trying to model the universe as a uniform distribution of matter. Again, thinking like a physicist Einstein asked what is the simplest possible universe I could have. It's one where they are the same density of stuff throughout and he thought this would be a good model for what the astronomers told him the real universe was like. Back in these days in the early part of the 20th century, we thought that our Milky Way galaxy was the universe. When you look at the Milky Way it's not changing very much over human time scales, it's essentially static.

Einstein tried to describe a static universe. If he had done that in Newtonian gravity it could have worked. The answer is actually a little ambiguous. It's sort of a mathematical puzzle, what would happen? But in general relativity the answer was completely unambiguous. If you had a uniform distribution of stuff filling the universe, the universe must either be expanding or contracting, it could not be static. To Einstein this was a puzzle. He had what his astronomer friends said were data, that the universe was static. He had his theory which he loved very much that said the universe would either be expanding or contracting so he had a chance to write a paper saying I don't believe the astronomers, I predict the universe is either expanding or contracting, but he didn't. Instead he blinked and he tried to change his theory to find a static solution. He ended up introducing the idea that there was energy filling empty space itself, what he called the cosmological constant or we would call vacuum energy.

It turns out that he was right to think that that is a possibility. We think that it is very possible there is a cosmological constant in the real world, but he was wrong to think that it balanced the force of matter and gave you a static solution to cosmology. He was very regretful later in life that he missed the chance to predict that we live in an expanding universe. Back in the 1920s, Edwin Hubble actually discovered that the universe is expanding which we will talk about in a later lecture. That was the prediction that Einstein should've made but he missed his chance. If the universe is expanding, which it is, we can run the movie forward and the universe will get more and more dilute, galaxies will move away from each other, but we can also play the movie backward. General relativity is a reversible theory. We can ask where we came from.

What we find is that in the past things were closer together and therefore the gravitational fields were stronger and if you keep going back into the past, the gravitational field just gets very, very strong. Winding the movie of the universe backward in time is very, very similar to following the formation of a black hole forward in time. The universe is like the time reverse of a black hole. Inside the black hole there is a singularity in your future because all the matter collects in one place. In the universe there is a singularity in our past, that is what we call the Big Bang. This was something that was theoretically predicted. People didn't like it. There was a small number of people who took it seriously and that's one of the consequences of this idea they called the Big Bang. It took many decades before we actually collected enough data to say that that idea was on the right track.

What this means is that there could be a beginning to the universe, that just like time ends inside a black hole the singularity in the future time ends at the Big Bang and the singularity in our past. Now we don't know for sure that that is right and we will talk about this in great detail. General relativity doesn't predict that there's a Big Bang, general relativity predicts its own downfall. It predicts that there's a singularity, but singularities aren't real, they are places where the equations fail to give us an answer. Nevertheless something is going on at that moment that we called the Big Bang, either it is something like the Big Bang or it's something else that we need to discover. For our purposes the relevant fact is that it completely changes the question of the past hypothesis.

Remember that Boltzmann, when he struggled to explain the past hypothesis, thought about an eternal universe with random fluctuations that turned out not to work, that makes predictions that turn out not to be true, but once general relativity comes on the scene there's a different option. You can simply say the universe doesn't last forever. Maybe there is a true initial condition at the Big Bang, a moment that we now know to be 13.7 billion years ago. That opens up a different door that we didn't have available before.

Now that there is general relativity, now that space and time are on the table as things that can change it might be that the Big Bang is a true initial condition with a very low entropy. There's still the question why was the entropy very, very low at that moment? But it is a different kind of question. It's not why did the universe go through a phase where the entropy was very, very low, it is why did the universe begin with low entropy? That may or may not be the answer. It is certainly one of the alternatives we will have to take very seriously.

Time Travel
Lecture 19

In this lecture, we'll look at the question: Is time travel possible? Like many of the other issues we've discussed, the answer gives us a kind of good news/bad news situation. The good news is that general relativity, Einstein's theory of gravity in which space and time are flexible, allows us to talk about the possibility of time travel in a scientific way. The bad news is that it's probably not possible, although interestingly, we can't say that it's definitely not possible.

Traveling to the Past?
- We travel through time every day, but we travel to the future. Special relativity allows us to talk about changing the rate at which we travel to the future. Because the amount of time that passes depends on our trajectory through the world, we can actually get to the future faster but not slower.

- The real question most people are interested in, however, is: Can we travel to the past? If Newton had been right that space and time were absolute, then the answer would be no. The laws according to Newton are that there are separate moments in the history of the universe, and you march forward through them; you cannot help but do that.

- In special relativity, the answer is also no. Special relativity replaces the Newtonian division of space and time with spacetime structured by light cones. You are forced to move into the interior of your light cone—that's the same as saying that you must move slower than light—and the light cones point into the future. They don't go into the past.

- In general relativity, the answer is probably not, but at least we can wonder whether we could tilt the light cones enough to travel into our personal futures and nevertheless end up in the past. In general relativity, a time machine would be a twisting of light cones, focusing them back on themselves so far that we could go into the past while still moving forward in time.

A Realistic Time Machine

- If time travel were possible, it would not be what we generally see in science fiction movies. It would not involve some kind of dematerialization in the present and rematerialization in a different time. Real time travel would be a journey through spacetime, and a true time machine would be some vehicle that moves you through space and time but in a spacetime that allows you to visit your past.

- The most realistic version of time travel we can imagine is not about building a machine but about building spacetime. Our goal as potential time travelers is to warp spacetime so much that we can personally move forward in time and nevertheless visit ourselves in the past.

- To think about spacetime, we need to think locally, in this case, about what is happening to you.

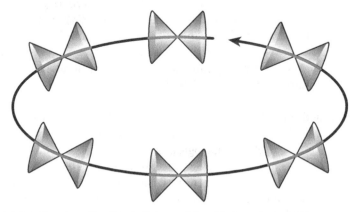

The trick to time travel isn't to build a machine; it's to warp spacetime so much that we can keep moving forward in time yet come back to where we started, that is, bending our light cones so much that we can move in a circle—a closed timelike curve.

- What's happening to you is that you're growing older; you are moving locally forward in time. In other words, you are staying inside your light cone.

- As we said, general relativity tells us that light cones can be twisted. Thus, we can imagine a light cone twisting so that you could locally move forward but visit your past self because the light cone had closed in on itself.

- This formation is called a closed timelike curve. A timelike curve is simply a path through spacetime that is moving slower than the speed of light. We ordinarily move on open timelike curves, but a time machine would be a closed timelike curve.

• There are spacetimes that allow closed timelike curves, but is that our universe? If we started in a universe that didn't have closed timelike curves, could we create them? Could we warp space and time so much that we were able to visit our own past? These questions remain open.

Kurt Gödel's Version of Spacetime

• The most famous example of a kind of spacetime that has the possibility of time travel built into it comes from Kurt Gödel, a German mathematician. Gödel dabbled in general relativity and was curious about Laplacian determinism, just as we are. He wondered whether it was possible to start with one moment in time, evolve it forward, and then evolve it backward.

• Neither quantum mechanics nor special relativity gets in the way of doing that, but what about general relativity?
  - In Gödel's cosmological answer to this question, instead of expanding, the universe is rotating. The stuff that sits inside Gödel's hypothetical universe is vacuum energy—the cosmological constant energy that is inherent in space itself—and swirling matter particles. The energy and particles cause the curvature of spacetime to be light cones that are tilting gradually as we travel through the universe.

- Every event in this universe sits on a closed timelike curve. Everywhere you start, you can travel through some trajectory in spacetime and eventually visit your past.

- It's not difficult to write solutions to Einstein's equation in general relativity that look like time machines, such as an infinite rotating cylinder or cosmic strings, but all these examples in Gödel's universe have the property that they are infinitely large.
    - If we ask whether we can start with a universe that doesn't have time travel built in and create a situation that resembles any of these, the answer seems to be no.
    - The naïve solutions require an infinite amount of energy. If we try to make a finite cylinder or finite cosmic strings or a finite amount of dust rotating, we don't seem to get closed timelike curves.
    - These solutions are curiosities, but they are not realistic ways to go about engineering a time machine.

Wormholes

- The most well-known way to construct a time machine in a finite region of space is to use wormholes. A wormhole is a tube through spacetime. It's as if you enter some sphere locally and you are spit out somewhere else arbitrarily far away. Wormholes can connect different regions of spacetime, and you can use them to travel in much shorter time periods than if you went the ordinary route.

- If it is possible to build a wormhole connecting two different regions of space, it is also possible to build a wormhole that connects two different moments in time. Again, it's relatively easy to write down the equations to qualify this as a solution to Einstein's theory of relativity.

- The problem here is that wormholes involve physics that we don't think works in our world. In particular, wormholes collapse instantly into black holes. To get around this problem, we need

Wormholes connect two distant parts of space; although this concept cannot be rendered graphically with complete accuracy, the physical distance could be much shorter than the usual distance between the two openings.

negative energy—something that supplies us with a repulsive gravitational force.

- o Everything we know about in the universe has the gravitational effect of pulling things toward it—positive energy—but to keep a wormhole from collapsing, negative energy is needed to push it apart.

- o The mathematical physicist and cosmologist Frank Tipler, as well as Stephen Hawking, have posited that manipulating matter and energy in such a way as to create any form of closed timelike curve will inevitably create some sort of singularity. The density of curvature and energy in the universe would go to infinity somewhere.

The Paradoxes of Time Travel

- The grandfather paradox is one of the most famous problems that arises from the idea of time travel: What stops me from traveling backward in time and killing my grandparents before they ever met so that neither my parents nor I were born? In that situation, who committed the murders?

- One problem with this scenario is that we can't pick the time we travel back to. The entrance to a wormhole is like a portal; you go in one end, and you come out somewhere else and some when else.

It's also true that if you can go backward, then someone else can come forward.

- Logic must still work even if there is time travel. You cannot kill your grandparents and then be born to go back in time and kill your grandparents. You can't change the present moment because you are in the present moment and you know what the present moment has. It has you, for example, so nothing you do can truly prevent you from coming into existence.

- If time travel were possible, the most likely scenario is that even if you made it into the past, something would prevent you from changing things that really happened.

- What is truly bothering us here is the arrow of time, which is absolutely built into how we think about the past, present, and future. As we said, we believe that we can make choices that affect the future but not choices that affect the past. The past is tied down in our epistemic knowledge because of the past hypothesis. If you have a memory of something happening and your memory is valid, then that is what happened and you can't change it.

- If you put the possibility of time travel into this situation, then your personal future becomes mixed up with the past of the universe. You personally always age into your future light cone, but you go off in a spaceship, zoom around a closed timelike curve, and come to the past. Now, something that you thought was fixed—the past—gets mixed up with something you thought was alterable—your personal future.

- It's likely that time travel isn't possible, but the many-worlds interpretation of quantum mechanics offers a tiny loophole to the impossibility of time travel.
    - It is conceivable that if we had a closed timelike curve, we could imagine going back into the past, truly changing the past, and by doing so, bringing into existence a new world, a new branch of the wave function of quantum mechanics.

- You could travel back in time from one branch of the wave function, in which your grandparents did exist and you were born, into another branch of the wave function, in which your grandparents were killed and you were never born.

Conceptual Implications of Time Travel
- For our purposes, the most significant implication of time travel is that it would destroy the universality of the arrow of time.

- When we have the possibility of time travel, we no longer have Laplace's demon. We cannot slice the universe into moments of time. The moments of time intersect with each other in complicated ways, so that we cannot record the data of the universe at any one moment and imagine running it forward and backward.

Suggested Reading

Carroll, *From Eternity to Here*, chapter 6.

Thorne, *Black Holes and Time Warps*.

Questions to Consider

1. Think of movies or stories that make use of time travel. Are they logically consistent? Can you think of an interesting plot for a movie that treats time travel seriously?

2. Stephen Hawking has joked that time travel must be impossible because we haven't been invaded by tourists from the future. Is this a good argument, as well as a clever joke?

Time Travel
Lecture 19—Transcript

In a previous lecture, when we were talking about the perception of time, I mentioned that the first question people have when you start talking about the mysteries of time is why am I always late? The second question people always have is can I travel through time, is time travel possible? This is the lecture in which we finally answer that question. You wouldn't actually be addressing the mysteries of time unless you at least talked about time travel a little bit.

Like many of our issues, there will be a good news, bad news situation. The good news is that general relativity, Einstein's theory of gravity that says that space and time are flexible, allows us to talk about the possibility of time travel in a scientific way. It's not just fantasizing, we can actually think about what time travel would really be like. The bad news as you might guess is that it is probably not possible. It's interesting that we can't say it's definitely not possible. We actually as scientist don't know the final answer. Time travel is on the table as something that's respectable to think about. The evidence is pointing against it so we probably don't need to worry about the paradoxes that time travel brings up.

The first obvious point is that, of course, we travel through time every day, but we travel to the future. Twenty-four hours ago I started traveling through time and a day later here I am. You can't help traveling through time in that very simple way. Again, as we talked about way back when it's debatable whether the word travel is the right way to think about this. I am now, I was yesterday, I will be tomorrow. That is all true, but if we want to think about a little me that is taking a journey through space and time I am traveling to the future. That is what people do.

A little bit more sophisticated says that special relativity allows us to talk about changing the rate at which we travel to the future. Because special relativity tells us that the amount of time that passes depends on our trajectory through the world, we can actually get to the future faster. We can hop in a spaceship, we can zoom around in circles if you want near the speed of light because this is a thought experiment, this is not legitimately possible right now, but if you were to do that, to you let's say a week passes back here on Earth, an arbitrarily long amount of time can pass, a year, a century, a millennium, if you travel close enough to the speed of light your personal clock can read much less than the clocks of the people who stayed back here on Earth.

On the other hand, this is not crucially fascinating. If you just went into suspended animation, the same thing would happen. It is probably easier to put a human being in suspended animation than it is to build a spaceship that can go in circles close to the speed of light for a long time. But still, going into the future is something we do and something we can do at different rates of time. We cannot go into the future more slowly. Remember if you zoom off and zoom back you always experience less time than if you just stayed still. Going into the future is something we are going to be forced to do. The real question is can we travel to the past, going to the future sounds like fun, but only if we can then come back and tell our friends what we saw.

When it comes to traveling to the past, the answer whether you can do that depends on your theoretical background. We haven't built a time machine yet so it's still a theoretical question. If Newton had been right, if Newtonian absolute space and time were the way we lived our lives, the answer would be simple. It's no. The laws according to Newton are that you move forward in time. There are separate moments in the history of the universe, which are moments of individual times, and you march forward through them and you cannot help but do that.

In special relativity, the answer is also no. Special relativity replaces the division of space and time that Newton gave spacetime into light cones, the idea that from any one event they point out all of these hypothetical trajectories that light would travel on and that tells you how fast you can move through spacetime. You are forced to move into the interior of your light cone; that's the same as saying that you must move slower than light and the light cones point into the future. They don't turn around and go into the past. In special relativity you cannot ever visit the past.

In general relativity, the answer becomes maybe or probably not to be more honest. But, it at least lets you wonder whether we could tilt the light cones enough to travel into our personal futures and nevertheless end up in the past. That is what a time machine would be like in general relativity, it would be a twisting of light cones, a focusing back on themselves so that as far as your personal wristwatch is concerned, you are moving forward in time, but as far as the universe is concerned, you have gone into the past. That is a very different idea of what time travel would be than we'd get from cheap science fiction stories. Even the cheap science fiction stories can be very intellectually challenging, but they rarely take time travel seriously.

What we're used to seeing in science fiction stories and movies is some sort of machine that we get into. This is our time machine. It could look like *H. G. Wells* or in *Dr. Who* it looks like a telephone box or it could look like a DeLorean in *Back to the Future*. The point is that you climb into the time machine and you push some buttons, lights flash, and your time machine disappears. It dematerializes and then it rematerializes at some other time, it pops back into existence in the past or the future. That is completely crazy. That is not what time travel would be like.

It's still more respectable than the even crazier science fiction stories that just have your consciousness travel, so you personally inhabit the body of your former self or your future self, that's right out of bounds, but even the dematerializing and rematerializing is not how time travel would work. To understand why this is true, let's think of a simple analogy. Think of a very primitive tribe that lives on an island so they have no technology whatsoever. They hunt and gather, but they cannot even build a boat. They see however in the distance there is another island across the ocean, they can still see it, they would like to visit, but they don't even have the materials or the know-how to build a boat to get there. They think about getting to this other island, what would it be like to travel through space to get to this other island and in their thought experiments they imagine it would be like teleportation, they would somehow disappear from where they are now and they would reappear on the other island.

Now you and I know that travelling from one island to another is possible. It does not involve teleportation; the reality is much more mundane. The reality involves you get in a boat and you actually traverse all of the points in between the island where you start and the island where you end up. The point is that if time travel were possible, it would be like that, it would not involve some weird kind of dematerialization which is a whole other next to impossible technological challenge. Real time travel doesn't skip over the intervening part of spacetime. Real time travel is a journey through spacetime; a true time machine would not be a telephone box with flashing lights and funny noises. It would be a spaceship. It would be some vehicle that moves you through space and time, but in a spacetime that allowed you to go visit your past.

In the most realistic version of time travel we can imagine, it's not about building a machine it's about building the spacetime. That's our goal as potential time travelers to warp spacetime so much that you can personally be moving forward in time and nevertheless visit yourself in the past. You might wonder, how is this possible? Remember my advice in thinking about spacetime was to think locally, to not try to compare what's going on here to what's going on faraway at the same time. Relativity says that's not the right way to think, think about what is happening to you.

What's happening to you is that you're growing older, your clock is ticking, you are moving locally forward in time, and the translation of that into jargon is you are staying inside your light cone. At any one event you have a place you can go which is slower than the speed of light. But, general relativity says that light cones can tilt, they can be twisted, that energy and momentum, and mass can bend spacetime itself. So, we can imagine that light cones actually twist around on themselves so that you can locally move forward in a light cone. You are always moving slower than the speed of light, but nevertheless you can come visit your past self because the light cones themselves have closed in on each other.

The jargon term for this is a closed timelike curve, a timelike curve is just a path through spacetime that is moving slower than the speed of light so you and I move on timelike curves and we ordinarily move on open timelike curves, just going to the future. A time machine is a closed timelike curve. We locally travel slower than light, but our path to the universe eventually visits itself in the past. The only reason I am even introducing this jargon is because, as a physicist, I use this jargon all the time. I cannot help but use those words so it would be more helpful for the rest of the lecture to think about closed timelike curves, Can we build them? Can we make a spacetime that has closed timelike curves in it?

The short answer is probably not, as I alluded to before, but we're not sure. There's a process we go through in our theories to ask the question are there solutions to general relativity. Are there spacetimes that fit into Einstein's equation that tells us how matter and energy bend space and time that have closed timelike curves; that have the possibility of time travel inherent in them? The simple answer is yes, there absolutely are solutions to Einstein's equation which have time machines in them, which would let you go backward in time so then the next level question is, is that our universe. If we started in a universe that didn't have closed timelike curves could we create them? Could we, essentially, build a time machine? Could we warp space and time so much that we were able to visit our own past?

This is the open question, we don't know whether or not that's possible or not. We try and when we try we seem to run into obstacles. There always seems to be problems when we try to create a closed timelike curve to build a time machine. An interestingly there are two different types of time machines we can imagine building and two different types of obstacles that get in our way. One kind of time machine is what you might call built into the universe. Essentially this kind of closed timelike curve takes advantage of the whole universe all at once. You can write down on your piece of paper a solution to Einstein's equation, a particular kind of spacetime with a certain kind of curvature of spacetime that has the possibility of time travel build in from the start. It was not created it was just always there.

The most famous example comes from Kurt Gödel, a mathematician and magician who was from Germany, but he lived for a long time in Princeton at the Institute for Advanced Study where he was a colleague of Einstein's. Gödel is most famous for his incompleteness theorem that says that no sufficiently powerful logical system has statements that are only provable and true, there are always some statements that are either unprovable and true or provable, but not true. In his spare time, Gödel dabbled in general relativity and he was very curious about this question of Laplacian determinism, just like we are, can you start with one moment in time, evolve it forward, and then evolve it backward.

We said that quantum mechanics did not get in the way of doing that. We said that special relativity did not get in the way of doing that, but what about general relativity. Gödel came up with a solution that is very astonishing. He was one of those geniuses just like Einstein. It's hard to know how he thought of this, but he came up with a cosmological solution, but one in which the universe is not expanding. Instead of getting bigger the universe is rotating. The stuff that sits inside Gödel's hypothetical universe is vacuum energy, the cosmological constant energy that is inherent in space itself and matter particles that are swirling, that are like vortices rotating around each other. What they cause the curvature of spacetime to be is light cones that are tilting gradually as you travel through the universe.

It turns out that in Gödel's solution to general relativity every event sits on a closed timelike curve. Everywhere you start you can travel through some trajectory in spacetime and eventually visit your past. Gödel is fascinated by this solution as was Einstein. He shared it with his pal Einstein, they were both a little bit worried about it. Many, many years later when the physicists Kip Thorne and John Wheeler and Charles Misner were writing a giant textbook on general relativity, they went to talk to Gödel and what Gödel wanted to know was there any evidence that galaxies have some uniform rotation in the universe. He was worried that our real universe might have time machines build into it.

Another idea along similar lines is a rotating cylinder, so not a finite cylinder, an infinitely long cylinder and you start it rotating except of course you can't start it rotating, it would take an infinite amount of energy to rotate an infinite cylinder. But, if you just happen to have an infinite cylinder lying around, which is rotating, and it has enough energy and a fast enough rotation then long ago it was shown that as you get very, very close to the cylinder there are closed timelike curves. You can actually spiral around into the past as far as you want.

A similar scheme come up with more recently is from Richard Gott, an astrophysicist at Princeton, who talked about cosmic strings, these are much more plausible, but still similar sounding infinitely long straight tubes of energy. We think that non-infinite cosmic strings could actually be created in the early universe, but infinitely long ones are easier to solve the equations for. What Gott showed is that two cosmic strings can pass by each other and if they're passing by each other fast enough you can do a loop in the opposite sense and go backwards in time. It's not hard to write down solutions to Einstein's equation in general relativity that look like time machines, but all of these examples in Gödel's universe, the infinite rotating cylinder, Richard Gott's cosmic strings, have the property they are infinitely big.

If you ask can you make any of these, can you start with a universe that doesn't have time travel built in and create a situation that resembles any of these, for example, could you have a finite cylinder that you start spinning faster and faster, would something go wrong to stop that? The answer seems to be yes. The naïve solutions require an infinite amount of energy. If you tried to make finite ones with a finite cylinder or finite cosmic strings or a finite amount of dust rotating you don't seem to get the closed timelike curves. It seems that these solutions are curiosities, but they are not realistic ways to go about engineering a time machine.

The other option people look at are trying to make a time machine in a finite region of space. The most famous way to do this is to use wormholes. I'm not sure how famous it is. In the circles I move, this is a very famous way to try to make a time machine. A wormhole remember, it sounds like a black hole, but it's a very different thing. In a black hole you enter and you will eventually hit the singularity and be crushed. A wormhole is a tube through spacetime. It's as if you enter some sphere locally and you are spit out somewhere else arbitrarily far away. The nice thing about wormholes is they can connect different regions of spacetime and you can use them to travel in much shorter time periods than if you just went the ordinary route.

It's like a super fast spacetime subway. You go in and then you pop out very, very far away, but it doesn't take you any time to do it, or at least it takes very little time to do it. The distance from one end of the wormhole to the other through the wormhole, can be much, much less than the distance from one end to the other outside the wormhole. So, this is a very straightforward way to make a time machine because remember the different ends of the wormhole don't just connect different points in space, they connect different events in spacetime.

If it is possible to build a wormhole, a curved tube through space that connects two different regions of space, it is also possible to build a wormhole that connects two different moments in time. You could just climb in a wormhole on one side you pop out the other side, but it's many years in the past or in the future. That is again something where it's easy to write down the equations to qualify this as a solution to Einstein's theory of relativity.

The problem with this is it involves physics that we don't think works in our real world. In particular, wormholes don't want to exist in the first place. They instantly collapse and you can't get through them. You can prove that with ordinary conventional kinds of matter and energy there cannot be a wormhole that takes longer to collapse than to travel through. In other words, there's the wormhole you try to get through it, but it collapses into a black hole before you can get to the other side. There is a loophole there, you can get around that, but you need nonconventional matter and energy. By that I mean not just forms of matter and energy we haven't yet found, but you need negative energy. You need something that would supply you with a repulsive gravitational force.

Everything we know about in the universe has the gravitational effect of pulling things toward it, positive energy. You are pulled toward the Earth, you are pulled toward the Sun. To keep that wormhole from collapsing you need negative energy to push it apart. We don't know how to do that so the simple answer is maybe you just can't build a wormhole. There's a more subtle answer which says that even if you try to build a wormhole, even if you had negative energy, you would still fail. There is actually a theorem due to Frank Tipler, and an update of it from Stephen Hawking that says that if you try to make a time machine in the local region of space, if you start in a universe that has no time machines in it, no closed-timelike curves, and you push around matter and energy in just the right way to create any form of closed-timelike curve at all whether it is with a wormhole or without, you will inevitably create some sort of singularity.

The density of curvature and energy in the universe would go to infinity somewhere. Now, it's not quite enough to say therefore you can't do it. We have a belief that if you make a singularity you make a black hole, it is inside an event horizon and you can't get there, but that's not a proof. We can't quite make it into a theorem and we don't understand gravity perfectly well. Einstein's theory of general relativity is the best idea we have, but as we will discuss it doesn't play well with quantum mechanics. Who knows whether or not in quantum gravity you might actually be able to build a time machine? That is why it is difficult for us to right now say for sure whether or not time travel is truly possible.

But, let's put aside the technical, the engineering difficulties with time machines. We don't think we know how to build one from scratch, but let's imagine that we did. The real worry is not that we don't know how to build it, it's that if you built it trouble would follow, paradoxes in particular. The famous one is the grandfather paradox. What stops me from travelling backward in time, killing my grandparents before they ever met so that my parents are not born and therefore I am not born. In that case, who did the killing. If you actually had a wormhole that you could go back through that would send you a hundred years in the past, how would that really work?

There's a couple of things to note about how it might work. One is that you don't get to pick when you go back to. If you have a wormhole you have an entrance, it's like a portal, you go in one end, you come out somewhere else, and some when else, at some particular point in time. It's not like *HG Wells* where you dial the dials on your time machine and decide where you want to go. If your wormhole sends you back a hundred years in time and you go through it in 2012, you will come out in 1912. If you go through that same wormhole in 2013, you will come out in 1913. There's nothing you can do to go in in 2012 and come out in 1950 or in 2035. The wormhole tells you where you're going to go.

Another thing is, of course, if you can go then someone else can come the other way. If you can go backwards a hundred years, someone else can go forward a hundred years. If people can go into the past they can also come to the future. That's helpful in some sense. If you go backward in time through this wormhole, you will eventually get to come home.

The way to think about a possible wormhole once we have figured out how to build it is as a portal, a subway as we said, that connects to moments in time rather than to places in space. If that existed could you use it to change the past, could you use it to kill Hitler or stop J.F.K. from being assassinated or for some reason you want to kill your grandparents and not let yourself exist. The honest answer is, of course, that we don't know because we don't have any experimental data about what time travel would be like. We don't even think that it is possible, but we do still have logic and physics on our side. We can ask ourselves the question could there be a version of reality where you could travel into the past and change it to somehow make it better, if you're not quite so ambitious as to kill Hitler or save J.F.K., maybe you want to stop an embarrassing incident that happened to you at the senior prom.

There's one thing that we know that even with time travel there cannot be paradoxes. There cannot be A and not A. this is the thing that gets you into trouble when you are thinking about how time travel would work. How can you kill your grandparents and then be born to go back in time and kill your grandparents. We know for sure that logic still works even if there is time travel. You can't change the present moment because you are in the present moment and you know what the present moment has. It has you, for example, so nothing you can do can truly prevent you from coming into existence. If time travel were possible, the best way to think about it, again we don't have data, we don't know for sure, the most likely scenario is you go into the past, you attempt to kill your grandparents and even if you make it into the past something prevents you.

This is the scenario that all the good time travel movies have. In *12 Monkeys*, for example, the time travel is perfectly consistent. As they said in the TV show *Lost* whatever happened, happened. If something occurred at some particular event, some particular location in space at some particular date and time, then that thing really happened. J.F.K. really was assassinated, you really were born. You go back, you try to change the past, you will fail. What stops you? We don't know. You trip on a banana, you get sick, someone gets in your way, we don't know what will stop you, but we know you will be stopped because physics and logic are still consistent even in the presence of time travel.

Now, this tells us what is really bugging us about time travel and the answer is the arrow of time. The reason why we are talking about time travel in this lecture is not because time travel helps illuminate the question of the arrow of time, it's because the arrow of time helps illuminate the question of time travel. The arrow of time is absolutely built in to how we think about past, present, future. We discussed how we have this strong feeling that we can make choices that affect the future; that we have free will. Neuroscientists and philosophers can debate about the reality of free will, but one way or the other we have a feeling that we can make choices. We are not predestined. We can choose to have one kind of food for dinner tonight or some other kind of food.

We don't think that we can make choices that affect the past. The past is something that is tied down in our epistemic knowledge because of the past hypothesis. If you have a memory of something happening and your memory is valid then that is what happened and you can't change it. What happens if you put into this situation the possibility of time travel? Then your personal future becomes mixed up with the past of the universe. You personally always age into your future light cone, your clock keeps ticking in the future direction, but you go off in a spaceship, zoom around a closed-timelike curve and come to the past. The point is that now something that you thought was fixed, namely the past, gets mixed up with something you thought was alterable, namely your personal future.

But, the simplest way of making time travel work is to just say that the past really is fixed, just like our future really is fixed if we were Laplace's demon, if we knew the state of the universe and all the laws of physics we would know what happens in the future. Time travel makes that real. If we have a memory that is reliable of what happened at our senior prom then that really happened. If you go back and try to change it, there is predestination. You will fail, we don't know how you will fail, but we know you will not change your senior prom because we know that whatever happened, happened. We don't like that, it violates our sense of free will, but that's the price you pay when you try to mess with the arrow of time. Your personal arrow of time is doing something different than the arrow of time of the universe.

If this bugs you, if you don't like the idea that your free will is being compromised, the simplest way out is just to say that time travel probably isn't possible. You probably can't get into the past and do anything about it. If you live in a universe that doesn't have closed-timelike curves then the arrow of time is universal. You can slice the universe into moments of time. Einstein says that it is not universal, but you can still do it, and then the arrow of time moves in the same direction for everyone.

Now I have to say there's a loophole, a very tiny loophole, called quantum mechanics. Remember quantum mechanics has this problem of interpretation. We don't really know what happens when you make a quantum observation and one of the possibilities on the table is the many-worlds interpretation that says that every time I observe a quantum system there is one universe, one world in which I get one answer, another universe, another world in which I get another answer. It is conceivable that you could combine that theory with time travel. If you had a closed-timelike curve you could imagine going back into the past, truly changing the past, and by doing that bringing into existence a new world, a new branch of the wave function of quantum mechanics.

In other words, you could kill your grandparents and stop yourself from being born. But the interpretation of that would be that you had traveled from one branch of the wave function in which your grandparents did exist and you were born into another branch of the wave function in which your grandparents were killed and you were never born, but the point is that the other branch doesn't go away. If you bring into existence a new reality in which the only you in that reality is the time traveler from the future, there is no you that is ever born, then that reality exists and the old one in which your grandparents were fine and you were born also exists. So, have you really improved anything by changing the sum total of all branches of the wave function?

This is an ethics and morality question, not a physics question, I cannot provide you with the answer, but this is the kind of conundrum that you get into when you try to make time travel respectable. For our purposes, the important feature of time travel is not whether it is possible, but what its conceptual implications would be. The biggest one would be that, as we mentioned, it destroys the universality of the arrow of time. When you have the possibility of time travel you do not have Laplace's demon anymore. You cannot slice the universe, not only absolutely, but even arbitrarily, you cannot slice the universe into moments of time. The moments of time intersect with each other in complicated ways so you cannot put down the data of the universe at any one moment and imagine running it forward and backward in a reversible way. The question of the arrow of time would be a much harder question to answer if time travel were possible. Since, it is probably not we are going to go back to thinking about the real world and explaining the arrow of time there.

Black Hole Entropy
Lecture 20

Gravity was of crucial importance in the early universe, and the Big Bang clearly plays some crucially important role in understanding why the early universe had such low entropy. Unfortunately, we don't understand the Big Bang very well. We don't understand gravity at the microscopic level, but such an understanding is exactly what we need when it comes to the Big Bang and entropy. The good news for us in our quest to understand entropy is that black holes are places where gravity is incredibly important, and in this lecture, we'll see that we do know the entropy of black holes.

Classical Black Holes and Black Hole Mechanics

- In classical general relativity, black holes are actually very simple things. We can completely specify a black hole by knowing its mass, its electric charge, and how fast it's spinning. The idea that black holes possess only these properties is known as the no-hair theorem: All black holes are bald.

- We think of a black hole as a place from which nothing can emerge. It was therefore extremely surprising when, around 1970, Roger Penrose showed that energy can be extracted from a rapidly spinning black hole if it is slowed down. The mass of a black hole actually decreases as it spins more and more slowly.

- Of course, there's only a finite amount of energy that can be extracted. The black hole is spinning at some rate. We can get energy out of it, but we slow it down in the process. Once it is slowed down to the point that it's no longer spinning, there is no more useful work we can extract from the black hole.

- This is exactly the situation we saw when we were talking about a thermodynamic system approaching thermal equilibrium. Just like the gas in our piston, if it's in a low-entropy state, we can extract useful work from it, but once it reaches equilibrium, there's nothing more we can do. This idea was developed in the 1970s into the subfield of black hole mechanics.

- The idea here is that there is an analogy between the behavior of black holes and thermodynamics. In thermodynamics, we have the energy of a system; for a black hole, we have its mass, and $E = mc^2$ tells us that mass is just a type of energy, so the analogy is obvious.

- The analogy to temperature in thermodynamics is surface gravity in black holes. Thermodynamic entropy is analogous to the area of the event horizon, which increases as energy is extracted from the spinning black hole.

- The zeroth law of thermodynamics, which states that systems in thermal equilibrium will all have the same temperature, is analogous to the zeroth law of black hole mechanics, which states that the surface gravity is the same everywhere on the horizon, whether the black hole is spinning, charged, and so on.

- The first law of thermodynamics says that energy is conserved, and the first law of black hole mechanics says that mass is conserved.

- The second law of thermodynamics says that the entropy of a closed system only increases; the analogy in black holes is that the area of the event horizon only increases. In fact, this is a theorem that was proven by Stephen Hawking. The area of the event horizon only increases; its shape can be changed, but it cannot be made smaller.

- The third law of thermodynamics says that zero temperature cannot be reached; the third law of black hole mechanics says that zero surface gravity cannot be reached.

Jacob Bekenstein v. Stephen Hawking

- The set of rules in the thermodynamics regime is spookily analogous to the set in the black hole mechanics regime. This prompted a graduate student named Jacob Bekenstein to make the radical suggestion that the laws of thermodynamics and the laws of black hole mechanics aren't separate.

- Bekenstein said that there is one form of generalized thermodynamics that includes black holes in it and that what we call the area of the event horizon isn't analogous to entropy but is entropy. He further said that the entropy of a black hole is proportional to the area of its event horizon.

- Boltzmann had said that entropy is found by counting the number of microstates that are macroscopically indistinguishable. If Bekenstein was right, black holes must have many indistinguishable microstates because they have a large degree of entropy.

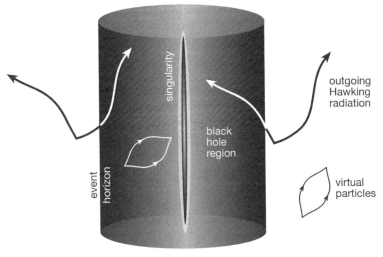

Hawking showed that black holes do radiate—in precisely the fashion to make the black hole mechanics/thermodynamics analogy perfect!

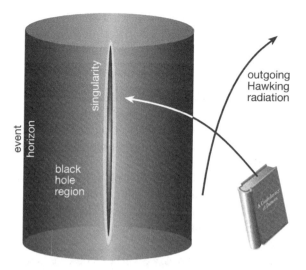

Information falls into a black hole; meanwhile, the radiation should be conveyed outward, but how can it be in two places at once?

- But in the classical general relativistic description, those microstates are nowhere to be found, just as they're nowhere to be found in thermodynamics. A black hole is specified by its mass, charge, and spin and nothing else. Every mass, charge, and spin equal black hole should be the same as every other one with the same values. How can it be that black holes secretly have a huge number of states to accommodate all the entropy they have?

- Further, thermodynamics tells us that a system with a temperature will radiate. If black holes are ordinary thermodynamic systems, then the surface gravity of a black hole should radiate, but we knew that black holes don't radiate.

• Stephen Hawking set about proving Bekenstein wrong by applying quantum mechanics to black holes. He found that in the vicinity of black holes, virtual pairs of particles could be created, one of which could fall into the black hole, but the other could escape. In this scenario, the black hole is actually losing mass to its environment because the escaping particle carries away mass, while the particle that falls into the black hole has a negative mass. The black hole gradually shrinks.

- In particular, Hawking derived a formula for the temperature of a black hole, $T = x/m$, which basically says that the temperature is inversely proportional to the mass of the black hole. He showed that the temperature of a black hole is not zero, but it is smaller for large black holes.
  - A tiny black hole gives off a furious amount of radiation, but a large black hole is very cold. A black hole with 1 solar mass has a temperature of about one-billionth of the background radiation that suffuses the universe.

  - In ordinary black holes, we will never observe Hawking radiation.

- If black holes do give off radiation, then they lose mass. As a black hole gets smaller, its mass goes down, and when it gets smaller, a black hole gets hotter. In a finite period of time, the black hole completely evaporates and disappears. The lifetime of a black hole is given by $t = yM^3$, a constant times the mass of the black hole cubed.

The Entropy of Black Holes

- If black holes radiate—if the analogy between thermodynamics and black hole mechanics is good—that means that black holes have entropy. This is our single most important handle on how entropy works when gravity is relevant.

- As Bekenstein said, the entropy of a black hole is proportional to the area of its event horizon. Why does the entropy go in conjunction with the area, not the volume? That question inspired something called the holographic principle: When quantum gravity is important, physics is no longer strictly local; there are correlations between unknown quantum states that all macroscopically look alike and exist on the horizon.

- Hawking gave us a formula for black hole entropy, and the short story is that it is huge. The entropy of a 1 million–solar mass black hole is 10^{90}. For comparison, the entropy of all the non–black hole particles in the observable universe is 10^{88}.

- We think there are something like 10^{12} or 10^{13} million-solar-mass black holes in the observable universe. This tells us that the entropy in our universe today is overwhelmingly in the form of black holes. This is a crucial clue to us when we talk about the evolution of entropy through time.
    o In the early universe, there were no black holes; there was just plasma and gas spread uniformly throughout the universe. Ordinarily, we think of that as a high-entropy state, but when gravity is important, things change, and in the early universe, gravity was very important. There was a huge amount of stuff that was pulling on all the other stuff.

    o This stuff that was pulling could easily have been in the form of black holes, and had it been, the entropy would have been much, much higher. This is a reminder that the entropy of the early universe was actually very low. It's difficult to keep so much stuff in a perfectly smooth configuration.

    o We also see this in the real evolution of the universe. The universe expands and cools, and gravity pulls things together and makes the universe lumpier. That perfectly smooth configuration was not stable. Entropy goes up as structure forms in the universe.

    o Gravity is important in cosmology, yet when we look back at the early universe, we see a very smooth configuration. That's a reflection of the fact that it had very low entropy. Black holes give us the one clue we have about how to relate gravity and entropy. We will have to see if we can use that clue to explain why the early universe was so smooth.

Suggested Reading

Carroll, *From Eternity to Here*, chapter 12.

Hawking, *A Brief History of Time*.

Susskind, *The Black Hole War*.

Questions to Consider

1. Physicists have long debated whether information is conserved or destroyed when black holes evaporate. What do you think? Should it bother us if it turns out that the laws of physics don't actually conserve information?

2. We have never observed Hawking radiation from a black hole. How confident should we be that it really exists?

Black Hole Entropy
Lecture 20—Transcript

The last lecture on time travel was a bit of fun, something of a palate cleanser. It did not actually address our main concern, which is the arrow of time. What it did really is to say that we are on the right track in asking that kind of question. We said that time travel to the future is possible, time travel to the past doesn't happen. We don't think that there're any closed-timelike curves in our universe.

The good news for us with our concern for the arrow of time is that the question is therefore well-posed. If there were closed-timelike curves in the universe than there is no division of spacetime into past, present, future so how could the past and the future be different if they get all mixed up? But, in a world without closed-timelike curves we know which is the past, which is the future, we can ask why they are different.

From now on, for the rest of the course, we are focused more or less like laser beams on this question of why the early universe had a low entropy, why is the past hypothesis true and useful in our universe. We're given both an opportunity and a bit of a puzzle. Both come from the fact that gravity is of crucial importance in the early universe. We said that we live in a universe that is expanding. If you go backwards in time all the galaxies, all the stuff in the universe were crunched together and if you go 13.7 billion years into the past you hit the Big Bang, the moment of infinite curvature, infinite density, that is the place when our equations break down and we don't know what to talk about.

The opportunity that is presented is that the Big Bang clearly plays some crucially important role. If we were to understand it better we would have a chance of knowing why it has such a low entropy. Before there was any such thing as the Big Bang, in Boltzmann's time, before Einstein came along, there was no special moment in the history of the universe. Newton said that time just goes on forever and any moment should be predictable from any other moment. It might be that what we call the Big Bang is truly special for some reason, either it was the real beginning of the universe or it came out of some very different process. Either way it makes sense to us that there is something there to be explained. If we understood the Big Bang better maybe we would understand why the entropy of the universe was so low at that time.

The puzzling aspect to us is that we don't understand the Big Bang very well. The Big Bang is clearly a moment in the history of the universe where gravity is of fundamental importance. All that matter so close together, spacetime expanding, so quickly you can't ignore gravity when you get to the Big Bang. But, as we mentioned, gravity and quantum mechanics don't play well together. We don't have a sensible definitive theory of quantum gravity. Gravity is the only force of nature for which that's the case. We have all the other forces put into a framework of quantum field theory, which is based on quantum mechanics.

Gravity is the only force of nature where we don't understand what is going on at the microscopic level. That is exactly what we do need to understand when it comes to the Big Bang and entropy. We want to be able to calculate what the entropy was in order to explain why it was so small and that's hard to do if we don't understand how general relativity and quantum mechanics are ultimately reconciled. One way to think about this is to go back to Boltzmann's understanding of entropy. The equation engraved on Boltzmann's tombstone is $S = k \, log \, W$. The entropy is a constant in k times the logarithm of the number of microstates that are macroscopically indistinguishable.

Our problem as cosmologists, as people who are trying to understand the Big Bang, is that we don't know what W is. We don't know the whole space of possible states because we don't understand that until we understand the quantum basis for gravity. If we don't understand the whole set of states, we don't certainly understand how to chunk up those states into macrostates that look the same to us. From some point of view, it's like we're Clausius before Boltzmann came along. We understand something about entropy, but we're not dealing with the fundamental theory. Boltzmann had atoms, Clausius just had the phenomenological theory of thermodynamics.

Our present understanding of gravity comes from the classical theory of general relativity. That's kind of like thermodynamics. It's a very, very good theory within its domain of applicability, but in the domain we care about when gravity becomes very, very strong and quantum mechanics becomes important it's not good enough and we need to do a little bit better. We need to understand how entropy works, what the formula for entropy is when gravity is very important and right now we don't understand that except for one shining counterexample and that is black holes.

We talked about black holes and what they are. The great news for us in our quest to understand entropy is that black holes are places where gravity is incredibly important. They are the places where gravity is more important than anywhere in the universe and we do know what their entropy is. It turns out that if you have some complicated messy situation like the solar system where you have the Sun, made of gas, and the Earth and the other planets, made of different things, and they have a gravitational field and the gravity can change, and maybe there's gravitational waves and radiation, we don't have a formula that tells us what the entropy is in that situation. But in the black hole situation, where everything is maximally simple, we actually have a formula that was given to us by Stephen Hawking.

To get to that formula, we should start thinking about classical black holes, black holes as Einstein would have thought about them even though the name black hole was not invented until long after Einstein had passed away. It turns out that black holes and classical general relativity are actually very simple things. It might be hard to understand what's going on, but they are not difficult to characterize which black hole we are talking about. Once you tell me the mass of a black hole, the electric charge that it has, and how fast it's spinning, there is nothing more to tell me about it. You have completely specified the black hole.

This is completely unlike a planet. You tell me the mass of a planet, the electrical charge of a planet, and the spin of a planet you are nowhere close. You need to tell me what it's made of, what the atmosphere is, what kind of mountains and oceans it might have, black holes have none of those things. This is called the No-hair Theorem. All black holes are bald. They all look the same. Once you tell me the mass, charge, and spin there's nothing more to tell me about it.

We think of such a black hole as a place from which nothing can emerge. You go into the black hole, you can never come out. It was therefore extremely surprising when back around 1970 Roger Penrose, who we talked about earlier because he has been emphasizing the need for a cosmological explanation for the past hypothesis. Penrose showed and you can actually extract energy from a black hole, not any old black hole, but some black holes are spinning very rapidly, some are not spinning at all. Penrose showed that if a black hole is spinning rapidly there is something we can do to slow it down and in the process extract some of its energy. The mass of a black hole actually decreases as it spins more and more slowly.

But of course, there's only a finite amount of energy that you can get out. The black hole is spinning at some rate. You can get energy out of it, but you slow it down in the process. Once you are done, once it's no longer spinning, there is no more useful work you can extract from that black hole. Those words should sound familiar. You have a situation where you can extract useful work from a system, once you extract a certain amount of useful work it reaches a state where there's no more useful work to get out. This is exactly the situation when we are talking about a thermodynamic system approaching thermal equilibrium. Just like the gas in our piston, if it's in a low entropy state, we can extract useful work from it, but once you reach equilibrium there's nothing more we can do.

Penrose is saying that there is a black hole that you can extract useful work from, extract energy that you can use to do things, but there's a point of no return where the black hole stops being useful to us. This idea was developed over the early 1970s into a whole little sub-field of study called black hole mechanics. The idea was there was an analogy between how black holes behave and thermodynamics. In thermodynamics we have the energy of a system, for a black hole we had its mass $E = mc^2$ tells us that mass is just a type of energy so that's an obvious analogy.

In thermodynamics, we have a temperature, in black holes we have the surface gravity. Now you might think of the surface gravity of a black hole should be infinitely big. Once you go inside isn't it true that the force is so strong you can never come out? That is true, but surface gravity is a technical term, which means if you were very, very, very far away from the black hole and you got a little weight and you lowered into the black hole on a string, when the weight approaches the event horizon how strongly is it tugging on you. It's true that if you were standing next to the weight the tug would be infinitely strong, but it dilutes as you go up the string and so from a point infinitely far away there's a finite strength of force that the black hole exerts on you. That's called the surface gravity. That is analogous to temperature in the rules of black hole mechanics.

Finally, in thermodynamics we have the idea of entropy, of course. In a black hole, the idea is that this is the area of the event horizon. Even though as you extract energy from a spinning black hole its mass goes down, you are extracting useful work from it, the area of the event horizon goes up. That's because a spinning black hole is kind of squashed, it becomes more spherical as you spin it down so the entropy keeps going up. Just like in a thermodynamic system, entropy keeps going up. In fact, there is a whole set of rules that are completely analogous between thermodynamics and black hole mechanics.

In thermo, we have the zeroth law that says that if you put two systems together in another system and they're all in thermal equilibrium they will all have the same temperature. Likewise, the zeroth law of black hole mechanics says that the surface gravity is the same everywhere on the horizon of a black hole whether it's spinning or charged or what have you. You can measure a surface gravity that's going to be the same everywhere, just like the temperature is the same everywhere in equilibrium.

The first law of thermodynamics says that energy is conserved and so does the first law of black hole mechanics. It says that mass is conserved. There are different ways to give the black hole energy with spin and charge, but it doesn't change over time. It's the second law of course that is most interesting. The second law of thermodynamics says that the entropy of a closed system only increases so the analogy in black holes is that the area of the event horizon only every increases. In fact, this is a theorem that was proven by Stephen Hawking, the famous area theorem of black holes. The area of the event horizon only goes up, you can change its shape, but you cannot ever make it smaller.

The third law of thermodynamics says that you cannot reach zero temperature, the third law of black hole mechanics says that you cannot reach zero surface gravity. We have a whole set of rules that are spookily analogous in the thermodynamics regime and the black hole mechanics regime. The question is, is this an accident? Does this mean something? Is it deep and profound or is it just a nice little analogy between two very, very different ways of thinking about the universe.

There's actually a graduate student named Jacob Bekenstein who was working with John Wheeler, the Princeton physicist who coined the term black hole. Bekenstein actually made the radical suggestion that these laws of thermodynamics and these laws of black hole mechanics aren't separate; that there is one form of generalized thermodynamics that includes black holes in it and they are what we call the area of the event horizon isn't just like the entropy or analogous to entropy it is the entropy. Bekenstein says that the entropy of a black hole is proportional to the area of its event horizon. This suggestion did not go easy among the people who think about black holes at the time.

Remember, Boltzmann said that the entropy is counting the number of microstates that are macroscopically indistinguishable. If you like, if you have a box of gas the number of ways we can rearrange the atoms of that gas so that it looks the same macroscopically. If Bekenstein was right, black holes must have many, many indistinguishable microstates because they have a large amount of entropy. In the classical general relativistic description, those microstates are nowhere to be found, just like they're nowhere to be found in thermodynamics. A black hole is specified by its mass, charge, and spin and that's it. Every mass, charge, and spin equal black hole should be the same as every other one with the same values.

How can it be that secretly these black holes have a huge number of states to accommodate all the entropy they have. What is even worse is that we know from thermodynamics that if you have a system with a temperature, it will radiate, it will give off light. This is black body radiation. You are a thermodynamic system yourself. You have a body temperature, you give off infrared light. You can't help it. Any object with a temperature gives off radiation. That is a consequence of how we think about thermodynamics. You heat up an oven, you see it begin to glow.

If you really, really, really took seriously the idea that black holes are ordinary thermodynamic systems, and you thought that the surface gravity of a black hole was like the temperature in thermodynamics then black holes should radiate. That seems to be the implication of Bekenstein's claim that black holes really are part of thermodynamics, not just an amusing analogy. This bugged a lot of people. People thought that it is clearly not true that black holes are just part of thermodynamics because they would radiate and black holes are black. They don't radiate.

In particular, Stephen Hawking got very, very annoyed. Hawking is a very opinionated guy. He was a little irked that this young, ambitious, graduate student in Princeton was saying that black holes are real thermodynamic systems. Like any good scientist, Hawking didn't just insult him or ignore him, Hawking set about proving Bekenstein wrong. He thought about what it would mean for a black hole to give off radiation and he set about showing that a black hole can't give off radiation.

What this means is that you have to apply quantum mechanics to black holes. Now we have already said why this is difficult because we don't have a theory of quantum gravity, but there is a little nice in between regime where you can hope to make progress. You treat the black hole as classical. The black hole's just sitting there. You are not quantizing the black hole itself, it's a classical object, but you look at the behavior of other fields in the vicinity of the black hole. You have photons and gravitons and electrons that are near the black hole and those can still obey the rules of quantum mechanics. Basically what you are studying is the field of quantum field theory in curved spacetime and that is actually perfectly well-defined even if true quantum gravity is not.

Everyone knows how this story ends. Hawking sat down very carefully to think about quantum field theory in the vicinity of a black hole. It wasn't by the way his specialty. Hawking was the world's expert in classical general relativity. He proved the area theorem. He proved different singularity theorems along with Penrose and so forth, but he's also a brilliant guy so he studied quantum field theory. He would sit silently, other people would have to prop up a textbook for him and he would read about quantum field theory.

Then he studied that quantum field theory in the vicinity of a black hole. What he found was that quantum field theory says there are virtual particles. Just like in ordinary quantum mechanics if you have a single particle you can't pin it down to where exactly it is located. In quantum field theory, you can't pin down the number of particles in a certain region of space. Virtual particles keep popping in and out of existence.

When Hawking took this idea and applied it to black holes, of course he's not just waving his hands and talking about virtual particles, he has equations. He is solving the equations of quantum field theory in the vicinity of a black hole and what he found was that virtual pairs of particles could be created, one of them could fall into the black hole and another one could escape. From the point of view of someone standing very far outside, the one that falls into the black hole is irrelevant, you don't see it. The one that escapes you see. What it looks like to you is that the black hole is giving off radiation. The black hole is actually losing mass to its environment because those virtual particles carry away mass.

You might say, well, wait a minute. Doesn't the virtual particle that falls into the black hole increase its mass? It turns out, miracle of general relativity, the particle that falls into the black hole has a negative mass. The black hole gradually shrinks. "Black holes ain't completely black," as Hawking himself put it, they radiate and eventually they will evaporate. They lose energy to the radiation around them and so even black holes aren't forever. Like I said, Hawking had equations to back this up. we don't need to know all the details of the equations, but it is useful to get the results in terms of how the properties of a black hole are affected by how big it is, how much the mass of the black hole is.

In particular, Hawking got a formula for the temperature of a black hole, $T = x / m$, x is just a constant, and so all we are saying is that the temperature is inversely proportional to the mass of the black hole. That constant of proportionality, if you must know, is $\hbar$, Planck's constant, which tells us how important quantum mechanics is since this is fundamentally a quantum effect $\hbar c^3$, c is the speed of light, divided by $8\pi Gk$, G is Newton's constant of gravity. Gravity is going to be important, and k we know is Boltzmann's constant. You're connecting energy and temperature so Boltzmann's constant makes an appearance.

The important thing is that Hawking showed that the temperature of a black hole is not zero, but it is smaller for large black holes. The temperature goes as one over the mass. A tiny black hole is giving off a furious amount of radiation. A large black hole, a macroscopic one like we would make in the center of the galaxy or after a star died, is actually very, very cold. A black hole with one solar mass has a temperature about 1 billionth of the background radiation that suffuses the universe.

For real size black holes, you are never going to observe the Hawking radiation as it is now called. This is too bad for Hawking and it makes it very hard for him to win the Nobel Prize. It's very hard to test his theory that most people believe that he's right, black holes do give off radiation even if ordinary black holes are just too cold for us to observe it. If they do give off radiation then they lose mass. As they get smaller, their mass goes down and when they get smaller a black hole gets hotter so it's a runaway cycle. The black hole gives off energy, it shrinks, and then it gives off more energy faster, and in a finite period of time the black hole completely evaporates and disappears.

This lifetime is given by $t = yM^3$, the lifetime of the black hole is some constant times its mass cubed. Obviously the black hole lives longer it goes as the cube of its total mass. That constant, again, if you need to know it, it's $5120\pi G^2/\hbar c^4$. You don't need to know the constant, all you need to know is that the bigger black holes last longer even though they are colder. This was a big surprise of course when Hawking showed that black holes give off radiation. But, it only became more surprising with time.

Remember one of our most precious principles is Laplacian determinism, information is conserved. We think that if you know the state of the world at one point in time and if were Laplace's demon in principle you could predict the future with perfect fidelity. For example, if I have a book and I throw it into a fire and it burns, ordinarily you would say well now I can't read the book anymore, that information is gone, but if you were Laplace's demon you would keep track of every little particle of soot, every little photon of light that came out of the fire, and in principle you would know exactly what the words in the book were before you burned them. It's obviously a thought experiment and we think that's how physics works at a deep level.

Now the question becomes what about a black hole. We used to think, before we knew that black holes radiated, you would throw a book into a black hole and it's just lost. It's not that the information has disappeared, it's just that you can't get it any more. Now, Hawking says that the black holes radiate that they give off their mass to the surrounding environment. Therefore, if we wait long enough all the information that was in the book should come out in the Hawking radiation. Still to this day, we don't know how that happens. In fact, there are debates, some people think it doesn't happen. You can argue about how difficult it would be to make it happen.

Hawking, himself, was a skeptic. Hawking, himself, strongly believed that the information is truly lost that Laplace is defeated by evaporating black holes. But more recent developments in our half-hearted attempts to understand quantum gravity have convinced Hawking that he was wrong. He had made a bet with John Preskill and Kip Thorne, two physicists at CalTech, and Hawking bet that information was really destroyed, Preskill and Thorne bet that it was preserved. Hawking has conceded his bet, he's given up, and the funny thing is that neither John Preskill nor Kip Thorne agree that he should have conceded the bet. They still think that information does get out, but they don't think that the issue is completely settled.

Most importantly for us, if black holes radiate, if the analogy between thermodynamics and black hole mechanics is good that means that black holes have entropy. This is our single most important handle on how entropy works when gravity is relevant. You can calculate what the entropy is, just like Bekenstein predicted, the entropy of the black hole is proportional to the area of its event horizon. The proportionality constant is $C^3 k / 4\hbar G$, but the point is a bigger black hole has a bigger entropy. The area goes like the size squared, an area goes like a radius squared, and the size goes like the mass, so the entropy of a black hole goes up as its mass squared. It gets to be a very, very big number.

Even before we plug in the numbers, we are puzzled by the fact that the entropy goes like the area. Remember way back in Boltzmann's time, one of the crucial features of entropy is that if you take one box who has a certain entropy in another box with a certain entropy you add them together, their entropies add. It is as if entropy is proportional to the volume that you are dealing with. But in a black hole, according to Hawking, the entropy goes like the area, not the volume. This is a big puzzle. This opens up a whole can of worms that we don't have time to talk about in this course, but it inspired something that we call the holographic principle. The idea that when quantum gravity is important. Physics is no longer strictly local in the sense that, in ordinary physics and special relativity and quantum field theory to describe what happens in the world, you describe what happens here and then what happens there and then what happens somewhere else. Different places in the universe have different things going on and if you describe them everywhere you would describe the whole world.

You can describe them everywhere separately. To know what happens here, I don't need to know what happens somewhere else. The holographic principle says there are secret correlations, secret connections, between what happens in the world at different places exactly in the right way that the entropy of a black hole goes like the area of its horizon. That is to say the states, those secret unknown quantum states that all macroscopically look alike. To give us Boltzmann's formula for the entropy of a black hole, those states don't live inside the black hole, in some sense they actually live on the horizon. This is a dramatic idea, the holographic principle. Basically it's saying that three-dimensional space can be described by a two-dimensional theory, but since it was proposed we've gotten more examples of how it works in different contexts outside black holes.

The holographic principle is an absolute guiding principle in future developments of quantum gravity. It doesn't really help us understand the arrow time. Let's plug in some numbers. Hawking gave us a formula for the black hole entropy and the short story is it is huge. If you take a 1 million-solar-mass black hole, we think we have a 4 million-solar-mass black hole at the center of our galaxy, but let's just take a 1 million-solar-mass black hole. It has an entropy according to Hawking of 10^{90}. For comparison purposes let's take all of the non-black hole particles in the observable universe, every galaxy, every atom, every star, every bit of dark matter, every photon, every neutrino, and so forth, add them all up, ignore gravity and calculate the entropy. The answer we get is 10^{88}. The entropy of all the non-black hole stuff in the observable universe is 10^{88}. The entropy of a single solar-mass black hole is 10^{90}.

We think there is something like 10^{12} or 10^{13} million-solar-mass black holes in our observable universe. What this is telling us is that the entropy in our universe today is overwhelmingly in the form of black holes. Most of the entropy is black hole entropy. It's not photons, stars, burning pieces of paper or anything like that. This is a crucial clue to us when we talk about the evolution of entropy through time. When you think about it in the early universe there were no black holes. As far as we know the early universe was smooth. There were not only no black holes, there were no stars or planets, there was just plasma and gas spread uniformly throughout the universe. Ordinarily, we think of that as a high entropy state, when everything is smooth the entropy is high. But, when gravity is important things change.

In the early universe, gravity was very important. There's a huge amount of stuff that was all pulling on all the other stuff. It could easily have been in the form of black holes. Had it been in the form of black holes the entropy would've been much, much, much higher. This is a reminder that the entropy of the early universe was actually very low. It's difficult to keep so much stuff in that perfectly smooth configuration. You also see this in the real evolution of the universe. The universe expands and cools, gravity pulls things together, and makes the universe lumpier. That perfectly smooth configuration was not stable. The entropy goes up as structure forms in the universe.

This is our task now. Gravity wants to make things not smooth. Gravity is very important in cosmology and yet when we look back at the early universe we see a very smooth configuration. We will talk about this in more detail, but we have data, we look with our telescopes at the early universe and we see that it is very smooth. That's a reflection of the fact that it had a very low entropy. Black holes give us the one clue we have about how to relate gravity and entropy. We will have to see if we can use that clue to explain why the early universe was so smooth.

Evolution of the Universe
Lecture 21

In this lecture, we're going to follow the evolution of entropy through the history of the universe. We start near the Big Bang, 13.7 billion years ago, and follow the physical processes that occurred as the universe expanded and cooled. Along the way, we'll track what the entropy was at each stage. Not surprisingly, we will learn that entropy has been increasing—the second law of thermodynamics is correct. We will then follow the evolution of the universe past the present moment into the future.

The Present Universe
- There are about 100 billion galaxies in the observable universe, each with about 100 billion stars. On very large scales, the universe seems to be uniform.

- As we know, the universe is expanding. Distant galaxies are moving away from us, and the farther away they are, the faster they are receding. The amount of space between us and the galaxies is becoming greater.

- Once we know that the universe is expanding, we can play the movie backward. If galaxies are moving away from each other now, in the past, they were closer together.

The Moment of the Big Bang
- If we keep tracing backward in time, 13.7 billion years ago, we hit a point where all the galaxies were on top of each other. That's the Big Bang, the point at which the density of matter was infinite.

- Einstein's general relativity gives us an equation that relates the rate at which the universe is expanding to the amount of stuff in the universe. This gives us a quantitative theoretical understanding of what the universe should have done in the past, and we can compare

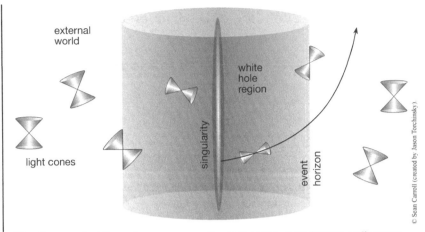

We often say that the universe was smaller in the past, but what we really mean is that galaxies were closer together in the past and they will be farther apart in the future.

that to the data. Remarkably, the data we have go all the way back to the first 3 minutes in the history of the universe.

- When all the matter in the universe was much more densely packed, its temperature was extremely high.
  - If we go far enough back, it was so hot that atoms could not exist; electrons could not stick to atomic nuclei.

  - If we go more than a few seconds after the Big Bang, even nuclei could not exist; protons and neutrons could not stick together.

  - In the period between a few seconds and a few minutes after the Big Bang, the universe expanded and cooled, and protons and neutrons could finally join.

  - Protons and neutrons want to fuse together to make heavier elements, but because the universe was expanding so quickly, they didn't have time. Thus, the density of the universe decreased rapidly.

- A certain fraction of the protons and neutrons became helium, lithium, and deuterium. Observations of the abundances of these elements in primordial regions of the universe match scientific predictions about what they should be just a few seconds after the Big Bang.

- That process, called Big Bang nucleosynthesis, occurred when the universe cooled down enough to allow nuclei to form. About 380,000 years after the Big Bang, it cooled enough to allow atoms to form.
  - This process, when electrons can finally stick to atomic nuclei to make atoms, is called recombination.

  - Before recombination—when electrons were free to roam—the universe was opaque. Once recombination occurred, the universe became transparent, and the photons from that period can move all the way across the universe to be detected in our telescopes.

  - In 1964, Arno Penzias and Robert Wilson, two scientists working for Bell Labs, first detected this cosmic microwave background radiation from the Big Bang.

The Formation of Large-Scale Structure

- As we've said, our current universe seems uniform on very large scales, but at the moment that the cosmic microwave background was formed, the universe was uniform on all scales.

- As the universe expanded and cooled, gravity turned up the contrast knob. A region of space with slightly more matter would have a slightly stronger gravitational field than a region with less matter; it would pull nearby atoms toward it. A region that was slightly less dense than average would lose atoms to the rest of the universe. This process is called the formation of large-scale structure in the universe.

- Some small perturbations in space became planets, stars, and black holes. Larger ones formed galaxies, and even larger ones formed large-scale structure. As this process progressed, entropy increased.

- Recall that when gravity is important, high entropy does not necessarily mean smooth. High entropy can be very lumpy. The highest entropy of all would be to have all matter in one large black hole.

- It's not true that the early universe had to be smooth. It could have been much denser in one region than another. That would have been a much higher-entropy configuration.

- If we took a high-entropy version of the current universe and played the movie backward, there's no reason why entropy would have had to go down. The early universe having low entropy is a fact that still needs to be explained.

- Of course, if the entropy of the early universe had been as high as it could possibly be, we would not live in a universe with the second law of thermodynamics. The reason there's a second law is because of the past hypothesis, the fact that the early universe had such a tiny entropy.

Dark Energy

- In 1998, astronomers made the surprising discovery that the expansion of the universe is accelerating. But ordinary stuff—matter, radiation, atoms, even dark matter—does not speed up the expansion of the universe.

- The fact that the universe is accelerating requires some nonordinary stuff: dark energy, which doesn't dilute away as the universe expands.

- In an expanding universe containing nothing but matter and radiation, the number of particles doesn't change in any one volume of the universe. In any fixed cubic centimeter of the universe, the density of particles is going down.

- But dark energy hypothesizes that in every cubic centimeter, some energy doesn't dilute away; some amount of energy is present that provides a perpetual impulse to the expansion of the universe. That impulse actually builds up as space itself gets larger, and that's what makes the universe seem to accelerate.

- The best candidate for dark energy is vacuum energy, a specific example of dark energy in which the dilution is exactly zero and the density of energy is exactly the same in every point in space. Vacuum energy is also called the cosmological constant.

- If vacuum energy is the right answer, it has a profound implication for the future of the universe, namely, that the expansion will continue forever.

The Future Universe

- Right now, the age of the universe is, roughly speaking, 10^{10} years. What will happen in the future if the universe does not recollapse?

- Stars in the galaxies of the universe are burning their fuel. The smallest stars, which are the ones that last the longest, will last about 10^{15} years before they finally burn out.

- Those dead stars will ultimately fall into black holes, which themselves will eventually evaporate. This process will take about 10^{100} years, about 1 googol years.

- The radiation of the black holes will be diluted away as the universe expands and accelerates, so it will be imperceptibly different from having nothing at all. And that, if the dark energy is completely constant, is the end of the story. After a googol years from now, we will have empty space and that empty space will last forever.

- In other words, the highest-entropy state the universe can be in is nothing but empty space. Right now, the universe is not high entropy, but entropy is growing. When entropy hits its maximum value, equilibrium is reached and evolution stops.

- That will be the heat death of the universe. Rather than recollapsing and experiencing a Big Crunch, the universe will calm down and become colder and colder until it reaches equilibrium.

An Inside-Out Black Hole

- There is, however, a fascinating extra fact that is forced on us by the idea of vacuum energy: If the universe is truly accelerating and vacuum energy never goes away, it turns out that the accelerating universe is like living inside an inside-out black hole.

- A black hole is defined by the fact that it has a horizon. If you go inside, you can never escape to the outside. There's a region of the universe from which information can never reach us.

- Likewise, in an accelerating universe, a galaxy that is far away from us now will eventually be moving away from us at a speed faster than light. Any light that galaxy emits will never reach us. In other words, that galaxy is past a horizon.

- Just as Hawking said that black hole horizons give off radiation, our universe, which has a horizon around us, has a temperature and gives off radiation. In the far, far future of our purportedly empty universe, there will still be a nonzero temperature, about $10^{-30°}$ Kelvin.

- What this scenario means is that we have an eternal universe that lasts forever. There's a fixed volume that we can see inside our horizon, and inside that volume, there is a temperature that remains constant forever.

- This should remind you of Boltzmann's scenario for understanding why the early universe had low entropy.
    o Boltzmann imagined an eternal universe that had some fluctuations and, every once in a while, would form a low-entropy configuration—a planet, a galaxy, or even the whole universe.

- o We said that couldn't be right, but it turns out that our real world seems headed for exactly that situation. It might take an enormously long time to fluctuate back into the whole universe, but we have forever to wait.

- If this is the right cosmological scenario, then we should be Boltzmann brains or the universe around us should have just fluctuated randomly into existence. We don't believe that's true, and therefore, there must be something wrong with our current best understanding of cosmology.

Calculating Entropy across Time

- At very early times in the universe, when there was no structure, the entropy of the universe was about 10^{88}.

- Today, most of the universe has black holes at the centers of galaxies, and most of the entropy of the universe is in those black holes. Adding up the entropy in black holes, we get about 10^{103}.

- In the future, the maximum entropy that we can possibly imagine from all the stuff we see in the universe today will be about 10^{120}. That would be the equilibrium version of the universe, the configuration we could be in with maximal entropy.

Suggested Reading

Carroll, *From Eternity to Here*, chapter 13.

Guth, *The Inflationary Universe*.

Penrose, *The Road to Reality*.

Questions to Consider

1. Before the discovery of the cosmic microwave background, many astronomers took seriously the steady-state theory. In that model, the universe was expanding, but it was constantly creating new matter so

that the average density remained the same. What would the status of the arrow of time be in such a universe?

2. Is it surprising that the predicted future of the universe stretches out for a much longer time than its known past? Is this something that science should be working to explain or just a random fact we should accept?

Evolution of the Universe
Lecture 21—Transcript

By this point, you're probably tired of hearing me say that the entropy of the early universe was small. In this lecture, I'm going to say that again and again, but we're also going to follow the evolution of the entropy through the history of the universe. We're going to start near the Big Bang, 13.7 billion years ago, and follow the physical processes that occurred as the universe expanded and cooled, and then we're going to track what the entropy was at each stage along the way. Unsurprisingly, we will learn that the entropy has been going up. The Second Law of Thermodynamics is correct, and then we will follow the evolution of the universe past the present moment into the future.

Most discussions of the evolution of the universe talk about between the Big Bang and today, but we know that that's just a prejudice. There's nothing special about the past versus the future. If you're going to talk about the whole history of the universe, you need to do the future just as well as the past.

We can begin with the present. We look outside into the universe and we see galaxies. If you go out on a clear night, the first things you see are stars. Galaxies are just collections of stars that are orbiting each other under their own mutual gravitational field. We live in a galaxy, the Milky Way Galaxy.

If you're blessed enough to have very clear skies and dark nights, then you can see the Milky Way stretching at nighttime from one horizon to the other. We're embedded in the disc-like structure of the Milky Way. It's a Galaxy with about 100 billion stars. We're not giving exact numbers here. We're just trying to give you orders of magnitude. It's a good number to remember. A hundred billion is about the number of stars in the Milky Way or any other typical or any other typical galaxy. There are smaller galaxies and larger ones, but 100 billion stars is about the average.

If we look out at the whole universe, we see that it is filled with many galaxies. In fact, there are about 100 billion galaxies in the observable universe, a similar number as there are stars per galaxy. Now you might wonder about that observable universe bit. You might say, well, as our observations become better and better, will we discover more galaxies? When I talk about the observable universe, I'm not referring to our present technological capabilities. I'm talking about a matter of principle.

Because the Big Bang was a finite amount of time in the past and because the speed of light is a finite number, there's only so much universe that we can possibly see. There's a finite part of the universe outside of which to get to us light would need to go faster than the speed of light.

The observable universe is what we call the part of the universe that we can, in principle, ever observe. It's not that our technological capabilities are not better. It's just that that's all the universe that we can possibly see. We haven't seen anything like it, yet. When I say there are 100 billion galaxies, I'm not trying to imply we've counted 100 billion galaxies. We count very tiny slices of the universe and then we extrapolate through the whole thing.

The nice thing about our universe on very, very large scales is that it seems to be uniform. If you look at large enough sizes of the universe, what you see is the same as if you looked at any other equivalent size somewhere else.

What NASA has done with its space telescope, is to basically take the telescope and point it at a blank region of the sky, a region where it doesn't look like there's anything. It left the shutter open on the telescope for quite a while and just let the photons accumulate in the camera and what you see when you do that is what they the Hubble Ultra Deep Field. The Hubble Space Telescope has taken an image of the furthest reaches of the universe and even though we didn't know there was anything there, we predicted that we would see lots of galaxies and that's exactly what you do.

When you look at this image you see a tremendous number of little splotches. Each one of them is a galaxy in its own right, comparable in size to the Milky Way. So our present universe is an essentially uniform collection of galaxies, 100 billion of them spread out amongst the universe with about 100 billion stars each.

The most important other fact about the universe is that not only is it big, it is getting bigger. The universe is expanding. When we say the universe is expanding, what we mean operationally is that distant galaxies are moving away from us and the further they are away the faster they are receding. That's known as Hubble's Law, named after Edwin Hubble, the astronomer who worked at Mount Wilson Observatory in California who discovered this in the 1920s. It's interesting to think about the history of this discovery. It wasn't Hubble who found that distant galaxies are moving away from us. That was actually a different astronomer, Vesto Slipher. Though what Hubble was able to do was to measure the distances to the galaxy.

Before Hubble came along, people weren't even sure that we now call galaxies are, in fact, separate collections of stars. They thought that they might be fuzzy nebulae here inside our galaxy. Hubble first showed that galaxies truly are distant, and then he was able to plot the speed at which they were moving away versus the distance to the galaxy. What he found was that nearby galaxies are moving away slowly, further away galaxies are moving away more quickly.

That is what we mean by the expanding universe. It's not just that all galaxies are moving away from us. It's as if space itself is expanding and that's exactly, of course, what general relativity predicts. Space-time is dynamical. It's space itself that is getting bigger. When the galaxies move away, it's not that the galaxies are moving through space, it's that the amount of space between us and the galaxies is actually growing.

Even though we see all the galaxies moving away from us, it's not because we're special. It's not because we're sitting at the center. If there's another galaxy you pick out far away, we are moving away from them and the galaxies on the other side are moving away from them even faster, so they see the same thing we do. They see all the galaxies in the universe moving away from them. Every person in every galaxy thinks the universe is expanding, thinks that everything is moving away. That's the expansion of the universe.

It's also interesting that Hubble never won the Nobel Prize. He really wanted to win the Nobel Prize, but back in the '20s when he made his discoveries and afterward, Astronomy was not really counted among Physics, which is what the Nobel Prize is given for. In fact, it wasn't until 2011 that the first Nobel Prize in Physics was given for work in Optical Astronomy.

Once we know that the universe is expanding, we can play the movie backwards. If galaxies are moving away from each other now, in the past they were closer together. It's not right to say the universe was smaller. We often say that but if you do say that the universe was smaller in the past, or will be bigger in the future, just be careful to understand what exactly you mean. What you really mean is that galaxies were closer together in the past and they will be further apart in the future.

The point is that we don't know the size of the universe. We know this relative size of different parts of the universe and we know how that size changes over time. If you pick some fixed collection of galaxies, the region that is drawn by those galaxies was smaller in the past and will be larger in the future, but the universe itself might very well be infinitely big. It could be finite; it could infinite. We don't know because we can only see a finite part of it. But even if the universe is infinitely big with the spatial extent of the universe goes forever, we can still say it's expanding or contracting because what we mean by that is that the galaxies were closer together or further apart.

Now if you keep tracing backwards in time, of course, 13.7 billion years ago you hit a point where all the galaxies were on top of each other. That's the Big Bang. That's the point where the density of matter was infinitely big and we need to think about what that means. Of course, this extrapolation isn't done naively. It's not just, well, the galaxies are moving away now, we naively extrapolate them back at the same speed.

We use the general theory of relativity. Einstein gave us an equation that relates the rate at which the universe expands to the amount of stuff in the universe so we have a very quantitative theoretical understanding of what the universe should have done in the past and we can actually compare that to the data. Remarkably, the data that we have goes all the way back to the first three minutes in the history of the universe. From about a few seconds after the Big Bang to a few minutes after the Big Bang the universe was a nuclear reactor.

When all the matter in the universe was much more densely packed, it kept banging into each other and the universe was a lot hotter, the temperature was extremely high. If you go far enough back it was so hot that atoms could not exist, electrons could not stick to the atomic nuclei inside and if you go more than a few seconds after the Big Bang, even nuclei could not exist. Protons and neutrons could not stick together. As the universe expands and cools, it's the period between a few seconds and a few minutes where the protons and neutrons can finally stick together.

In fact, there's a race. What the protons and neutrons want to do is fuse together to make heavier and heavier elements, but because the universe is expanding so quickly, they don't have time to do it. The density of the universe goes down very rapidly.

Instead, a certain fraction of the protons and the neutrons get turned into Helium, Lithium, and Deuterium. Deuterium is just Hydrogen, one proton with an extra neutron. Deuterium is basically heavy Hydrogen. These three other elements, Helium, Lithium, and Deuterium, are produced in the first seconds after the Big Bang and if you have a theoretical understanding for what the density was, what the nuclear physics is, and how fast the universe was expanding, you could make a prediction. You can say, there should be this much Hydrogen, this much Helium, this much Lithium, this much Deuterium and then you can try to check that against observation.

It's not easy, of course, because in the later universe where we live stars and other processes in the universe have changed the elemental abundances, but you can look for regions of the universe where that change has been minimal. You can look for regions of the universe where the abundances of different elements are close to primordial.

The wonderful fact is that we've done that. We've made the observations. They match the predictions beautifully. In other words, we have data that tells us we know what the universe was doing just a few seconds after the Big Bang. We can, of course, go further than that.

That process, called Big Bang Nucleosynthesis, was when the universe cooled down enough to let nuclei form. It wasn't until hundreds of thousands of years after the Big Bang that it cooled down enough to let atoms form. That's about 380,000 years after the universe got started. This process is called recombination, when electrons can finally stick to atomic nuclei to make atoms. After recombination it is atoms that are in the universe, Hydrogen and Helium; before recombination it's nuclei and electrons.

The reason why that's important is because when electrons are free to roam, when they're not stuck to atoms, the universe is opaque. The light that is in the universe, the light is filling the universe because everything is so hot that it is glowing, but you can't see your hand in front of your face if you were living during that period of the universe's history because the light keeps bumping into the electrons. Once recombination happens, once the electrons get together with the nuclei to form atoms, the universe becomes transparent and the photons from that period can move all the way across the universe, to be detected in our telescopes.

That was done in 1964 by Arno Penzias and Robert Wilson. They had a radio telescope at Bell Labs. They were trying to see how quiet the universe was because they were interested in communications. They were working for Bell Labs. What they found was there was some noise they could not get rid of in their radio telescope. Other physicists eventually let them know that what they had discovered was the relic radiation from the Big Bang, from just a few hundred thousand years after the universe started.

This is now known as the Cosmic Microwave Background radiation because the wavelength is in the microwave regime. It's basically a snapshot of what the universe looked like at that moment in time when atoms were first born. What we see is a universe that is pretty smooth. It's not a universe that has galaxies, stars, planets or anything like that. There are tiny fluctuations in the density of the universe from place to place. Basically, the early universe was much, much smoother than our current universe.

Our current universe seems uniform on very, very large scales but at the moment that the Microwave Background was formed the universe was uniform on all scales. In fact, when Penzias and Wilson discovered it and for decades thereafter, we couldn't detect any variation in the density of the Microwave Background from place to place.

It wasn't until the Cosmic Background Explorer satellite, which was launched in 1989, that we finally were able to detect very tiny fluctuations in the temperature of the microwave radiation. Just one part in 100,000, one in 10^5 is the difference in temperature in the Microwave Background from point to point. This got the Nobel Prize for George Smoot and John Mather in 2006, the masterminds behind the Cosmic Background Explorer satellite.

Later in 2001, NASA launched the WMAP satellite, which gave us a much better view of this Cosmic Microwave Background. Basically the highly focused image of what the universe was like, the snapshot just 380,000 years after the Big Bang.

The early universe was fairly smooth but there were tiny differences. In one region of space, there might be 100,000 atoms. In the region right next to it, there's 100,001 and in the region next to that there's 99,999. There's fluctuations about one part in 10^5, but as the universe expands and cools gravity turns up the contrast knob on the universe.

If you have a region with a little bit more matter, that region has a slightly stronger gravitational field so it pulls atoms that are nearby toward it. If you have a region that is slightly less dense than average then it loses atoms to the rest of the universe, which is more dense. As the universe expands and cools you go from a very, very smooth universe to one where structure is forming. This is called the formation of large-scale structure in the universe.

Some of those perturbations are fairly tiny. They grow into planets and stars and black holes. Larger ones form galaxies and even larger ones form large-scale structure in the universe. As this process happens, as the initially smooth early universe forms structure, entropy increases. Remember we talked about the fact that when gravity is important, high entropy does not mean smooth. High entropy can be very, very lumpy. The highest entropy of all would be to take all the matter you have and put it in one big black hole.

There's a misapprehension that sometimes gets passed around, that I want to undo just in case you've been exposed to it which says that in the early universe because it was smaller, there was less room for entropy and as the universe expanded there was more and more entropy you could have. That way of thinking is just false. It's assuming that the early universe had to be smooth, but that's not true. The early universe could've been wild, roiling chaos. It could have been much more dense in one region than in another region. That would have been a much higher entropy configuration.

If you took a high entropy version of the current universe and played the movie backward there's no reason why the entropy would've had to go down. The early universe having a low entropy is a fact that needs to be explained. It is not at all automatic. Of course, if the entropy had been higher, if the entropy had been as high as it could possibly be, we would not live in a universe with the Second Law of Thermodynamics. The reason why there's a Second Law is because of the past hypothesis, the fact that the early universe had such a tiny entropy. That is what we would like to explain.

Recently, relatively recently, we've learned a little bit about the future of the universe. In 1998, astronomers discovered that not only is the universe expanding but the expansion is accelerating. What we mean by that is that if you look at one galaxy today, it looks like it is moving away from you, as Hubble discovered. What was discovered in 1998 was that if you came back a billion years from now and measured the velocity of that galaxy again it would be moving away from you faster.

Individual galaxies are moving away faster and faster as the universe expands. This was a complete surprise to everyone who tried to do these measurements. They thought they were going to be measuring the deceleration of the universe. What they actually discovered was the acceleration of the universe.

The reason why this was a surprise is because, if the universe is full of ordinary stuff, matter, radiation, atoms, dark matter, even dark matter counts as ordinary for these purposes, as the universe expands everything pulls on everything else, you would naturally expect the expansion rate to slow down. Ordinary stuff does not speed up the expansion of the universe. To account for the fact that the universe is accelerating, you need some non-ordinary stuff. The astronomers who discovered this are the ones who won that 2011 Nobel Prize for the acceleration of the universe.

The question is what is the non-ordinary stuff that is doing it, that is pushing things apart? Well we have a name for it. We call it dark energy and we even know some of the features of dark energy. The most important feature is that it doesn't dilute away as the universe expands.

If you had a universe with nothing but matter and radiation in it, then as the universe expands the number of particles, either matter particles or photons in radiation, doesn't really change in any one volume of the universe that is expanding. It's not that particles are being created or destroyed, so in any fixed cubic centimeter the density of particles is going down. There's the same number of particles but an increasing volume of space. That's why in the ordinary way of doing things the expansion of the universe slows down. There's less and less stuff in it to drive the expansion. But dark energy hypothesizes that in every cubic centimeter there's some energy that doesn't dilute away, some amount of energy that is just stuck there that provides a perpetual impulse to the expansion of the universe. That impulse actually builds up as space itself gets bigger and that's what makes the universe seem to accelerate.

The best candidate for dark energy is called vacuum energy. So dark energy is just whatever it is that doesn't dilute away, that makes the universe accelerate. Vacuum energy is a specific example of dark energy where the dilution is exactly zero and the density of energy is exactly the same in every point in space. It's also called the cosmological constant. It's what Einstein invented to make the universe static way back when. Now we think that the cosmological constant could be the best explanation for why the universe is accelerating.

It's not the only explanation, of course. We can also have kinds of dark energy that are slowly varying, so gently that we haven't detected it yet, and even better than that we could have no energy at all but a modification of general relativity. After all our expectation that the universe should be decelerating is because that's what general relativity says, but maybe in the late universe on very, very large scales, general relativity isn't right. Maybe we need a modification of gravity.

It is unlikely that that's true. It's a long-shot idea, but it's well worth considering. That's what theoretical cosmologists like myself do for a living. We try to propose models that explain why the universe is accelerating and then we compare them against the data, but we have to admit that the simplest model is the one that is most likely to be right, vacuum energy, the cosmological constant, a strictly constant amount of energy in every cubic centimeter of space.

If that's the right answer, it has a profound implication for the future of the universe, namely that the expansion of the universe will continue forever. We're not going to recollapse because there's nothing pulling the universe back together. We'll just expand and expand and expand and the future history of the universe will last forever. This idea gives us a plausible future scenario for the universe. We don't know as much about the future as we know about the past because of the asymmetry of epistemic access. We can't see the future yet, but we can at least extrapolate our current theories.

Right now the age of the universe is roughly speaking, 10^{10} years. 10^{10} years would be 10 billion years. The real age is 13.7, but roughly speaking let's call it 10^{10} years since the Big Bang. What will happen in the future if the universe does not recollapse? We have stars that live in galaxies that light up the universe but they are burning their fuel. There's only a finite amount of fuel per star. If you look at the smallest stars, which are the ones that last the longest, they will last about 10^{15} years before they finally burn out. So we have 100,000 times the current age of the universe before it becomes completely dark. That will eventually happen.

Even after the universe becomes dark, that's not the end of the story. These dead stars will not be radiating but they will still exist and they're going to gradually fall into black holes. Remember the centers of galaxies have giant black holes in them. If you wait long enough all of the dead stars in the galaxy will fall into that black hole but even that black hole doesn't last forever. Black holes, remember, evaporate so you can calculate how long will it take before all the mass of the galaxies falls into the black holes and then those black holes radiate it all into empty space. The answer is about 1 googol years.

In the old-fashioned sense of the word googol, 10^{100}. So 10^{100} years after the Big Bang there'll be nothing left but empty space. The black holes will have radiated but that radiation is diluted away as the universe expands and accelerates, so it is imperceptibly different from having nothing at all and that, if the actual dark energy is completely constant, is the end of the story. After a googol years from now we will have empty space and that empty space will last forever.

In other words, the highest entropy state the universe can be in is nothing but empty space. The universe is right now not high entropy. The entropy is growing. What happens when the entropy hits its maximum value, you reach equilibrium and evolution stops. In our case, that's when all of the black holes have evaporated away and you're left with nothing but empty space. Empty space is simple. Remember our cup of coffee example that starts simple before you mix it with the milk, becomes complicated and then becomes simple again when it hits equilibrium. The universe does the same thing. It was simple at early times; it was very smooth. It's complicated now; here we are, complicated beings. The future of the universe is empty space, nothing going on, simple again and very high entropy.

That would be the heat death of the universe. Rather than recollapsing and having a big crunch, the universe just calms down and becomes colder and colder and colder until it reaches equilibrium. However, there's a fascinating little extra fact that is forced on us by the idea of vacuum energy. If the universe is truly accelerating and that vacuum energy never goes away, it turns out that the accelerating universe is like living inside an inside out black hole.

What I mean by that is a black hole is defined by the fact that there's a horizon. If you go inside, you can never escape to the outside. There's a region of the universe where information can never reach us from. Likewise, if we live in an accelerating universe, think about a galaxy that is very far away. It's moving away from us faster and faster, eventually be moving away from us so fast that it's moving faster than the speed of light.

This does not violate relativity. What relativity says is the two galaxies cannot pass by each other faster than the speed of light, but when space is expanding the rules are a little bit different. A very, very distant galaxy can appear to be moving faster than the speed of light. What that really means is that the galaxy is moving away from us so fast that any light it emits never gets to us. The space in between us and that galaxy is just expanding so quickly that the light will never reach us. In other words, that galaxy is past a horizon.

We live inside the horizon and we see it all around us and just like Hawking says that black hole horizons give off radiation, they have a temperature in a black body radiation, likewise our universe, which has a horizon around us, has a temperature and gives off radiation. When you live in this purportedly empty universe, the far, far future where all the stars have died and all the black holes have radiated away, you don't live at absolute zero. There's still a nonzero temperature. If you had a thermometer or a particle detector usually it would be perfectly quiet but every now and then a very, very low energy photon would come along and hit it. You would measure a nonzero temperature.

Now that temperature is very, very tiny. In numbers, it's about $10^{-30°}$ Kelvin. Our current microwave background, which is pretty cold, is about $3°$ Kelvin. So 10^{-30} is an incredibly tiny temperature but it's still not zero. What the scenario means is that we have an eternal universe that lasts forever. There's a fixed volume that you can see inside our horizon and inside that volume there is a temperature that never changes, a temperature that is just constant forever.

If this reminds you of something, it's Boltzmann's scenario for the understanding why the early universe had a low entropy. Remember Boltzmann, who didn't know anything about the Big Bang, dark energy or quantum mechanics, imagined an eternal universe that had some fluctuations, had some particles that were just moving around with some temperature and every once in a while those particles would form a low-entropy configuration. They would form a planet or, if you wait long enough, they would form a galaxy or, even longer than that, they would form the whole universe.

We said that that can't be right, but if it were right you could have all of these things happening. It turns out that our real world seems headed for exactly that situation. If you take seriously the idea that we have a vacuum energy that will not evaporate away we will live in an eternal universe with a temperature and in that temperature there will be fluctuations and those fluctuations will make planets and people and galaxies. It might take an enormously long time to fluctuate back into the whole universe. You would need 10 to the 10^{120} years to fluctuate back into a situation where you had 100 billion galaxies, but you have forever to wait. Compared to forever 10 to the 10^{120} years is really not that long.

Here's the depressing aspect of that story. If this is the right cosmological scenario we should be Boltzmann brains or the universe around us should've just fluctuated randomly into existence. We don't believe that is true and therefore there must be something wrong with our current best understanding of cosmology. This idea that I've been talking about, the universe expanding under the influence of vacuum energy, this isn't wild and speculative. This is our best fit to the current data, but if it's right something goes terribly wrong in the far, far future.

There is an escape hatch. If that vacuum energy isn't truly permanent, if the vacuum energy at some point in the future goes away, then you can avoid these disastrous consequences, but that's telling us something very, very interesting about how vacuum energy works. In the long run, what we will get is not just a universe with vacuum energy. Something has to happen. The vacuum energy has to decay or change into something else, otherwise we would be Boltzmann brains.

We can now put numbers on what the entropy of the universe does. Remember, we have an observable universe that is only so big. It's a finite sized thing so the entropy of it is not infinite. At very, very early times when there was no structure, no planets, no black holes, that entropy was about 10^{88}. Basically what we're doing is we're ignoring the effects of gravity simply because the self-gravity of the particles was not that important. The universe was smooth. There weren't any structures formed by gravity. You can just plug in 19th Century formulas for what the entropy is and you get about 10^{88}.

Today on the other hand, we have black holes in the universe. Most of the universe has black holes at the centers of galaxies and most of the entropy of the universe is in those black holes. So even though we don't have perfect formula for the total entropy, we can add up all the entropy in black holes and get a good answer, it's about 10^{103}, much larger than 10^{88}. In the future, you can get the maximum entropy that you could possibly imagine from all of the stuff that we see in the universe today. The answer is about 10^{120}. It's a much larger number yet than the 10^{103} that we have today. That would be the equilibrium version of the universe, the configuration we could be in where the entropy was maximal. We argue that it looks like empty space.

The real challenge for cosmology is simply why aren't we in equilibrium? Why don't we live in empty space? Why aren't we a random fluctuation in the thermal atmosphere that an empty universe would have? We clearly need to change our cosmological theories to explain why we see what is around us.

The Big Bang
Lecture 22

So far in these lectures, we have not shied away from speculating. We've talked about time travel, for instance, and Boltzmann brains. But these speculations have been grounded in theories we understand fairly well. In the next couple of lectures, we move into the realm of speculative speculations. We've reached a point where what we know about the universe isn't good enough; we don't know why the entropy of the early universe was so low, so we have to imagine what the possibilities might be.

Time Equals Zero

- If we extrapolate Einstein's equation for general relativity backward, given the conditions in the universe as we understand them, we hit a singularity, a moment when the expansion rate and the density of the universe were infinite. That's what we call the Big Bang.

- Notice that this is a time, not a place. It is a moment in the history of the universe, not a location in a preexisting universe. The Big Bang was not an explosion of stuff in an otherwise empty spacetime. It was all of spacetime as far as we know. The Big Bang is just like the singularity of a black hole except backward in time.

- We know that the early universe was very dense and very smooth. The matter in the universe about 1 second after the Big Bang was essentially in thermal equilibrium. It was as smooth as it could be and as hot as it could be, as long as we ignore gravity. But we can't ignore gravity, of course; that's why we have this mystery in front of us.

- Einstein's general relativity predicts that if we go backward in time, we hit this singularity, but a singularity doesn't mean a boundary to spacetime. What it means is that our equations blow up.

- o If general relativity were the correct theory of reality, the Big Bang would be the beginning—we cannot go past that singularity—but we know that general relativity is not right. It is not compatible with quantum mechanics. To reconcile general relativity with quantum mechanics, we need a theory of quantum gravity.

- o General relativity tells us that there is geometry to spacetime, a certain amount of curvature. Quantum gravity would say that there's a wave function of spacetime, a set of possibilities for what spacetime would look like if we were to observe it. That's the theory we would like to have, but we don't have it yet.

- Even though we don't have the right theory, we know what quantum gravity should be like in certain regimes. It should, for example, reduce to classical general relativity in our solar system. We know that 1 second after the Big Bang, the universe was doing more or less what general relativity says it should do. We can try to see what a full theory of quantum gravity would tell us and use that to understand the Big Bang.

Creation from Nothing
- One plausible scenario is that, with quantum gravity, the Big Bang is the beginning, just as our classical intuition tells us. There is no such thing as before the Big Bang; time doesn't go forever.
  - o In all the ordinary quantum mechanical theories we understand, time goes forever, but gravity is not ordinary in this sense.
  - o We have to take seriously the possibility that gravity added to quantum mechanics makes time come to an end at the beginning of the universe. That would be a universe that essentially comes into existence out of nothing.
  - o This nothingness is not an absence of anything. It's not even a thing, not a state of the universe, not even a quantum mechanical possibility.

- This kind of scenario is favored by Stephen Hawking. He tries to solve equations for the wave function of the universe that would describe the universe that looks like ours today and came into existence at the Big Bang.

- That is a plausible scenario given what we understand today. Does it help us explain the arrow of time? That depends on what the wave function of the universe is. It could be just a new law of nature that according to this wave function, the universe came into existence at a certain moment, and when it came into existence, the entropy was very low.

A Bouncing Universe

- An alternative to creation from nothing is a bouncing universe, a universe in existence before the Big Bang that essentially was the time reverse of the Big Bang. It was a universe in which matter and energy were contracting and getting denser, and somehow, the magic of quantum gravity, rather than collapsing it to a singularity, caused it to bounce back and create what we call the Big Bang.

- In some sense, it seems natural that entropy would be decreasing as it approached the bounce. If the bounce that we now call the Big Bang is a moment of symmetry in the history of the universe, then the entropy goes up in both directions away from it, both toward the past and the future.

- That's an aesthetically appealing universe, but it still leaves us with the question: Why was entropy so low at that point? What is the principle of nature that says that when the universe bounces, the entropy should be low?

The Multiverse and the Theory of Inflation

- A third possibility is the multiverse. According to this idea, our Big Bang was an event that is really quite small in the history of a much larger universe. We see only a finite bit of the universe; perhaps farther away than what we can see, the universe looks very different. The theory of inflation gives some credibility to this idea.

- To understand how inflation works, we need to go back to the idea of dark energy. Remember, dark energy doesn't dilute away as the universe expands. In a universe dominated by dark energy, expansion accelerates because space gets bigger and bigger. Inflation posits much higher energies at the earlier time when the universe began.
    - What would happen if, instead of ordinary matter and radiation, the early universe was filled with nothing but super–dark energy?
    - The rate at which something expands depends on the energy density, so if the energy density was very high, the universe would expand very quickly.
    - At the same time, if the universe was filled with dark energy rather than ordinary matter, it would smooth out.
    - Thus, after a very short time, the universe would go from whatever state it was in to a much larger configuration, very smooth over very large distances, and full of this super–dark energy.
- In the process known as reheating, this super–dark energy would be converted into ordinary matter and radiation. And, in fact, the quantum field that was responsible for inflation could decay after a certain amount of time, eventually becoming a hot, dense gas.
- What inflation gives us, then, is a universe dominated by super–dark energy that expands and smoothes out by a huge amount, and then suddenly, throughout the universe, all that energy is converted into ordinary matter and radiation. This looks exactly like our Big Bang.

The Benefits of Inflation
- In modern cosmology, inflation has become the dominant paradigm, and there are numerous benefits to believing that inflation happened. One of these is that the universe can start very tiny.
    - In conventional cosmology, the actual size of the observable universe when the universe started is about 1 centimeter. This

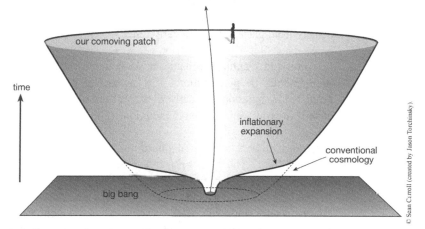

Inflation expands a tiny patch of space to a tremendous size in a fraction of a second (representation not to scale).

seems small to us, but by the standards of particle physics, it's huge.

- o The question is: Why would the universe be so incredibly smooth over this 1-centimeter size?

- o In inflation, if we trace our currently observable universe backward in time, it can start with a region that's about 10^{-30} centimeters across, and it's much more plausible to particle physicists that the universe started in exactly that state.

- Inflation also explains the spatial geometry of the universe, in particular, that its large-scale spatial geometry appears to be flat. The universe could have had substantial curvature at early times, but the process of inflating would have flattened it out.

- Further, inflation explains the fact that the universe is smooth over great distances. Again, the process of inflating would have smoothed out small-scale perturbations.

- o If there was just inflation in a classical universe, there would be no way of understanding why there was any deviation in density from place to place at early times.

- o But if we add quantum mechanics to the mix, we find that there will, in fact, be perturbations. The pattern of perturbations predicted by inflation is exactly what has been observed in the cosmic microwave background radiation.

- Inflation does raise a conceptual question: Where does all that energy come from? The short answer is that with the expanding universe, energy is not conserved.
    - o In the expanding universe, there is energy in stuff—dark energy, matter, radiation, dark matter, and so on—but there is also energy in spacetime itself, the energy of the curvature of spacetime.

    - o The energy of stuff is positive, while the energy of spacetime is negative. In a compact universe or a universe that is perfectly smooth everywhere, the total energy is exactly zero.

    - o As this universe expands, tremendously more energy is created in stuff that is compensated by the energy of spacetime itself.

    - o Energy in stuff does not remain the same as the universe expands. We can get as much universe as we like from an arbitrarily small amount of energy.

Inflation and Entropy

- Inflation tries to give us a natural explanation for certain features of the universe: the fact that it is spatially flat, that the density was so smooth, and so on. The theory tells us that we shouldn't be surprised by these facts, because if we begin with a small region ready to inflate, that is what we will get. However, the question is: Why did we begin with that small region ready to inflate?

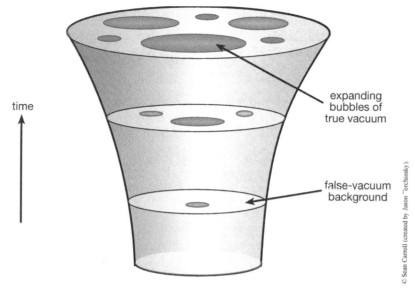

The expanding bubbles never completely collide. During the false-vacuum phase, the amount of space grows forever; inflation never truly ends.

- One way of attacking that question is to think about entropy and calculate the entropy of that proto-inflationary region. This entropy is about 10^{10}, much, much lower than the entropy after reheating.

- Remember, we said that the entropy of the early universe was something like the 10^{88}, and now we're saying that the entropy of the inflating region was something like 10^{10}. This follows the second law of thermodynamics, but it doesn't explain why entropy was so low. We've simply started at an even lower-entropy configuration.

Suggested Reading

Carroll, *From Eternity to Here*, chapter 14.

Guth, *The Inflationary Universe*.

Questions to Consider

1. The Big Bang might be a true beginning, a bounce, or part of a larger multiverse. What's your favorite option? What are the advantages and disadvantages of each?

2. Should we be surprised that energy is not conserved in an expanding universe? Can you think of good analogies for this phenomenon?

The Big Bang
Lecture 22—Transcript

So far, in these lectures, we have not shied away from speculating. We've talked about time travel. We've talked about Boltzmann's Brains. We've talked about some of the frontiers of neuroscience. But even when we've been speculating, they've been grounded speculations. In other words, we haven't ever seen a Boltzmann brain, but that's a prediction of a theory that we understand pretty well.

We've been talking about speculations within the context of a theoretical framework that seems to be right to us. When we talked about time machines, we asked, what does general relativity have to say about time machines? For the next couple of lectures we're going to be doing speculative speculations. We're reaching a point now where what we know about the universe is not good enough. We've discussed what the universe is, how it has evolved and we've come up against the fact that the entropy of the early universe was really, really tiny.

We don't know why and so we're going to have imagine what the possibilities are. So from now on, for the next couple of lectures we're going to be putting out scenarios, none of which are backed up by data or even fully flushed out theories. That's okay. That's what life is like at the edge of research. We're trying to understand the universe and trying to ask the questions which we don't yet have answered.

In particular, in this lecture we're going to talk about the Big Bang. Not the Big Bang model that says we began in a hot, dense, early state, but the actual moment of time, the initial singularity. Sometimes people say the Big Bang and they mean the early universe. They mean that moment really early on when we think the entropy of the observable universe was about 10^{88}. It was very smooth. It was very hot. It was very dense. It was rapidly expanding. But today we want to go right down to time equals zero.

If we extrapolate backward, Einstein's equation for general relativity, given the conditions in the universe as we understand them, we hit a singularity, a moment when the expansion rate was infinity, the density of the universe was infinity. That's what we call the Big Bang.

Notice that it is a time, not a place. It is a moment in the history of the universe, not a location in a preexisting universe. The Big Bang was not an explosion of stuff in an otherwise empty spacetime. It was all of spacetime as far as we know. The Big Bang is just like the singularity of a black hole except backwards in time.

The singularity of a black hole is a moment you reach in the future once you cross the event horizon, a singularity where the curvature of spacetime is infinitely big. The Big Bang is a moment you hit as you go to the past. It's a singular moment but it's not one that you could possibly avoid. If we imagine hypothetical travels backward in time, it's not like you can deviate from your path and miss the Big Bang. It's not a place. It's a time. It's a time when everything blows up and we need to do a better job of understanding what was really happening.

We know that in the early universe, it was very, very dense. It was very smooth. The matter in the early universe, the part that we understand about one second after the Big Bang when there was nucleosynthesis that gives us data about what was going on, the matter was essentially in thermal equilibrium. It was as smooth as it could be. It was as hot as it could be as long as we ignore gravity. We can't ignore gravity, of course. That's why we have this mystery in front of us. Maybe the answer to that mystery can be obtained by going all the way back to time equals zero.

You will sometimes hear cosmologists talk as if they know what happened at time equals zero, at the moment of the Big Bang. You will hear them say that there is no such thing as time before the Big Bang, that that's the moment when the universe came into existence, that asking what is before the Big Bang is like asking what is north of the North Pole. It's not that we don't know yet. It's that there is no answer to that question, but that's a little bit misleading to talk that way.

It is a prediction of Einstein's general relativity, that if you go backward in time you hit this singularity, but a singularity doesn't mean a boundary to spacetime when you take it seriously. What it means is our equations blow up. If general relativity, as Einstein wrote it down in 1916, were the correct theory of reality, the Big Bang would be the beginning. You cannot go past that singularity, but we know that general relativity is not right. We don't know it on the basis of data. Every experiment we've ever done, every astronomical observation we've ever made is perfectly compatible with general relativity, but we know that general relativity can't be right because as a theory it is classical.

It is not compatible with what we know to be quantum mechanics. The right theory of how the world works as far as our experiments tell us. To reconcile general relativity with quantum mechanics we need a theory of quantum gravity. Just like quantum mechanics says if you have an orange, it can be here or there, in fact, it truly is in a super position of both possibilities until you look at it. Whereas a classical theory would say the orange is here or there whether you know, classical general relativity says there is geometry to spacetime. There's a certain amount of curvature. There is something spacetime is doing.

Quantum gravity would say there's a wave function of spacetime. There's a set of possibilities for what spacetime would look like if we were to observe it. That is the theory we want to have, but we don't yet have it. People say this over and over again and it's true. We don't yet have a theory of quantum gravity, but I don't want you to think that we know nothing at all about quantum gravity. We know how quantum mechanics works. We know a lot about how classical gravity works, so it's not that we are completely clueless. We actually know what quantum gravity should be like in certain regimes even though we don't have the right theory.

We know that quantum gravity should reduce to classical general relativity in our solar system for example. We know that one second after the Big Bang the universe was doing more or less what classical general relativity says that it should do. We are not completely without guidance. We can try to see what a full theory of quantum gravity would tell us and to use that to understand the Big Bang.

From the point of view of this course, we can also say we know the early universe had a low entropy. Maybe that fact about the universe is a clue that should help us understand how quantum gravity works. The right theory of quantum gravity is one that will help us understand why the early universe had a low entropy.

What does such a theory say about the Big Bang? Again, since we don't have it, we don't know the final answers, but what we can do is list the possibilities that people have thought of. We have more than one plausible scenario about what could have occurred at the Big Bang. One scenario is that the Big Bang really is the beginning, that our classical intuition, from Einstein's general relativity, that there is no such thing as before the Big Bang, maybe that's just as true in quantum gravity as it is in classical gravity. Maybe time doesn't go forever.

This would be a surprise from one point of view because in all of the ordinary quantum mechanical theories we understand, time does go forever, but gravity is not ordinary in this sense. We have to take seriously the possibility that gravity added to quantum mechanics makes time come to an end at the beginning of the universe and end in our past. That would be a universe that essentially comes into existence out of nothing.

It's very difficult to describe this accurately because our human vocabulary doesn't have the right words. We say the universe could be created from nothing. That sounds as if there was nothing and then for a little while nothing lingered on and then the universe came into existence. That's not the right way to think about it. This nothingness we're talking about is not a way of existing. It's not an absence of anything. It is not even a thing, not even a state of the universe, not even a quantum mechanical possibility.

When we say creation from nothing that's really just a vocabulary term for saying time comes to an end. There is no moment before the Big Bang. This kind of scenario is what is favored by Stephen Hawking, for example. He talks about the wave function of the universe and tries to solve equations for the wave function of the universe that would describe the universe that looks like ours today and came into existence at the Big Bang. Not in a moment of time, but time itself came into existence along with the universe.

That is an absolutely plausible scenario given what we understand today. The question is does it help us explain the arrow of time and the answer is it depends on what that wave function of the universe is. It could be just the new law of nature or a new feature of the world that the wave function of the universe says the universe is created, comes into existence at this moment and when it comes into existence the entropy was very low. That might be a way the universe works.

In some sense it would be disappointing if that were true because we're not learning anything new over and above the fact that the early universe had a low entropy. This encourages some of us to try to think even a little bit closer about what the alternatives are.

One alternative is a bouncing universe. So maybe there was a universe before the Big Bang that essentially was the time reverse of what we call the Big Bang. At least some sort of universe that was contracting. There was matter and energy that was getting closer and closer, more and more dense and somehow the magic of quantum gravity, rather than collapsing it to a true singularity, caused it to bounce back and create what we call the Big Bang. There are different ways to play this game.

The question from our point-of-view is in the preexisting universe, in the universe that was contracting and getting more dense, was entropy going up or going down? In some sense, the most natural thing would be for the entropy to be going down as it approached this bounce. After all, it goes up after the bounce so if the bounce that we now call the Big Bang is a moment of symmetry in the history of the universe, then the entropy goes up in both directions away from the bounce, both toward the past and the future.

In some sense, that's an aesthetically appealing universe. It looks symmetric. The far, far past of that universe would look like the far, far future and this Big Bang moment is the one special time. This big bounce is the one time in the history of the universe when entropy was low.

On the other hand it's a little bit disappointing because we still have the question why was the entropy so low at that point? What is the principle of nature that says that when the universe bounces, the entropy should be low? There could be some principle. It could be that we need to look for it, we need to do a better job so that it's absolutely on the table, but we're still trying to do even better to try to find a dynamical reason why the entropy had to be low at the earliest times.

The third possibility on the table is what we call the multiverse. The multiverse is not the same as a bounce where the whole universe was contracting, bounces and then creates our Big Bang. The multiverse says that our Big Bang was just some kind of event that is really quite tiny in the history of a much larger universe. The multiverse theory says that look, we only see a finite bit of the universe. What right do we have to extrapolate what we see infinitely far? Maybe further away, much, much further than what we can actually see, the universe looks very, very different in different places. What cosmologists call the multiverse is really a mundane thing in some sense. It's just the idea that far away from us, things look very different.

It's not truly separated universes disconnected from us. It's just regions of space that we don't see because light can't get to us from them. That is a way of thinking about how our Big Bang could've come into existence. Clearly, to make it more sensible we're going to need some physics that would bring the multiverse into existence.

There is a theory on the market that was not invented because of the multiverse, but sort of gives some credibility to the multiverse idea. That theory is called inflation. Inflationary cosmology was a scenario invented by Alan Guth, a physicist at MIT back in 1980, which purports to explain why the Big Bang looked the way it did. We are dealing with the past hypothesis, the idea that the early universe had a very, very low entropy. The low entropy of the early universe is only one of the interesting features of the early time.

Inflation was invented, not to explain why the entropy was low but to explain certain other curious features of our universe on very large scales that can be traced back to the beginning. The nice thing about inflation is that it is based on sensible physics. In other words, even though we don't yet know whether inflation is true or false the physics underlying it seems very, very plausible. We don't know that it's right, but it doesn't require great leaps of faith, maybe physics works in a certain way, given the way that we do know physics works. It seems natural for something like inflation to happen.

To understand how inflation works, you can go back to the idea of dark energy. Remember dark energy is something that doesn't dilute away as the universe expands. Einstein tells us that the rate at which the universe expands depends on the energy density of stuff. If you have a universe full of ordinary matter and radiation, stuff dilutes away as the universe expands and the expansion rate slows down as time goes by.

If you have a universe that is dominated by dark energy, something that doesn't dilute away or dilutes away only very, very slowly, then the expansion of the universe actually accelerates because space gets bigger and bigger. Every cubic centimeter of space has the same amount of energy, so that pushes on the universe a little bit and we observe that as an accelerated expansion. The idea behind inflation is to basically play exactly that game, but play it at much higher energies at the much earlier time when the universe began.

Imagine a temporary phase of super dark energy, not energy that is super dark, but dark super energy. Dark energy with a really, really high energy density. Remember, at early times in the history of the universe, matter was compacted to a much greater density, so to start the universe inflation says maybe there's dark energy that has an enormous amount of energy in every cubic centimeter.

What would you have if instead of ordinary matter and radiation, the early universe had nothing but this dark super energy filling it? Well, what would happen is you get an accelerated expansion and the universe would grow by an enormous amount. The rate at which you expand depends on the energy density so if the energy density is very high, the universe expands very, very quickly.

Now while that happens, unlike ordinary evolution with matter in it with perturbations grow, where small variations in density get increased by the force of gravity, if the universe is filled with dark energy, the universe smoothes out. If you have an initial universe with super dark energy in it and variations from place to place, as inflation occurs, as this expansion and acceleration with the super dark energy happens, any perturbations get smoothed out. It's like you're pulling on the edges of a sheet, smoothing out the wrinkles inside it.

So after just a very, very short while, the universe goes from whatever state it was in to a much larger configuration, very smooth, over very large distances full of this dark super energy. What you need to have happen then is this dark energy needs to go away. We don't live in a universe full of this kind of dark energy. You need to convert that dark energy into ordinary matter and radiation. That's a process known as reheating. It is something we know how to do. In other words, this is not just words. We right down theories with equations. We invent fields that would have given us this dark super energy in the early universe and the physics of these fields is like that of a particle that decays.

Remember we talked about muons or for that matter, neutrons, particles that don't last forever. They last for a little while and then they decay into other particles. Likewise the quantum field that was responsible for inflation could decay after a certain amount of time and when it decays it basically turns into other particles. They decay and interact and they turn into a hot, dense, gas.

What inflation gives you is you start wherever you start, but with the universe dominated by this dark energy, you expand by a huge amount, you smooth everything out and then suddenly all throughout the universe you convert all of that energy into ordinary matter and radiation. What are you left with? A universe that is very hot, very dense, very smooth and very rapidly expanding. In other words, something that looks exactly like our Big Bang.

The wonderful thing about inflation is it starts with a very specific initial point, but one that is not implausible. You can imagine that for some reason the universe had this dark energy in it and after that very, very plausible physics leads you to something that looks exactly like our early universe, looks like the Big Bang with small deviations in density from place to place, but basically a hot, dense, rapidly expanding universe.

When inflation came along in 1980, people were initially a little bit surprised and skeptical, especially astronomers who like to see data and this seemed like a very wild, far out idea. More and more over time inflation has come to be accepted, even though we still don't know for sure whether it happened, it is the predominant theory among early universe cosmologists.

In modern cosmology, all discussions of the early universe begin by imagining that there is inflation. You can still imagine alternatives and people talk about them a lot but inflation is a dominant paradigm right now. There are a lot of benefits to believing that inflation happened. Let me just name one of them, namely, that the universe can start very tiny. Now you might not imagine that that is something necessary. After all, the universe was smaller at earlier times so it was tiny whether or not there was inflation, but there's a quantitative difference.

If you imagine the good old Big Bang cosmology, a universe filled with matter, radiation, dark matter, and so forth, if you take our current observable universe, so the 100 billion galaxies that we see with 100 billion stars inside them, no dark energy, no inflation or anything like that, just play the movie backwards and ask when does the density of matter become so high that we would call that the Big Bang. And ask, how big is the physical size that expanded into our universe?

The answer is about one centimeter across. The actual size of the whole observable universe, when the universe started in conventional cosmology is about one centimeter. Now that's very small. From one point of view, it's something that we need to try to imagine, stuffing 100 billion galaxies worth of stuff into that one centimeter across, but by particle physics standards, one centimeter is huge. It's much, much bigger than an atom or an elementary particle.

The question is why would the universe be so incredibly smooth over this one-centimeter size? That's basically a restatement of the past hypothesis. Why was the entropy of that one centimeter region so small compared to all the fluctuations, all the ripples and inhomogeneities that we could have fit into that little one-centimeter sphere?

In inflation, on the other hand, if you trace our currently observable universe backward in time it can start with a region that is in size about 10^{-30} centimeters across. In other words, the same amount of stuff that we see today when you trace it back in time to the Big Bang could've all began in a region 10^{-30} centimeters across. That's much, much smaller than one centimeter across. It is much, much more plausible to many people that there is some particle physics reason why a 10^{-30} centimeter universe started in exactly that state. We don't know what that would be, but that's one of the benefits of inflation. To start the universe going you needed a much smaller region than in conventional cosmology.

There are other benefits of inflation that are more popular, the actual benefits that people talked about when Guth invented the idea. One is it explains the spatial geometry of the universe. Remember we talked about the fact that there's Euclidean geometry when you have parallel lines that are initially parallel they remain exactly the same distance apart. When you take a triangle and add up the angles inside, they always add up to 180°. That is a flat geometry. That's the geometry of the tabletop.

There is also spherical geometry with positive curvature where initially parallel lines come together. There is satellite geometry where initially parallel lines move apart from each other. So you might ask when you look at the universe on very, very large scales, if you drew a triangle between three galaxies billions of light-years apart would the angles inside add up to 180°? Is the large-scale spatial geometry of the universe flat or curved?

The answer is it seems to be flat. We have observations that say that the overall geometry of the uniform universe on very large scales is indistinguishable from Euclidean geometry. We don't see any obvious deviations, any spatial curvature, either positive or negative.

Why is that? We don't know exactly, but if you believe in inflation, you could've had quite a substantial curvature at early times and that process of inflating would've flattened it out. Inflation predicted the universe would be flat and it predicted it back in 1980. It was not observed to be flat until the 1990s. This was a prediction that was made before the observation. Inflation helps explain the spatial geometry of the universe.

It also explains the fact that the universe is smooth over great distances. When we look at the radiation from the Cosmic Microwave Background, it's more or less the same temperature in every direction. We're looking, when we do that, at radiation that came to us from different ends of the universe. There's a question you can ask, how does it know to be the same temperature even though it's millions of light-years away at a time when the universe was only hundreds of thousands of years old?

Well inflation explains that because this process of inflating smoothes everything out, both the overall spatial geometry and these small-scale perturbations. At the same time, because of quantum mechanics, inflation says you can't make it perfectly smooth. If there was just inflation in a classical universe, there'd be no way of understanding why there was any deviation in density from place to place in the early universe, but you add quantum mechanics to the mix and you predict that there will, in fact, be perturbations, that some regions are slightly more dense than others. The pattern of perturbations predicted by inflation is exactly what we observed in the Cosmic Microwave Background.

This is why, even though we're not sure that inflation is happening, it's a very good theory that a lot of people like to think about. It made predictions that fit the data.

There is a conceptual question that inflation raises which is, where does all that energy come from? You start with a tiny little region and you expand it into 100 billion galaxies. The short answer is energy is not conserved. I know that we've said in other context that energy is conserved, but the expanding universe is different. One way of thinking about it is that in the expanding universe there is energy in stuff, that means dark energy, matter, radiation, dark matter, whatever. There's also energy in spacetime itself, the energy of the curvature of spacetime.

The energy of stuff is positive. Real particles weigh a certain positive number. The energy of spacetime is negative. In a compact universe or a universe that is perfectly smooth everywhere, the total energy is exactly zero. This is just a remarkable fact about general relativity. The total energy of the whole universe is zero because the energy of the stuff, of the matter and radiation, including dark energy, is exactly matched by the negative energy of the curvature of spacetime.

So what that means is as this universe expands, you create a tremendous amount of more energy in stuff, which is why I said energy is not conserved. That's because it's compensated by the energy of spacetime itself. The moral here is that it costs nothing to make a universe. Alan Guth calls inflation the ultimate free lunch. You can get a whole universe by starting with essentially nothing. This is a feature of dark energy, but it was always a feature of general relativity. It's not something that is new to inflation or dark energy or any new-fangled modern theory.

Think about radiation. If you have a universe full of photons, every photon loses energy as the universe expands. The total energy density in a bunch of photons like the Cosmic Microwave Background decreases as the universe grows. Decreasing is just as much not being conserved as increasing is. The simple statement is that the energy in stuff does not remain the same as the universe expands. You can get as much universe as you like from an arbitrarily small amount of universe, amount of energy.

Inflation is a good theory. It's a very promising theory. It's the predominant way of thinking among modern theoretical and observational cosmologists. It says you can start with a tiny universe. It can grow to exactly the kind of universe that we see. For our purposes, what we want to know is does it help explain why the early universe had a low entropy?

That's what inflation is trying to do. It's trying to account for why our early universe looks the way it does. Remember you don't need inflation to explain any feature of the universe whatsoever. It's not that the universe disagrees with a theory, it's just that our early universe seems finely tuned to us. Why was the universe spatially flat? Why were the density ripples so small? Why is the density approximately the same in different parts of the universe?

These are all facts that simply could be true. The answer might just be well that's the way it is. What inflation tries to do is give a natural explanation for these features. It says you shouldn't be surprised at all these facts because if you begin with a little region ready to inflate, that is what you will get.

However, the question is why did you begin with that little region ready to inflate?

So one way of attacking that question is to think about entropy. Just like we can calculate the entropy of the hot, dense universe at the time of nucleosynthesis, we can calculate the entropy of that little patch ready to inflate. The answer is about 10^{10}. We don't have an exact number. It depends on when inflation started and what the energy density was. The point is that the entropy of that little region, that little proto-inflationary region of space is much, much lower than the density after reheating, than the entropy, sorry, after reheating once you convert it into matter and radiation.

Remember we said the entropy of the early universe, our observable part of it was something like the 10^{88} and now we're saying that the entropy of the inflating region is something like 10^{10}. On the one hand it follows the second law of thermodynamics. If you start with inflation and end with the conventional early universe, entropy goes up.

On the other hand, you haven't explained why the entropy is low. You've simply started at an even lower entropy configuration. This is the puzzle that inflation leaves us with. It does not explain why the entropy of the early universe was low. It pushes that problem further into the past. It says the entropy of the early universe was low because it used to be even lower. Just like the past hypothesis says the entropy of yesterday was lower than the entropy today because the entropy was even lower the day before yesterday, inflation says the entropy of the early universe was low because it was even lower before that.

Now we want to do better. We don't want to simply say that's the way it was, inflation started and everything happened. We want to know why inflation started so that means we need to go beyond this more or less well-defined theory of inflation and ask why the universe started like that at all.

The Multiverse
Lecture 23

We're beginning to put together a comprehensive theory to explain the origin of the difference between the past and the future. We've said many times that the entropy was lower yesterday, and the reason the entropy was lower yesterday is that it started low near the Big Bang. In the last lecture, we looked at three different ways of thinking about the Big Bang: as the actual beginning of the universe, as a bounce, or as an event that occurred in just one universe of a very large multiverse. In this lecture, we'll dig a little deeper into the possibility of the multiverse.

A Scientific Parable

- To think about the multiverse and how it could be part of science, consider this parable: Imagine there is a planet that is much like ours, except that it's always completely overcast. Scientists on this planet wouldn't know that there are other stars in the sky.
    - To these scientists, their planet is the whole universe because that is all that they can see. At some point, a philosopher wonders whether there might be many similar planets that can't be seen because of the opaque atmosphere.
    - Of course, the other philosophers and scientists say, that idea is nonsense. We can't invoke other parts of the universe that can't be seen. That's not how science or philosophy works.

- The point here is that when we talk about the multiverse, we're not invoking a new kind of thing. Instead, we're observing that there is a point in our universe's past that we cannot see. Is there a finite amount of stuff out there? Is there an infinite amount of stuff that works exactly like the stuff we can see? Or is there an infinite amount of stuff and conditions that are very different from place to place?

- In some sense, assuming that the entire rest of the universe is infinitely large and just like the region we see is just as presumptuous as assuming that there's an infinitely large universe where conditions are very different from place to place. The cosmic microwave background in our universe functions exactly like the opaque atmosphere on our hypothetical planet: It's a barrier past which we cannot see.

The Multiverse and the Anthropic Principle

- When we consider a multiverse, we necessarily invoke the anthropic principle: If there are many different conditions throughout our universe, intelligent beings like ourselves will find ourselves only in those conditions that are compatible with us existing.

- The anthropic principle may be the explanation for the vacuum energy that makes the universe accelerate.
    - In ordinary general relativity, there is a classical contribution to the vacuum energy that is present as an unknown constant of nature, but there are also quantum mechanical contributions to vacuum energy.

    - All the particles in the universe vibrate in hidden ways in the virtual particle sense. That is to say, even in empty space, these particles are providing energy. If we add up these contributions to the vacuum energy, we get a huge number.

    - The real puzzle of vacuum energy is not why it exists but why it is so small compared to its natural value. And the answer might be the anthropic principle.

    - If the vacuum energy were huge, space would accelerate very, very quickly. In fact, if the vacuum energy had its natural value, we couldn't have atoms or even nuclei. The vacuum energy would rip things apart so effectively that no molecules could form. Under those conditions, it's unlikely that life could develop.

- o We could also have a large negative vacuum energy. Instead of making the universe accelerate, it would make the universe recollapse in a tiny fraction of a second. Even if the local physics allowed for the existence of life, there wouldn't be time for it to evolve.

- o However, if the vacuum energy is small but not zero, then there's plenty of time for life to evolve and there's nothing to stop molecules from forming.

- In 1988, Steven Weinberg, a Nobel Prize–winning physicist, made the following prediction: If the vacuum energy is not a constant of nature, if we live in a multiverse where the vacuum energy takes on different values from place to place, then it can't be too large, and it can't be too negative. If we live in a multiverse, we should observe a nonzero vacuum energy with a certain typical value. In 1998, Weinberg's prediction of the energy density was confirmed.

- We cannot, however, explain the entropy of the early universe using just the anthropic principle. If we try to make the same kind of prediction with entropy that Weinberg did with the cosmological constant, we get a prediction that the entropy of the early universe was much, much larger than it actually was.

Inflation and Quantum Mechanics
- According to quantum mechanics, things fluctuate, and we can't predict exactly what will happen; we can predict only different probability distributions.

- As we saw in the last lecture, inflation involves super–dark energy expanding to make a large region of space and then decaying into ordinary matter and radiation; that's the process called reheating. What happens if we add quantum mechanics to that process?

- The answer is that space can expand and grow, but we can't predict that it certainly turns into matter and radiation; we can predict only that this process takes place with a certain probability.

- It's possible that a small region of space expands to a large region and perhaps 90% of that larger region reheats. In the aftermath, we get what we call the Big Bang, but that means that in 10% of the space, inflation does not stop and reheating does not take place.

- That 10% continues to expand, and after a certain period of time, 90% of that region reheats, resulting in a Big Bang. But again, 10% continues to inflate. In other words, we create more and more regions of space where inflation is still going on, even though, in any one region, 90% will stop inflating soon.

- This 10% that keeps inflating grows in size by a tremendous amount. In actual cubic centimeters, the region of space that is still inflating grows without bound. Inflation never ends in this scenario. This is called eternal inflation. It's a natural marriage between inflation and quantum mechanics.

- This theory of eternal inflation tells us that we can get a very different universe from place to place.

String Theory

- Eternal inflation could create many regions of universe just like our own, or it could create many regions, each of which has different local laws of physics. This is a consequence of string theory, the leading candidate to reconcile quantum mechanics and gravity.

- According to string theory, spacetime has at least 10 or 11 dimensions; this seems to be an obvious conflict with experiment because we have only 4 dimensions of spacetime. But it turns out that in general relativity, it's not difficult to hide extra dimensions of space. In fact, there could be as many as 10^{500} ways to hide extra spatial dimensions.

- Every different way of curling up extra dimensions affects the mass and charges of elementary particles, the different forces through which they interact, and the ranges over which those forces stretch.

Basically, every one of these possibilities is a different way the universe can be.

- If we combine string theory with eternal inflation, we bring a wildly varying multiverse to life. Inflation says that we can have different regions of space. String theory says that the local physics in all those regions can be different.

Testing the Multiverse Hypothesis
- The biggest problem with the multiverse hypothesis is that it is difficult to test, given that other regions of the universe are too far away to visit.

- It's possible to look for direct evidence of other universes, that is, instances where another universe with different local physics bumps into our universe.
    o If another universe came into existence because of inflation and reheating, then that universe would leave a tiny relic of radiation and energy heating up our cosmic microwave background.

    o The data we have on the microwave background right now aren't good enough to rule out this idea, but they're also not good enough to confirm that other universes are present.

- The more likely scenario is that we will get indirect evidence in favor of the multiverse. Here, we use the anthropic principle and the multiverse to make statistical predictions for certain parameters of particle physics that haven't yet been measured.
    o Could we use this combination to make predictions about dark matter, the Higgs boson, supersymmetry, or other aspects of particle physics?

    o This is very difficult to do, but it is the more likely way we will discover that the multiverse is on the right track.

The Measure Problem

- The measure problem is a puzzle associated with the multiverse: If we take the multiverse seriously, everything that can happen will happen, and it will happen an infinite number of times.
    - One of the things we want to know, for example, is how many observers would observe a small vacuum energy? How many observers would observe a large vacuum energy?

    - Again, if we take the multiverse at face value, the answer is: An infinite number of observers measures a certain value and an infinite number measures another value. How can we compare these? How can we say that one infinite number is greater than another one?

- This may be a profound problem with the multiverse idea. It may be that even if the multiverse is real, we will never be able to use it to make predictions. As of now, the multiverse is a speculative idea that is not yet developed well enough by theorists to make firm predictions.

The Future of the Multiverse Hypothesis

- The reason we think about the multiverse is not that we know it's there, and it's not even that if we knew it was there, it would make a definite prediction about the universe. It's that when we think about the puzzles we face in explaining the universe we see, whether or not there is a multiverse changes the way we think.

- The way to move forward with the multiverse idea is to understand how gravity, quantum mechanics, and particle physics work together. We need to develop at least one well-defined, unambiguous theory, and right now, when it comes to explaining the past hypothesis, we have no compelling theories.

Suggested Reading

Carroll, *From Eternity to Here*, chapter 14.

Susskind, *The Cosmic Landscape*.

Vilenkin, *Many Worlds in One*.

Questions to Consider

1. Do you think that speculation about the multiverse is a respectable part of science? If not, how should we deal with the possibility that the multiverse is real? And if so, what if we are never able to test the idea?

2. The anthropic principle relies on some idea about what kind of life can possibly exist in the universe (or some other universe). What do you think are the necessary conditions for the existence of life? What is the form of life you can imagine that is least like our own?

3. In 1998, Stephen Weinberg used the anthropic principle to predict the value of the vacuum energy, and his prediction turned out to be correct. How much weight should we give to this success?

The Multiverse
Lecture 23—Transcript

By this point we're beginning to put together a comprehensive theory. Remember our goal is to basically understand, why do I remember yesterday and not tomorrow? What's the origin of the difference between the past and the future? Well we said that the origin is that the entropy was lower yesterday and the reason the entropy was lower yesterday is because it started low near the Big Bang.

That's the next question and in the last lecture we gave you three different ways of thinking about the Big Bang. One is that it could be the actual beginning to the whole universe. Another is that it could be a bounce. There could be a symmetric version of the universe on the other side of the Big Bang. The third way is that the Big Bang could just be one of the things that happens from time to time in the history of the universe and we could live in a very large multiverse.

The problem with the first two ideas is that even if we have those theories of the Big Bang, making the entropy low at that point seems arbitrary. There seems to be no good reason why the entropy should be low at the moment we call the origin of our observable universe. The third scenario, the multiverse, it's at least possible to imagine explanations for why the entropy was so low. In this lecture we're going to dig a little bit deeper into that possibility of a multiverse.

To do that we have to start off with some sort of general philosophical statements. The idea of the multiverse, if you hang around with the kind of people I hang around with, gets people very worked up. You get an emotional response to the scientific hypothesis that the universe we see is not all there is. Some people love the idea. They think that it truly helps you understand all sorts of features of the universe we do see that would otherwise be mysterious. Other people think that the multiverse is a capitulation to the forces of anti-science, that because you're invoking all these things that can never be observed, this is not how true science should be done.

I want to take a middle ground. I'm actually very willing to consider the multiverse as a serious possibility but I don't think that we know that it's right. We're very, very far away from saying that the multiverse makes predictions that fit data we have. Nevertheless, it's also wrong to think of the multiverse as nonscientific. The ultimate goal of science is to find the truth and according to the modern physics that we understand, the idea that there is a multiverse is very, very possible. It is something that could be the truth. It's our duty to figure out what we can learn about what the multiverse would teach us and if we can make any progress on it. If not, then we should abandon the idea. But we're nowhere near the point, yet, where that statement could be made.

To think about the multiverse and how it could be part of science, I like to tell a little parable. Imagine that there was a world that was much like ours. The conditions were very similar except that it was always cloudy. It was always 100% completely overcast and the temperature down on Earth was always exactly the same, on this planet, I should say. So what would the scientists on this planet think? If they were at a primitive stage of development, they were not able to develop yet satellites or ways to go above their atmosphere, they would think about the universe in terms that did not involve what we think of as the night sky. They wouldn't know there were other stars in the sky.

To them, their planet was the whole universe. That is all that they can see, and at some point in their intellectual development one of their philosophers comes along and says, well, maybe there are lots of planets. Maybe we live in a universe that has many other planets just like ours, it's just that we can't see them because our atmosphere is opaque. Of course, his philosophical collaborators would say that is nonsense. You're invoking all of these parts of the universe that can't be seen. That's not how science or philosophy works. We should stick to observable things.

The point is that when we talk about the multiverse, we're not invoking a new kind of thing. We're just saying that there is the universe that we live in. There's only part of it that we can see. Because there is a point past which we can't see, what is out there will never be directly observable to us. That is not what we're imagining, that someday we'll build a multiverse telescope.

But there's still the question, is there only a finite amount of stuff out there? Is there an infinite amount of stuff out there that goes on exactly like the stuff inside? Or is there an infinite amount of stuff out there and conditions are very different from place to place?

In some sense, assuming that the entire rest of the universe is infinitely big and just like the region we see is just as presumptuous as assuming that there's an infinitely big universe where conditions are very different from place to place. Both options are absolutely on the table. The question is, does choosing one over the other matter to how we think about the universe that we do see?

The point is that imagining that there is a multiverse changes what we consider to be interesting questions to ask about the universe in which we live. Think about our foggy planet. Think about the planet where you can't see other parts of the universe. The scientists on that planet might have the question, why did we get lucky enough that the universe has exactly the right temperature and conditions for our civilization to arrive? They live on a very temperate planet where the temperature is the same everywhere and it is very hospitable to life arising and they asked themselves the question why is that the case. Why are the laws of the universe such that we can live here on our planet? How lucky, how finely tuned was that?

Of course, we know better. We know that it's not lucky or weird or fortunate at all that they could live on their planet because there are many planets. There are 100 billion stars in our galaxy and the best current estimates are that there are probably more planets than stars. There are billions and billions of planets just in our galaxy and 100 billion galaxies in the universe. Eventually you will get one where the conditions are just like that.

Knowing that they lived in a universe where there are other planets changes our hypothetical scientists way of thinking about what is an interesting question versus a not interesting question. If you live in an ensemble, if you live in a universe where conditions are very, very different from place to place, there are selection effects. You will only find yourself in the parts of that universe that are hospitable for you being there. If you live in a unique universe, then you have a challenge. How can you once and for all explain the conditions of this unique universe in which you live?

Our universe could be either way. We don't know. The Cosmic Microwave Background in our universe functions exactly like the opaque atmosphere in our hypothetical planet. It's a barrier past which we cannot see. We look out into the sky. We look with light, with visible light, with radio waves, and so forth, but the early universe was opaque. We will never be able to see beyond it using light so there is literally a barrier past which we can't see. There is the rest of the universe that might be like ours or might not. We need to take both possibilities very, very seriously.

Now when we do that we necessarily start invoking the anthropic principle. We already mentioned the anthropic principle. There are different versions of it and some of them sound more profound than others. I like the sort of simple version of the anthropic principle that simply says, if there are many different conditions throughout our universe, our multiverse, whatever you want to call it, intelligent beings like ourselves will only find ourselves in the middle of those conditions that are compatible with us existing.

Now we don't know what the conditions are for intelligent life, generally. That's why applying the anthropic principle in practice is very, very hard. My personal opinion is that people assume way too much about what we know about life and intelligence and consciousness when they actually try to put the anthropic principle to work. I think there are many, many different ways that you could get complexity and intelligent life. Until we have a much more developed theory of complexity, consciousness and intelligence, we can't use the anthropic principle in too quantitative a way. But we can still use it in a rough qualitative way.

The example that I gave before was we're not surprised that we live on the surface of the Earth rather than the surface of the Sun or in between the planets in the solar system, even though there's a lot more area and volume outside the Earth, the Earth is a very hospitable place to live. Nobody considers it a profound question why are we living here rather than on the Sun.

As a more specific cosmological question it might be that the anthropic principle is the explanation for the vacuum energy that makes the universe accelerate. We talked about the idea that the universe is accelerating, there's an explanation for it, that empty space itself has an energy density, the vacuum energy, the cosmological constant. Now you might be surprised that there can be vacuum energy at all but it turns out that's the wrong direction for your surprise to go. Once you understand that there can be such a thing as vacuum energy, as a physicist you sit down, you do a back of the envelop calculation saying, well, how big should the vacuum energy be?

It turns out that you can have a classical contribution to the vacuum energy that would be there as an unknown constant of nature, even in ordinary general relativity. That's what Einstein discovered, but there are also quantum mechanical contributions. All of the particles in the universe vibrate in hidden ways in the virtual particle sense. That is to say, even in empty space these particles are there providing energy. These are contributions to the vacuum energy.

You add them all up and you get a huge number. The real puzzle of the vacuum energy is not why there is vacuum energy, but why is it so small compared to its natural value. The answer might be the anthropic principle. Think about what the world would be like if the vacuum energy were huge. It would cause space to accelerate very, very quickly. In fact, you can easily do the calculation and say that if the vacuum energy had its natural value you couldn't have atoms. You couldn't even have a nucleus of an atom. The vacuum energy would rip things apart so effectively, you couldn't make a single molecule.

So imagine that under those circumstances you couldn't make life. This is not 100% because we don't know what life is. We don't know under what conditions it could exist but if you can't make atoms, it's very plausible you can't make life.

You could also have a large vacuum energy, but a negative vacuum energy. A lot of vacuum energy but with a different sign. Instead of making the universe accelerate it would make the universe recollapse. And then again you can easily do the calculation and you can show that the magnitude were the natural value, the universe would recollapse in a tiny fraction of a second, even if the local physics allowed for the existence of life, there wouldn't be time for it to evolve.

Whereas if the vacuum energy is small but not zero then there's plenty of time for life to evolve and there's nothing to stop molecules from forming. Steven Weinberg, who won the Nobel Prize for Particle Physics a long time ago, in 1988 made a prediction. He said that if the vacuum energy is not a constant of nature, if we live in a multiverse where the vacuum energy takes on very different values from place to place, then it can't be too large. It can't be too negative. He predicted that if that were the universe, we should observe a nonzero vacuum energy with a certain typical value.

In 1998 when we discovered that energy density, when we discovered that the universe was accelerating it was exactly the same to within the quite substantial error bars as Weinberg's prediction. So in other words, under this huge set of assumptions that there is a multiverse, that we understand how life can form, and so forth, the anthropic principle made a prediction for the vacuum energy and it came out right.

The question is can we explain other things about the universe? People are trying. Can you explain the mass of the Higgs Boson, for example? But what we can't do is explain the entropy of the early universe just using the anthropic principle. Remember in Boltzmann's scenario where you have an infinitely big universe with fluctuations you can fluctuate to lower entropy and then bounce back but we only need a very, very tiny fluctuation of entropy to make a person, or even to make the Earth compared to the huge fluctuation of entropy that our actual universe represents.

In other words, if you tried to do the same thing that Weinberg did but with entropy rather than the cosmological constant, you would predict that the entropy of the early universe is much, much larger than it actually is. So even if the anthropic principle can explain the vacuum energy, it can't explain the entropy. We need to do better.

Nevertheless, even if the anthropic principle is not by itself explaining the low entropy of the early universe, the multiverse might still be part of the explanation. It's important to emphasize that the multiverse is not a theory by itself. The multiverse is a prediction of other theories. It's not that a bunch of cosmologists sat around and started saying, well, wouldn't it be amazing if there were lots of different universes. That would change how we think about everything. That's not the way people became interested in the multiverse idea. It started from other theories that we believe for other reasons and we realized that those theories predict the existence of a multiverse.

The most basic of those theories is inflation. We thought that inflation is a good idea because inflation helps explain things we observe. It explains why that spatial geometry of the universe is flat. It explains why there are tiny ripples, but only tiny ones in the universe from place to place. It starts with a very tiny region, expands it tremendously and then you reheat. It's plausible physics. We think we understand how inflation works. But almost as soon as inflation was invented, people started investigating it a little bit more carefully.

In particular, they started taking this story and putting it in the context of quantum mechanics. Quantum mechanics says that things fluctuate. You don't predict exactly what will happen. You predict different probability distributions. With inflation you had this super dark energy that expands to make a big region of space and then that dark energy decays into ordinary matter and radiation. That's the process called reheating.

What happens if you add quantum mechanics to that process? What happens is space can expand and grow but you don't predict that it certainly turns into matter and radiation at a certain place. You predict that there's a probability. You have this dark energy field that makes inflation go. It reheats with a certain probability. It's just like a decaying particle, remember. A muon decays but we can't predict the exact lifetime of any one muon. We can predict the probability per second or per microsecond that the muon will decay.

Likewise for inflation, we can predict the probability that any one region of space converts from dark super energy into ordinary matter and radiation. So maybe what happens is you take a little region of space, it expands to a big region and perhaps 90% of that larger region reheats. Its dark energy converts into ordinary matter and radiation. In the aftermath you get something that we would call the Big Bang, but that means that in 10% of the space inflation does not stop. It does not reheat. It keeps on being dark super energy so that 10% keeps inflated. That is to say even though the rest of the universe is still expanding, it is decelerating, it's slowing down because now it's ordinary. Now it's matter and radiation and that 10% where quantum mechanics says it did not turn into ordinary matter and radiation, inflation keeps going so the size of that universe expands much more quickly.

Then after a certain period of time, 90% of that region can reheat, turn into ordinary matter and radiation and become a Big Bang, but 10% keeps inflating and then again keeps growing. What happens is you keep creating more and more regions of space where inflation is still going on, even though, in any one region 90% of it will stop inflating pretty soon. This 10% that keeps inflating, keeps growing in size by a tremendous amount. So in actual cubic centimeters, the region of space that is still inflating grows without bound. Inflation never ends, in this scenario. This is called for obvious reasons eternal inflation. It's a very natural marriage between inflation and quantum mechanics.

Remember, inflation is the ultimate free lunch. Once you start it you get as big a universe as you want. The theory of eternal inflation, which is just quantum mechanics combined with inflation itself, says that you can get a very different universe from place to place. You get one little region where you've reheated and you've created a Big Bang, other regions where inflation keeps going.

That is the simplest way to get a multiverse. Remember it didn't come out of a desire to get a multiverse. It just came out of taking inflation seriously. We think that quantum mechanics is the right way of thinking about the world. If you think that inflation is also the right way of thinking about the early universe, eternal inflation is very, very natural.

Now if it is the case that every time inflation ends you create ordinary particles like ours, protons and neutrons and electrons and whatever the dark matter is, then what eternal inflation would give you is a big collection of universes just like our own. We call them universes. In fact, Guth calls them pocket universes, little regions of space that have their individual Big Bangs where inflation ended, but they're just separated in space itself. We call them different universes even though they're connected. They're in between the regions that are still inflating, but it could be that there are different kinds of regions so this is another variation on the same theme.

One way that eternal inflation can happen is to create many, many regions of universe just like ours. Another way eternal inflation can happen is to create many, many regions, each of which have different local laws of physics. This is not, again, completely made up. This is a consequence of other theories that we know and love, in particular in this case, string theory which is the leading candidate to reconcile quantum mechanics and gravity. Just like inflation, string theory is a theory that we don't know whether it's right or not. We don't have direct experimental evidence, but it seems to be on the right track.

One of the major problems with string theory is that it says that spacetime has at least 10 or 11 dimensions. That seems to be an obvious conflict with experiment because we only have four dimensions of spacetime, but it turns out that in general relativity it's not difficult to hide extra dimensions of space. You can curl them up into a little ball, a little geometry of some shape and size, so small that you and I would never notice them. Once you realize you can do this, it's actually very easy to do this. The problem is it's not unique. There's not one way to hide the extra dimensions of space; there are a gazillion ways of hiding the extra dimensions of space.

We try to estimate how many ways there are. We get numbers something like 10^{500}. That's obviously not a very precise number, but it gives you a feeling for the fact that there are many, many ways to hide all these extra dimensions that string theory predicts.

The reason why that matters to our current discussion is that every different way of curling up the extra dimensions affects, what we call, particle physics, what we would call the mass of the elementary particles, the charges of the elementary particles, the different forces through which they interact. The ranges over which those forces stretch and even the number of dimensions. Maybe you curl up six dimensions, but maybe you only curl up five or four or three of them. So basically every one of these possibilities is a different way the universe can be. It's a different phase of the universe. Just like water can be liquid water, water vapor, solid ice, string theory says that spacetime can have different phases corresponding to each different way you can curl up the extra dimensions.

If we combine string theory with eternal inflation we bring a very wildly varying multiverse to life. Inflation says you can make different regions of space. String theory says that the local physics in all those regions can be very different. For example, you can have different kinds of elementary particles, so life could be different. Maybe in some of these regions you can't have any kind of life at all. Maybe in other regions you get exactly our kind of life. In still other regions, you get a very different kind of life.

Another thing that can be different is the vacuum energy. Steven Weinberg just assumed that the vacuum energy was different from place to place when he made his prediction but string theory brings that to life. String theory says here is how you could actually get different amounts of vacuum energy in different regions of the universe. It's like combining inflation with string theory that we get the multiverse.

You could have a multiverse even without string theory but the different regions of space would all be different. The important thing is that we didn't just invent the multiverse from the start. It is a prediction of other theories, inflation and string theory, that we like for other reasons.

Still what we'd love to do is be able to test this prediction. It's one thing to have a prediction. It's another thing to know that it is right. The single biggest problem with the multiverse hypothesis is that it is difficult to test. We are never going to go visit one of these other regions, they're too far away. If they're there we need to look for some reason why we should believe them? Basically, there are two possibilities.

One possibility, which is the winning-the-lottery possibility, is that you directly look for evidence of other universes. Now I just said we can't go there and we can't see them, so how can you directly look? The answer is that another universe with different local physics could literally bump into ours. In the early moments of our universe as it stopped inflating, turned into ordinary matter and radiation, maybe nearby another universe came into existence because inflation stopped in another region. That universe also grew and literally bumps into ours. It would leave a tiny relic of the radiation and the energy from that universe heating up our Cosmic Microwave Background.

What you look for in the data are little circles in the microwave background where other universes bumped into ours. People have looked for such evidence and they have found that maybe they're there. In other words, the data that we have on the microwave background right now aren't good enough to rule out the idea, but it's also not good enough to say that these other universes really are there. To be very, very honest, this is a long shot. It's probably not going to happen but it'd be incredibly, amazingly exciting if you actually saw evidence for a different universe in the Cosmic Microwave Background.

The more likely scenario is that we get indirect evidence in favor of the multiverse. One thing to do is to make statistical predictions. Just like Weinberg did for the cosmological constant, make predictions for other parameters of particle physics that we haven't yet measures. We have, for example, dark matter. We know that there is dark matter. We see the evidence for it in astrophysics and cosmology, but we haven't detected it yet in the laboratory. Could we use the anthropic principle and the multiverse to explain that a certain kind of dark matter is more likely than another kind? Could we use the anthropic principle in the multiverse to make predictions for the Higgs Boson that hasn't been discovered yet, for super symmetry or other hypothetical aspects of particle physics?

This is very, very difficult to do but this is the more likely way that we'll think that the multiverse is on the right track. If we can use it to make predictions which then later come true. The problem with this hope is that the multiverse idea might be right, but that doesn't mean that we understand it.

In particular, there is a puzzle associated with the multiverse called the measure problem. This is different than the measurement problem of quantum mechanics. The measure problem says that if you take the multiverse seriously everything that can happen will happen and it will happen an infinite number of times. So what we want to know, for example, is how many observers would observe a small vacuum energy? How many observers would observe a large vacuum energy? We can calculate some probability. You should probably observe the vacuum energy to be a certain number.

The problem is if you take the multiverse at face value, the answer an infinite number of observers measure a certain value and an infinite number of observers measure this other value. How can you compare them? How can you say that one infinite number is bigger than another one?

This is not just annoying. This could be a really profound problem with the multiverse idea. It might be that even if the multiverse is real, we will never be able to use it to make predictions. This is an honest problem with the multiverse that even its advocates need to admit. Now there are people trying. There are certainly many attempts, many models, many theoretical proposals on the market for how to make believable predictions in the context of the multiverse, but it's very hard to know whether you're on the right track. We don't have data that says, oh, one proposal right or one proposal is wrong. The real lesson here is not that the multiverse is useless or that it's a panacea.

The lesson is that the multiverse is a speculative idea that is not yet developed well enough by theorists to make firm predictions. Maybe it will be developed well enough or maybe it won't, but it just takes time to do it.

The reason we think about the multiverse is not that we know it's there, and it's not even that if we knew it was there, it would make a definite 100% prediction about the universe. It's that when we think about the puzzles we face in explaining the universe we do see, whether or not there is a multiverse changes the way we think about the universe that we actually do observe. If we are part of an ensemble, then the kinds of questions we ask and the kinds of answers we would give, are different than if we are a unique universe. Just like our planet with an atmosphere, knowing that they are just one of billions and trillions of planets would change the questions they ask about the conditions they find themselves in.

In particular we care about the past hypothesis. Why did the early universe have such a low entropy? The question is could we address that by imagining that our universe is not unique? If the universe we see, if the local observable universe is the once and for all universe, then we have a real puzzle. Why in this one appearance of the universe did the entropy start out so small?

If there is a multiverse, if there are many, many individual sub-universes then we have a different kind of question. Can we make a model? Can we make a theoretical understanding of the multiverse in which almost all universes have strong arrows of time? Is it natural for a universe to begin with very low entropy? That's the kind of idea that we can try to develop

The way that we're going to do this is not by waiting for the multiverse to show up. That would be great, but it's unlikely for this to happen. What we want to do is to have better ideas about fundamental physics. The real way to move forward with the multiverse idea is to understand how gravity and quantum mechanics and particle physics play well together. We need to develop at least one well-defined unambiguous theory. The lack of data, the lack of direct experimental input would be really a problem if we had more than one theory and we were trying to decide which one was right. But right now, when it comes to explaining the past hypothesis, we have zero compelling theories. We have not even one theory that is absolutely clear and consistent and predicts the data that we see.

Therefore, it's not a matter, yet, of deciding between different theories on the basis of experiments. It's a matter of understanding the questions we're asking and developing that kind of theory. That's work which we can, in principle, do even without experimental input. It's very, very hard and we might fail. It might turn out that our human brains are just not quite large enough to figure out how the universe or the multiverse work without direct experimental input. We might have to wait 1000 years for some brilliant insights to come along, but we might have to wait six months. We don't know until we try. It's a matter of the experimentalists and the theorists and for that matter philosophers and mathematicians working together to try to build a theory that will explain the universe we see.

Approaches to the Arrow of Time
Lecture 24

Throughout these lectures, our challenge has been to connect two realities: the deep underlying reality that there is no difference between past and future and our phenomenological reality, in which it is obvious that there is an arrow of time. The reconciliation seems to be that we are a complicated system, embedded in an environment that is far from equilibrium. There is something called entropy that characterizes the organization or disorganization of us and our environment. The amazing fact about our world is that all the manifestations of the arrow of time seem to be explained by increasing entropy. But the question remains: Why did the universe start with such low entropy? In this lecture, we'll look at three possibilities for answering that question.

An Intrinsic Arrow of Time?
- The first possibility for addressing our central question is that despite everything we think we know about the fundamental laws of physics, maybe there is an intrinsic arrow of time.
    - Ever since Galileo and Newton, our attempts to develop deep laws of physics have been time symmetric and have conserved information, but maybe that's wrong.
    - Perhaps laws that are yet to be discovered are not reversible and do not conserve information. Maybe entropy is something that has an intrinsic arrow of time.
- If we do have dynamical laws that pick out a direction of time, they need to be laws that make entropy spontaneously decrease. We need to start with a high-entropy universe, apply the laws of physics to it, watch the entropy decrease, and then imagine that we live in that universe, but we perceive it in the other direction.
- It's not hard to imagine laws of physics that are like that.

- We can conceive of a billiard table that is not frictionless but sticky along the walls.

  - The billiard balls could start in a high-entropy configuration (with the balls scattered around the table) and end with a low-entropy configuration (with the balls "stuck" to one wall of the table).

  - This is an irreversible process, but it doesn't seem to be our world. It seems intuitively implausible that we could start with an empty universe, expanding with dark energy, and then gradually form galaxies all the way back to a reverse-time Big Bang and that the result would be our universe.

- It's worth emphasizing that the past hypothesis, this idea that our early universe started with low entropy, is both necessary and sufficient for explaining the arrow of time. Any new theories about the arrow of time must explain why our early universe was hot, dense, and smooth because these are the conditions that obtained in the early universe.

- The idea of an intrinsic arrow built into the laws of physics doesn't seem to explain the past hypothesis.

The Simplest Possible Quantum State?

- The second possible explanation for understanding why entropy was so low in the early universe is that there's just a reason that initial conditions were like that. Maybe it's just a fact that conditions near the Big Bang had low entropy.

- Stephen Hawking and other physicists favor an approach that says the quantum state of the universe is the simplest possible. The Big Bang was the beginning. The entropy was low, and there's some principle of nature that explains why that makes sense.

- That principle of nature can't be that most possible universes look like that because that is simply false. The actual configuration

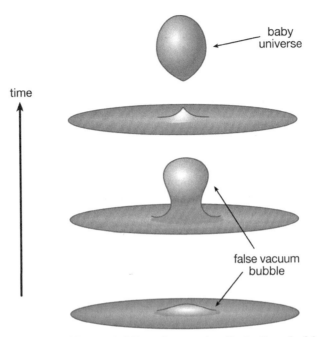

A baby universe could be created through a quantum fluctuation of a false-vacuum bubble.

of stuff in our early universe is a tiny fraction of all the different possible ways it could have been.

- It might be simpler to start with a low-entropy early universe, but we don't seem to have a principle that says the early universe had to be simple for that reason. It seems fairly arbitrary to simply say that the early universe was just like that; further, this approach doesn't seem to fit the universe we see.

A Spontaneous Arrow of Time?
- The third possibility for answering our question is that the arrow of time arises spontaneously out of the natural, reversible evolution of the universe. In other words, the universe does not have an arrow of

time in the laws of physics, but the solutions to the equations of the laws of physics always have an arrow of time.

- This is actually a common phenomenon in physics, where the symmetries of specific solutions to equations are not the same as the symmetries of the equations.
  - As a very simple example, think of Newtonian mechanics, which is perfectly reversible; it has no arrow of time built in. Laplace told us that if we know the state of the universe in Newtonian mechanics at one time, we know it at all other times, but it's not hard to think of systems obeying the rules of Newtonian mechanics in which an arrow of time automatically arises.

  - For example, think of a ball rolling on an infinite hill. The history of this ball is that for an infinite amount of time, it rolls up the hill; it reaches a turning point and then spends an infinite amount of time rolling down the hill. This is the only thing the ball can do in this system; if the hill always has a slope, the ball can never remain stationary.

  - This is a physical system obeying Newton's laws, which are perfectly reversible, but it doesn't have an equilibrium.

  - Could the universe be like that? Could we live in a universe where the fundamental laws of physics say that entropy doesn't have a maximum value? If entropy doesn't have a maximum value, the reason we are not in the maximum-entropy configuration becomes much less mysterious: There is no maximum-entropy configuration. Entropy can always grow without bound.

- Even if entropy can grow forever, that certainly doesn't explain why our local region of the universe once had such low entropy. We need to do much more work to create a physical system in which regions where the entropy had the low value it had at the Big Bang are created naturally.

Inflation and the Multiverse

- What we really want to explain is why we're not in equilibrium, that is, why we don't already have the highest entropy we could possibly have. As we know, the entropy of our universe will continue to increase into the far future, until it becomes empty space.

- When we apply quantum mechanics to this empty space, there will be thermal fluctuations. Very rarely, there will be a random fluctuation of thermal energy that will result in an atom, a molecule, a person, a planet, or the universe. The larger the thing that will be created, the longer we need to wait for the random fluctuations, but they will all eventually happen.

- It sounds difficult to fluctuate the entire universe, which has much greater entropy than a planet or a person. As we said, it's easier to fluctuate entropy a little than a lot, so how can we think that it's easier to fluctuate a whole universe than just a person? It seems like this scenario gets us Boltzmann brains. The answer might lie in inflation.

- Although it might be very rare to fluctuate 100 billion galaxies, it may not be that difficult to fluctuate the tiny region of space that is dominated by super–dark energy and is ready to inflate.
    - Imagine we have empty space, nothing but vacuum energy and the very cold thermal radiation we expect because of quantum mechanics. Random thermal fluctuations occur all the time.

    - Because space itself is flexible, perhaps the shape of space can be fluctuated. Perhaps a small bubble can be pinched off that is full of dark energy and ready to inflate. That little bubble can inflate, expand, reheat, and look exactly like our Big Bang.

    - In this way, what we thought was equilibrium—a state of maximum entropy—could be shown not to be equilibrium. We can always increase the entropy of the universe, according to this way of thinking, by creating new universes that split off and go their own way, and this process continues forever.

- Just as we can evolve this scenario forward in time, we can evolve it backward in time and make more universes toward the past. Inside every one of those universes is an arrow of time that is pointed backward compared to ours. Just as the ball rolling down the hill comes from and goes to infinity, our universe does the same thing into the far past and the far future.

- This kind of scenario takes advantage of the flexibility of the universe according to the laws of general relativity and quantum mechanics to escape the trap of thermal equilibrium. Given the flexibility of space and time, given the quantum mechanical freedom to fluctuate into different things, maybe there is no maximum-entropy state the universe could be in.

The Future of Our Theories

- As of right now, our understanding of the multiverse and inflation isn't good enough to compare these ideas to the data we have. We need to push our understanding of cosmology, particle physics, gravity, and the evolution of complexity and entropy through time to be able to really understand what our theories predict. Once we get there, we can compare the theories to data and try to make a choice about which theory is right.

- We personally are organized systems; we have a great deal of complexity but low entropy. Even so, we are not flouting the second law; in fact, we're manifestations of it. We are side effects of the universe's increasing entropy. If the universe were in thermal equilibrium, intelligent life would not exist.

- In that sense, we are fortunate to live in a universe with such a pronounced arrow of time. The reason we have this arrow of time is because of the early universe. It's because of the conditions in the early universe that we have aging, metabolism, memory, and causality—everything that, to us, distinguishes the past from the future.

Suggested Reading

Albert, *Time and Chance*.

Carroll, *From Eternity to Here*, chapter 15.

Greene, *The Fabric of the Cosmos*.

Lockwood, *The Labyrinth of Time*.

Price, *Time's Arrow and Archimedes' Point*.

Questions to Consider

1. Despite what we currently know about the laws of physics, do you think it's plausible that the ultimate laws are actually not invariant under time reversal?

2. Do you think the multiverse is a promising route to explaining the arrow of time?

3. Would you be satisfied if the ultimate conclusion of cosmologists was that the low entropy of the early universe was simply a brute fact, without any deeper explanation?

Approaches to the Arrow of Time
Lecture 24—Transcript

We've come a long way. This is the last lecture. Just as there is an arrow of time, there's an arrow of lectures. If you have followed them in order, then a certain amount of time has passed since you watched Lecture 1 or listened to it to today. In a very real sense, you're a different person than you were when we started. At the very least, you are older. A certain number of moments have accumulated in your life. You have different memories than you had. You hopefully can remember some of the stuff that was in the previous lectures. Maybe you've thought about it, maybe it has affected the way that you have accumulated other memories when you were not listening to the lectures.

It's even possible you are wiser. Maybe the things that we've talked about have made you look at the world in a different way. Going toward the future it's possible that some of the things that you've heard about in these lectures will affect choices you make. It might affect which Great Courses you will listen to next. It might affect how you think about the nature of time, so again the arrow of time makes us who we are. We reach a certain state at the present moment as an accumulation of different moments. We live in a universe that doesn't just happen randomly. It moves in a certain direction.

Now that's true for the whole universe. It's true for every one of us. It's because of the arrow of time that we remember yesterday, we don't remember tomorrow. You've heard the lectures. You don't know what's going to happen next. And yet, at the same time you are made of individual pieces. You are made of atoms. Those atoms are made of particles. These individual pieces don't accumulate anything as time goes on. They cannot tell the difference between past and future. At the level of two particles interacting with each other, there is no arrow of time. The deep underlying laws of reality as we understand them today do not distinguish between one direction of time and the other.

It's the task of connecting these two realities that has been our challenge. The deep underlying reality we believe is true, which says that there is no difference between past and future, there's no intrinsic arrow of time. You can make a movie of particles and atoms, play it backward and no one would be able to notice versus our obvious phenomenological reality around us at all moments when it is obvious that there's an arrow of time.

You make a movie of an egg breaking or someone diving into a pool or mixing milk into coffee, you play it backwards and everyone will notice. The reconciliation seems to be that we are not alone in the universe. We are a complicated system. We're embedded in an environment. Our environment is very, very far from equilibrium. In other words, there's something called entropy that characterizes how organized or disorganized we are and our environment is.

High entropy is equilibrium. It's a configuration of stuff that will stay just like that. That's where it wants to be. That is the most natural place for a system to be in, but our universe is very, very far from equilibrium. We are experiencing an unmistakable arrow time. Entropy is increasing through time and that's why you remember the previous lectures. You don't remember what's going to happen next.

We know why entropy increases. It's not because the fundamental laws of physics say it will. It's not because of microscopic laws that govern individual atoms and particles say that there's an arrow time. It's because of our environment. There are just two very simple principles that completely explain why entropy is increasing.

First, there are more ways to be high entropy than to be low entropy. That explains why starting today going to the future, entropy is going to be increasing. Given whatever we're doing today and no more information than that, the overwhelmingly high probability is that entropy will be higher tomorrow. You let a physical system evolve with time, it will tend to move towards higher and higher entropy.

The second principle is that the universe started out with a very low entropy. That explains why the entropy was smaller yesterday, why it was lower in the past. The first principle, that there are more ways to be high entropy than to be high entropy, is absolutely symmetric with respects to time. There's no intrinsic arrow of time built into the definition of entropy.

The second principle is the past hypothesis. The universe started in a low entropy state of the right form to create a universe like ours, that is manifestly asymmetric with respects to time. It's a past hypothesis. It's not a future hypothesis. The amazing fact about our world, one of the unifying triumphs of physics is that all of the different manifestations of the arrow of time can be explained by the increasing entropy.

I don't want to over claim that seems to be true the actual explanations are in many cases yet to be constructed. We are working hard to explain memory, prediction, causality, evolution, all of those other time-directed processes in terms of entropy, but our deep down theoretical understanding seems to say that that is the way it is going to work. The question is why is the universe like this? Why did we start with such low entropy?

It's not surprising, the first part, where we say there are more ways to be high entropy than to be low entropy. That doesn't seem to need to be explained. What needs to be explained is why the universe started in such a small subset of all the different configurations it could have started in. We don't know the final answer, but at least we know enough to talk about the different alternatives intelligently to put them on the table and wonder which one is going to eventually turn out to be right.

Essentially, there are three options when it comes to trying to figure out why the early universe started with a low entropy. Some of them are better than others, but we'll discuss all of them and you can decide for yourself.

The first option is that despite everything we think we know about the fundamental laws of physics, maybe there is an intrinsic arrow of time there. Ever since Galileo and Newton, our attempts to develop deep down laws of physics have been time symmetric, have been reversible. They conserved information. They had the property if you go from one state to the next, you know exactly where you came from if you keep track of exactly where you are. All the information at one moment of time is preserved as you evolve through time, but maybe you could imagine that's just wrong. Maybe we have even better laws yet to be discovered that do not conserve information, that are not reversible. Maybe entropy is something that has an intrinsic arrow of time.

This is obviously something to think about. We're trying to reconcile the fact that the deep down laws don't have an arrow and the manifest laws do. Well, maybe the deep down laws do too. That's just an obvious thing to consider. The problem is you need to think very, very carefully about what that would mean. Remember it is easy to get entropy to increase with time. You don't need to come up with new laws of physics to get entropy to go up. What you would be trying to explain is the low entropy of the early universe. So what that means is if you have dynamical laws that pick out a direction of time, they need to be laws that make entropy spontaneously decrease. You need to start with a high entropy universe, apply the laws of physics to it, watch the entropy go down and down and down and then imagine that we live in that universe but we perceive it in the other direction. We remember the direction of time in which entropy was lower. We predict the direction of time in which entropy was higher, but in this hypothetical viewpoint the laws of physics would work backwards from that.

This is not hard to imagine laws of physics that are like that. For example, we talked about on a billiard table, it is low entropy when all the balls are sticking in one place. It is high entropy when all the billiard balls are scattered around the table. Imagine laws of physics which instead of an absolutely perfect frictionless billiard table, we had one end of the billiard table be sticky. We have four walls on a rectangular table. Imagine all the balls can bump into each other but whenever they hit one of the walls they just stick there, so in that case it would naturally happen that as the balls move around the table, occasionally they would hit that wall and they would stop. You could start with a high entropy configuration where all the balls are scattered around the table. You would end with a low entropy configuration with all the balls stuck to one wall. That is a decrease of entropy.

It is an irreversible process. Once you got to that wall, you wouldn't know where you came from. The thing is that it doesn't seem to be our world. There's nothing in the laws of physics that seems to act like that. It seems just sort of implausible, intuitively, that you could start with an empty universe, expanding with dark energy. Then form very, very gradually galaxies backward in time, undo everything that we know, have laws of physics that explain why scramble the eggs naturally turn into eggs, why glasses of water naturally form ice cubes all the way to a reverse time Big Bang and say that that is our universe.

It's worth emphasizing that the past hypothesis, this idea that our early universe started with low entropy is both necessary and sufficient for explaining the arrow of time. You will sometimes read papers in the physics journals or news articles that say a new theory explains why there is an arrow of time, and this new theory is based on quantum mechanics or gravity or something like that. You should not believe any of these purported explanations unless they tell you why our early universe was hot, dense and smooth because that is sufficient.

If you explain why our early universe is hot, dense and smooth, then you have explained the arrow of time. After that entropy increases which is the most natural thing in the world. It is also necessary because our early universe really is hot, dense and smooth. Therefore, it really does have low entropy.

So the thing to be explained is the conditions that we think actually obtained in our early universe. That's the only thing you need to explain, as far as we know, to account for the arrow of time. The intrinsic arrow that you might try to build into the laws of physics, that doesn't seem to explain the past hypothesis. It might be possible to invent laws like that but no one knows how to do it. There's no empirical evidence that that's the way physics works so this is actually not a very popular approach to explaining the arrow of time.

The second option is that maybe there's just a reason why the initial conditions were like that. All we need to do is to explain why conditions near the Big Bang had a very low entropy. Maybe it's just true. Maybe there's no bigger dynamical framework in which that comes to be like that. Maybe it's just a fact. Maybe even God made it that way. That could be true, but as physicists we would like to do better. We want a physics-based explanation for why the early universe was in a certain configuration and you can try to do that.

You can say maybe the quantum state of the universe is the simplest possible. This is the approach favored by Stephen Hawking and other people. They want to understand the Big Bang from a quantum gravity point-of-view. They don't want to imagine there was something before the Big Bang. They want to say, yes, this was the beginning. Yes, the entropy was low and there's some principle of nature that explains why that makes sense.

That principle of nature can't be that most possible universes look like that. That is simply false. The actual configuration of stuff in our actual early universe is a very tiny fraction of all the different possible ways it could have been. It might be simpler to start with a low entropy early universe, but we don't seem to have a principle that says the early universe had to be simple for that reason. It seems fairly arbitrary to simply say that early universe was just like that. Furthermore, it seems to not really fit to the universe that we see.

Our universe has 100 billion galaxies with 100 billion stars per galaxy. It's much more extravagant than just us. If you want to believe that the early universe simply had a low entropy, that is one thing, but why did it have a low entropy with so many particles? Why such a big universe in such a hugely low entropy state compared to how high the entropy could've been?

Again, this is very much on the table, you have to accept that this could be the way the universe is, but I think we need to think a little bit more deeply about why it would be like that. There's some philosophy question as well as physics question. What counts as an explanation? At what point do we just say this is how it is, versus at what point do we just say, we can do better? There needs to be some reason why it is like that rather than some other way.

The third option besides it is just like that or the laws of physics have an arrow of time is that the arrow of time arises spontaneously out of the natural, reversible evolution of the universe. In other words, the universe does not have an arrow of time in the laws of physics but the solutions to the equations of the laws of physics always do have an arrow of time. This is actually a very common phenomenon in physics where the symmetries of specific solutions to equations are not the same as the symmetries of the equations.

As a very simple example, think of Newtonian mechanics. Newtonian mechanics is the paradigmatic example of no arrow of time built in. It's perfectly reversible. Laplace told us that if you know the state of the universe in Newtonian mechanics at one time, you know it at all other times, but it's not hard to think of examples of systems obeying the rules of Newtonian mechanics where you automatically get an arrow of time.

Think for example of a ball rolling down a hill but because this is a thought experiment, think of an infinite hill. Think of a hill that never has a bottom, it just goes on forever. You imagine you see the ball at some point in time. It is rolling down the hill and there is an arrow of time and it's moving from left to right. It has a direction in which it is moving. Now if time is truly infinite we can imagine tracing the past of that ball so it's moving from left to right at the moment we see it but we can go to the past. It's reversible according to Newtonian mechanics. It would go up the hill, but if it's moving at a finite speed and the hill just grows higher and higher without bound then it only has a certain amount of energy, this ball.

So when we go backward in time it slows down as it's climbing the hill and we eventually reach a point where the ball was stationary. And then before that the ball was actually moving in the opposite direction. So the whole history of this ball rolling on the hill for all eternity is that it's been an infinite amount of time rolling up the hill, it reaches a turning point and then it spends an infinite amount of time rolling down the hill.

The reason we know that must be what it's doing is because that is the only thing you can do in this system. There is only one kind of trajectory. If the hill always has a slope, the ball can never remain stationary.

So this is a restless system. This is a physical system obeying Newton's laws which are perfectly reversible but one that doesn't have an equilibrium. It doesn't have a place it can sit. If you put the ball on the hill, it will start rolling. If you let the ball roll forever without disturbing it, it will roll forever except for that one moment when it was perfectly still.

The question is could the universe be like that? Could the entropy of our universe be like the position of the ball. It's just an analogy. It's not an exact physical equivalent. The question is could we live in a universe where the fundamental laws of physics say that entropy doesn't have a maximum value? If entropy doesn't have a maximum value it becomes much less mysterious why we are not in the maximum entropy configuration. There is no maximum entropy configuration. Entropy can always grow without bound.

The key point of this way of thinking is that there is infinitely far for the ball to roll and an infinite time to do it. There is an infinite amount of entropy that can be generated in our universe, so this theory would say, and there is also infinite time to do it. That's easy to imagine in the context of a toy model. That toy model creates an arrow of time spontaneously because the ball is almost always rolling in one direction or the other.

A difficult question is can we really make a realistic cosmology that makes our universe be like that? Can we invent a theory of physics where the entropy of the universe can grow forever, there is no equilibrium state? But more importantly and trickier, the way that entropy grow is by creating universes like ours. In other words, by creating regions of space that look like our hot Big Bang in a low entropy configuration where it expands and cools. That's not easy to do.

Even if entropy can grow forever and ever, that certainly doesn't automatically explain why our local region of the universe once had such a low entropy. We need to do much more work to create a physical system where you naturally create regions where the entropy looked like a low value that we had at the Big Bang.

This is where inflation and the multiverse might come in. One way of thinking about it is to say what we really want to do is explain why we're not in equilibrium, why we don't already have the highest entropy we could possibly have. So we argued before, when we're talking about the evolution of the universe that our entropy is increasing. It will continue to increase for quite a while.

Our far future, a googol years from now will be empty space. We're imagining that the dark energy is truly vacuum energy. It's truly constant.

It's going to stay there. That might not be right but it's the most plausible theory we have right now. So when we say high entropy, when we say thermal equilibrium what that means is empty space, a universe that has expanded, emptied out and cooled until there's nothing there. But we said that when you apply quantum mechanics to such a universe there are fluctuations, there is radiation. Just like Hawking says black holes radiate, the whole universe radiates in the presence of vacuum energy so there will be thermal fluctuations.

Occasionally, very, very rarely there will be so much thermal energy generated just as a random fluctuation that you can make an atom or you can make a molecule, a DNA molecule, or you can make a person or a planet or a solar system or a galaxy or the universe. The bigger thing you want to make the longer you need to wait for the random fluctuations but they will all eventually happen. If you want to explain our reality, you want to have that kind of system that we said could be real but instead of fluctuating into individual brains or individual solar systems, you want to fluctuate into the entire universe.

Now that sounds difficult. We said that the universe has a much larger entropy than just a little planet or just a little person and we also said that it's easier to fluctuate entropy a little bit than a lot so how in the world would you think that it's easier to fluctuate a whole universe like we live in than just a person. It sounds like this scenario just gets us Boltzmann brains not the universe in which we live. The answer might be in inflation.

Remember the idea of inflation is that you can start with a very, very tiny region of space, 10^{-30} centimeters or smaller and if that region of space has just the right conditions, if it's dominated in energy by this temporary dark super energy, then that tiny region will expand by an enormous amount. No matter how tiny it started it can grow into a universe as big as we see. The point is that maybe even though it would be very, very rare to fluctuate 100 billion galaxies, maybe it's not that hard to fluctuate into this little bubble that is ready to inflate.

It is true that the entropy in that region, we said that the entropy in the region ready to inflate is very, very low, but it's, still it's, that's not a disadvantage in this particular situation. The point is that this little bubble is not the whole universe. We're not fluctuating from the whole huge entropy of the universe, all if it into that little bubble. We're just saying that this little bubble of space is a tiny fluctuation. The rest of the universe remains unchanged. So it's actually quite a small cost to fluctuate into this little bubble ready inflate.

Now let's put this together. It's very, very speculative so I don't want to say that it's all settled, but let's ask what this would look like. What we're saying is that we have empty space. Nothing but vacuum energy and the very, very cold thermal radiation, 10^{-30} Kelvin, that you would expect just because of quantum mechanics. You get random thermal fluctuations all the time.

Occasionally, you fluctuate into Boltzmann brains and Boltzmann universes, and so forth, but we're saying that there is a new kind of fluctuation. We take advantage of the fact that space itself is flexible so maybe you can fluctuate not just electrons and photons and quarks and so forth. Maybe you can fluctuate the shape of space. Maybe space itself can bend and pinch off a little bubble that is full of dark energy that is ready to inflate, so that little bubble separates from the rest of the universe and becomes what we call a baby universe.

Through a fluctuation of spacetime combined with the fluctuation of the fields that live in spacetime, we can create a whole new universe in a very tiny region of space. The best thing about this theory is that it doesn't end. That little universe that we made, that little baby universe that is a bubble of spacetime that pinches off and goes its own way, costs exactly zero energy. Inflation is the ultimate free lunch. You have empty space. You evolve from empty space, which is in a pretty high entropy state to a similar looking universe, again, empty space in a high entropy state plus a little bubble.

That little bubble can inflate, expand, reheat and look exactly like our Big Bang. That is the way that what you thought was a high entropy state, what you thought was equilibrium could be shown to really not be equilibrium. This is a new way to increase the entropy when you thought you were at the maximum entropy. You were not correct. You can always increase the entropy of the universe, according to this way of thinking, by creating new universes that split off and go their own way. This little baby universe that you make can expand, make people like us, cool off and become high entropy itself, and then this process continues forever.

The original universe we started with can create more and more baby universes. Every baby universe can expand, cool off and become high entropy then create more baby universes itself. This is a way to make real the hope that our universe could have a potentially infinite entropy, that there is no thermal equilibrium state. The universe never stops evolving because you can always create more universe.

One of the interesting spins on this scenario is that you can play the same game backward in time. Imagine that there's one moment in the history of the universe when all you have is this big empty universe with high entropy. You evolve it, according to the laws of physics forward in time. You make baby universes and make more entropy. You also evolve it backward in time, according to the laws of physics and because the laws of physics are reversible you get the same answer. You make more universes toward the past as well as toward the future.

Inside every one of those universes there is an arrow of time that is pointing backwards compared to ours. Of course, our arrow of time is pointing backwards compared to them. The point is that just like the ball rolling down the hill comes from infinitely far away and goes to infinitely far away, it's doing the same thing in the far past as the far future. In this scenario the universe is doing the same thing in the far past and the far future. It's growing, expanding, creating more and more entropy.

Now this is one of my favorite theories. I've worked on this kind of proposal myself. I don't want you to buy it. I think it is an example of the kind of thing we are looking for. There's much unknown physics lurking here. We don't know whether, in fact, you can make baby universes. There are some who have argued very strongly that baby universes simply don't exist. Even if you could make baby universes, this scenario very much has the measure problem. There's an infinite number of things happening. Have you compared the probability for one thing to the probability for something else? The answer is we just don't know.

There is one question that it answers which is why do you get universes rather than brains? How do you solve the Boltzmann brain problem? The claim in this scenario is that it is actually easier to make a whole new universe than it is to make a brain. A brain, you know, has a few kilograms worth of stuff, 10^{25} particles in it. But because of inflation, a universe doesn't need 10^{25} particles. It can have almost no particles at all, just a really tiny bubble of space filled with dark energy. That will expand and cool and create as many particles as you'd like.

Entropy can always increase. The number of particles is not conserved. This kind of scenario tries to take advantage of the flexibility of the universe according to the laws of general relativity and quantum mechanics to escape the trap of thermal equilibrium. Given the flexibility of space and time, given the quantum mechanical freedom to fluctuate into different things, maybe there is no maximum entropy state the universe could be in. This relies on a whole bunch of assumptions that we don't know how to test, that entropy can always increase, that time goes on forever. There's other ways out, of course.

If time is finite, if the Big Bang was really the beginning, if there's really an end at some point in the future, then there's no Boltzmann brain problem. We shouldn't be accepting that one or another of these proposals is right. We should be investigating all of them. The real question is how will we know? How will we test whether one theory is right or another? The short answer is patience. It is absolutely true that, ultimately, science is always empirical. We will not decide which theory is right just on the basis of thinking about it. We will only ever decide which theory is right by comparing the different theories to data, but it might take a very long time to develop these theories before we can compare them to the data.

People are trying right now. My personal belief is that we're just not there yet. Our understanding of the multiverse, inflation, and so forth, isn't good enough to compare to them to the data we actually have. We need to push on our understanding of cosmology, particle physics, gravity, not to mention how complexity and entropy evolve through time and eventually you'll be able to really understand what our theories predict. Once we get there, we can compare them to the data and try to make a choice about which theory is right.

So that's where we are. That is where we find ourselves in our quest to understand the mysteries of time. We personally are organized systems. We are very low entropy. We have a lot of complexity and a very low entropy but we are not flouting the second law. We are not just ignoring the laws of physics. We are manifestations of the second law. We are side effects of the universe's increasing entropy. If the universe were in thermal equilibrium, intelligent life would not exist.

Life, consciousness, processing information, metabolizing, thinking, feeling, existing, talking, communicating, all of these require a departure from thermal equilibrium. In that sense we are very fortunate to live in a universe with such a pronounced arrow of time. The reason we have this arrow of time is because of the early universe. It's because of the conditions in the early universe that we have aging, metabolism, memory, causality, everything that to us distinguishes the past from the future.

One way of thinking about it is we are surfers riding a wave of growing entropy. The wave started at the Big Bang, 13.7 billion years ago. It's riding along and it will continue to do so for billions of years. Eventually, it will peter out. Eventually, it will crash upon the shore and at least locally we will be in equilibrium. We are riding that wave. We are side effects. We are parasitic upon the fact that entropy is increasing. Someday it will go away. In the meantime, we can enjoy the ride. Thank you for sharing some of the time you have with me.

Bibliography

Adams, F., and G. Laughlin. *The Five Ages of the Universe: Inside the Physics of Eternity*. New York: Free Press, 1999.

Albert, D. Z. *Time and Chance.* Cambridge: Harvard University Press, 2000.

Barnett, J. E. *Time's Pendulum: From Sundials to Atomic Clocks, the Fascinating History of Timekeeping and How Our Discoveries Changed the World*. New York: Harcourt Brace, 1999.

Callender, C. *Introducing Time*. Illustrated by Ralph Edney. Cambridge: Totem Books, 2005.

Carroll, S. *From Eternity to Here: The Quest for the Ultimate Theory of Time*. New York: Plume, 2010.

Davies, P. C. W. *About Time: Einstein's Unfinished Revolution*. New York: Simon & Schuster, 1995.

Falk, D. *In Search of Time: The Science of a Curious Dimension.* New York: Thomas Dunne Books, 2008.

Frank, A. *About Time: Cosmology and Culture at the Twilight of the Big Bang.* New York: Free Press, 2011.

Gell-Mann, M. *The Quark and the Jaguar: Adventures in the Simple and the Complex.* New York: St. Martin's Griffin, 1995.

Greene, B. *The Fabric of the Cosmos: Space, Time, and the Texture of Reality*. New York: Knopf, 2004.

Guth, A. H. *The Inflationary Universe: The Quest for a New Theory of Cosmic Origins.* Reading, MA: Addison-Wesley, 1997.

Hawking, S. W. *A Brief History of Time: From the Big Bang to Black Holes*. New York: Bantam, 1988.

Klein, E. *Chronos: How Time Shapes Our Universe*. Translated by Glenn Burney. New York: Thunder's Mouth Press, 2005.

Levine, R. *A Geography of Time: On Tempo, Culture, and the Pace of Life*. New York: Basic Books, 1998.

Lindley, D. *Boltzmann's Atom: The Great Debate That Launched a Revolution in Physics*. New York: Free Press, 2001.

Lockwood, M. *The Labyrinth of Time: Introducing the Universe*. Oxford: Oxford University Press, 2005.

Lucretius *De Rerum Natura (On the Nature of Things)*. Edited and translated by A. M. Esolen. Baltimore, MD: Johns Hopkins University Press, 1995.

Nahin, P. J. *Time Machines: Time Travel in Physics, Metaphysics, and Science Fiction*. New York: Springer-Verlag, 1999.

Penrose, R. *The Road to Reality: A Complete Guide to the Laws of the Universe*. New York: Knopf, 2005.

Price, H. *Time's Arrow and Archimedes' Point: New Directions for the Physics of Time*. New York: Oxford University Press, 1996.

Schacter, D. L. *The Seven Sins of Memory: How the Mind Forgets and Remembers*. New York: Houghton Mifflin, 2001.

Schrödinger, E. *What Is Life?* Cambridge: Cambridge University Press, 1944.

Susskind, L. *The Cosmic Landscape: String Theory and the Illusion of Intelligent Design*. New York: Little, Brown and Company, 2006.

———. *The Black Hole War: My Battle with Stephen Hawking to Make the World Safe for Quantum Mechanics.* New York: Little, Brown and Company, 2008.

Thorne, K. S. *Black Holes and Time Warps: Einstein's Outrageous Legacy.* New York: W. W. Norton, 1994.

Vilenkin, A. *Many Worlds in One: The Search for Other Universes.* New York: Hill and Wang, 2006.

Von Baeyer, H. C. *Warmth Disperses and Time Passes: The History of Heat.* New York: Modern Library, 1998.

Zimbardo, P., and J. Boyd. *The Time Paradox: The New Psychology of Time That Will Change Your Life.* New York: Free Press, 2008.

Notes

Notes

Notes

Notes